Brown Skins, White Coats

Brown Skins, White Coats

RACE SCIENCE IN INDIA, 1920–66

Projit Bihari Mukharji

The University of Chicago Press Chicago and London

The University of Chicago Press, Chicago 60637
The University of Chicago Press, Ltd., London

Published 2022

31 30 29 28 27 26 25 24 23 22 1 2 3 4 5

ISBN-13: 978-0-226-82299-0 (cloth)
ISBN-13: 978-0-226-82301-0 (paper)
ISBN-13: 978-0-226-82300-3 (e-book)
DOI: https://doi.org/10.7208/chicago/9780226823003.001.0001

Library of Congress Cataloging-in-Publication Data

Names: Mukharji, Projit Bihari, author. | Rāẏa, Hemendra Kumāra, 1888–1963.
Title: Brown skins, white coats : race science in India, 1920–66 / Projit Bihari Mukharji.
Other titles: Race science in India, 1920–66
Description: Chicago : The University of Chicago Press, 2022. | Includes bibliographical references and index.
Identifiers: LCCN 2022021774 | ISBN 9780226822990 (cloth) | ISBN 9780226823010 (paperback) | ISBN 9780226823003 (e-book)
Subjects: LCSH: Race—Research—India—History—20th century. | Scientific racism—India—History—20th century. | BISAC: SCIENCE / History | HISTORY / Asia / South / General
Classification: LCC DS430.M79 2022 | DDC 305.800954—dc23/eng/20220518
LC record available at https://lccn.loc.gov/2022021774

For
David Arnold, Gautam Bhadra, and Majid Hayat Siddiqi
In gratitude and as *gurudakshina*

CONTENTS

ILLUSTRATIONS

PARABLE OF BROWNNESS

Brown is a funny color. Left for too long in the desert of power, the white heat of colonialism scorches it into a deep, dark black. Put it to bed in the lap of capital and comfort for too long, and it begins to dream in white. Mostly it's an unstable color stretched taut between the black and the white, fleeting restlessly from one pole to the other.

It is also an uneven color. Unlike both white and black it does not spread smoothly. Shimmering shades of light and dark alternate throughout, breaking up its unity and questioning its homogeneity. Patches of it are deeper, darker, and blacker than the rest. Blotches within it develop a sickly, pale whiteness. The blotches and patches appear divorced from each other. They hate being yoked together.

Then there is the shine. The sickly white blotches emit a peculiar shininess. Glint, glamor, and newness cloak the white blotches with an aura of modernity. That aura appears to raise the white blotches above the rest of the brownness. It is mostly an optical illusion. But illusions are entertaining, spectacular, and lucrative. They flourish and overwhelm.

Beneath the glitter, the darker patches of brown subsist, persist, and resist. All the while slipping in and out of the sickly blotches that emit the white dreams. Occasionally finding succor from the purer, more homogenous blackness that exudes its historical destiny beyond the contradictory life of brownness.

Brownness, audaciously, had once attempted to tryst with destiny on its own terms by promising to remain nonaligned with either pole. But haunted as it always was by the sickly patches of whiteness spreading cancerously upon its epidermis, ere long it had been seduced by the shininess of white and descended into a Shining India.

This is a book about brownness. Its impossible place betwixt and between white and black, and its shimmering inner violence. Its highfalutin' national dreams, and the smoldering heaps of broken dreams of equality and respect upon which those national dreams stand tall. It is a book about the peculiar, plural, and persistent alienations that make up brownness.

AN ADVERTISEMENT FOR WHITE COATS

[Published somewhere in South Asia sometime in the mid-twentieth century]

Sale! Sale! Sale!

Once it hung in the closets of Sahibs

Now it can be YOURS!

The Magic White Coat allows you to dream new dreams!

Dreams of a better world! Dreams of a truer truth! Dreams of a plush, better paid job!

Price: Old ways of knowing and being

Special Deal: Old hierarchies upgraded free of charge!!

Statutory Warning: Wearing a white coat may
produce alienation from coatless neighbors,
intimate enemies, and nationless pasts.

ACKNOWLEDGMENTS

It is always difficult to determine in retrospect exactly when a book begins to take shape in the author's mind. Whichever date one arrives at then provides a rough outline for acknowledging the intellectual and practical debts that went into the writing of the book. In my case, however, there are three people whose scholarship and thinking have deeply permeated mine and to whom my own work has been so pervasively indebted that it transcends all such temporal determinations. Gautam Bhadra, Majid Hayat Siddiqi, and David Arnold have shaped my understanding of historical methods, stimulated my thematic interests, and sculpted my political investments. I dedicate this book to them.

Predictably, the more immediate inspiration for the book is the political climate in which we have been living over the past few years. The 2014 victory of Narendra Modi in India and the 2016 victory of Donald Trump in the USA have both attested to and authorized the new respectability of racism. This has forced us to rethink the platitudes about race merely being a "pseudoscientific" idea that would soon go the way of the dinosaurs. For me, as an upper-caste Hindu born in India and living in the United States, the moment also brought out the contradictions of contemporary articulations of race. Many of those who bellowed in full-throated support of the unfolding bigotry in India simultaneously opposed the racist prejudices in Trump's America. Somehow rampant Islamophobia or caste discrimination in India did not seem to them to be connected to the racism that nonwhite Americans are subjected to. This contradiction seems to permeate many Indian American settings and calls for a rigorous interrogation.

What allowed me to distill this zeitgeist and the contradictions it was throwing up into a scholarly project where my historical training would be useful were the many conversations, both formal and informal, with friends and colleagues. Among those whom I particularly wish to thank are Warwick Anderson, Elise Burton, Harald Fischer-Tine, Erika Milam, Marina Mogilner, Ishita Pande, Dorothy Roberts, Suman Seth, and Keith Wailoo. They have all been generous with their time and ideas, and these have helped me formulate my own ideas. Conversations with Yulia Egorova, James Poskett, Ricardo Roque, and Thomas Trautmann have similarly helped me hone my arguments. I am also hugely grateful to Christopher Fuller. Not only did our many exchanges prove extremely helpful, but I also benefited immensely from being able to read his unpublished book manuscript on Sir H. H. Risley. Risley remains the person mentioned most often in the scholarship on race science in colonial India, and having an exhaustive biography to hand was most instructive.

As the book took shape, I also realized that I was making forays way beyond the time period, region, and archives that I had been familiar with in my past work. This posed a challenge since there were complex scholarly traditions that I had only limited acquaintance with. Once again, this is where the help of generous colleagues proved essential. Barbara Gerke guided me through the complexities of Tibetan medical ideas, while Steven Weitzman helped me navigate intricacies of Jewish history. I remain ever grateful to both of them.

Rajat Kanti Sur graciously copied certain key sources I needed, while Subhashis Bhattacharya helped me access important Bengali sources. I am immensely thankful to both of them.

I owe an even bigger debt to Robert Aronowitz and Allegra Giovine, both of whom have read and commented upon multiple draft chapters. Their insights have helped me sharpen my arguments.

More generally, my ideas about history of science have been enriched by the passionate conversations in three editorial groups that I am fortunate to be part of. First among these is the *Isis* editorial team. Matt Lavine and Alix Hui have been excellent colleagues and I consider myself fortunate to work with and learn from them. Second, the *Osiris* editorial collective, comprising Elaine Leong, Ahmed Ragab, and Myrna Perez Sheldon, has helped me realize that hoping for a more just society, working to make our field more inclusive, and writing cutting-edge histories of science are all in fact interconnected projects. I deeply cherish our comradeship. Finally, the Cambridge Elements editorial team, comprising Marwa Elshakry, Pablo Gomez, Gabriela Soto Laveaga, Sean Hsiang-lin Lei, and Helen Tilley, has helped me realize that while the Indian experience might not be akin to the settler colonial experiences, it is far from unique in the majority world. It is inspiring to be a part of this team.

A wider set of interlocutors within the histories of science and medicine has also contributed toward my thinking more generally. I particularly want to thank Nicole Barnes, Alex Csiszar, Mary Fissell, Michael Gordin, Jeremy Greene, Marta Hanson, Lauren Kassell, Hisa Kuriyama, Eugenia Lean, Clapperton Chakanetsa Mavhunga, Lissa Roberts, Pierce Salguero, Victor Seow, John Tresch, and Simon Werrett.

While conversations with fellow historians of science have inspired me to take up new questions and methods, my own analysis has necessarily drawn upon the works of fellow South Asianists. Since the historical scholarship on the twentieth century is just beginning to emerge, conversations with fellow South Asianists have been crucial to developing my own analysis. I particularly thank Manan Ahmed, Daud Ali, Nikhil Anand, Sekhar Bandyopadhyay, Dwai Banerjee, Ishita Banerjee-Dube, Debjani Bhattacharyya, Dipesh Chakrabarty, Shefali Chandra, Partha Chatterjee, Prasun Chatterjee, Burton Cleetus, Lawrence Cohen, Sangeeta Dasgupta, Saurabh Dube, Ken George, Greg Goulding, David Hardiman, John Bosco Lourdusamy, Justin McDaniel, Durba Mitra, Rahul Mukherjee, Prabir Mukhopadhyay, Kirin Narayan, Shailaja Paik, Sarah Pinto, Gyan Prakash, Arvind Rajagopal, Bo Sax, Martha Selby, Terenjit Sevea, Mitra Sharafi, Charu Singh, Mrinalini Sinha, Kavita Sivaramakrishnan, Banu Subramaniam, Renny Thomas, and Bharat Venkat. I owe a special debt to Doug Haynes, Rochona Majumdar, and Amit Prasad for many long and enjoyable conversations about South Asia, its pasts, and its presents.

My longest conversations, however, have been with my students. I feel privileged to have been able to work with a wonderful group of graduate students, and every conversation with each of them has been educational in one way or another. I would like to thank Arnav Bhattacharya, Baishakh Chakrabarti, Nikhil Dharan, Christopher Fite, Allegra Giovine, Ngamlienlal Kipgen, Prashant Kumar, Claire Sabel, Rovel Sequeira, and Koyna Tomar.

The fellowship and intellectual vibrancy of my departmental colleagues have been another source of crucial support and inspiration. Andi Johnson, David Barnes, Etienne Benson, Meghan Crnic, Stephanie Dick, Steve Feierman, Sebastián Gil-Riaño, Harun Küçük, Susan Lindee, Beth Linker, Ramah McKay, Jonathan Moreno, Elly Truitt, and Adelheid Voskuhl have been the best colleagues one could ask for.

Courtney Brennan at UPenn and Marchelle Brain and Paige Estefan at the *Isis* office have also assisted me in numerous ways big and small. Without their help, I would not have been able to find the time to write this book.

This is my second book with the University of Chicago Press and I cannot thank my editor, Karen Merikangas Darling, enough. Her support for the project and her candid suggestions have helped me see the key interventions with greater clarity. The comments from the anonymous referees have also been extremely helpful. Lys Weiss has been a wonderful copyeditor while Leonard Rosenbaum was a perceptive indexer. I remain grateful to them.

Beyond the so-called ivory tower, a small circle of friends in the United States and India has sustained me through the process of writing and much more. I would particularly like to thank Robby Aronowitz and Jane Mathisen, Debjani Bhattacharyya and Adam Knowles, Srilata Gangulee, Shampa Chatterjee, Pushkar Sohoni, and Prabhat Kumar.

Toward the end of my work on this book Pronoy arrived in my life. Along with his feline siblings, Siraj and Mohan, every day he conspired to find new ways to keep me from finishing the manuscript. But I could not have asked for a more beautiful and fulfilling reason to look away from the book.

My deepest and most abiding gratitude, however, is to the one person who saturates everything I am and everything I write: my wife, Manjita. Even though my writing has often eaten into our shared time, she has not only unstintingly supported the project but also enriched it with her candid insights and creative suggestions. It is my singular good fortune to have the opportunity to share my life with her.

Introduction

As a South Asian immigrant to the United States I have access neither to white privilege nor to Black angst. Yet, like other Asian immigrants, I too have to navigate an inescapably racialized terrain. From overt differences in pay and promotion structures and the disproportionately high frequency of "random" searches at airports to the banality of so-called microaggressions like a colleague inquiring whether Indians really ate human flesh, race incessantly seeps into my social existence as an "Asian" in America.

Yet, as an upper-caste Hindu male born in postcolonial India, I have also—often unknowingly—benefited from another set of racial logics. The educational opportunities that enabled and motivated me to travel west were and unfortunately remain marred by racialized exclusions of so-called lower-caste and non-Hindu citizens.

While I parse apart these racialized regimes and their contexts, it is obvious, from the biographies of thousands of people, ideas, and objects over the course of the last two hundred years, that these regimes and contexts are never entirely insulated from each other. They are constantly interpellated into each other and in the process render "race" itself a far more nebulous and occasionally self-contradictory concept with a much more polymorphous political footprint than is usually granted by studies focused exclusively on Euro-American constructions of "race."[1]

A global history of non-European whiteness remains to be written.[2] We know that nineteenth-century discourses of Aryanism had a transcolonial and transregional career. Besides the European votaries of raciology, constituencies as distinct as Hindu revivalists and Maori nationalists laid claims to discourses of Aryanism.[3] But Aryanism was not always consistently pegged to whiteness. Ishita Pande, for instance, has described the "strength and flexibility" of nineteenth-century European raciology by mapping the emergence of the category of the "black Aryan."[4] By the early twentieth century this very "strength and flexibility" of European racial science permitted the "cultivation of whiteness" beyond European geographies.[5]

Besides the expected rearticulations of whiteness in settler colonial nations like Australia and Brazil, nationalized discourses on whiteness proliferated in emergent or reborn postcolonial nations like China, Iran, Lebanon, Turkey, Egypt, and the more ambiguously situated new state of Israel.[6] When Frantz Fanon, writing in the mid-twentieth century, famously spoke of people with "black skins" wearing "white masks," he was not referring merely to the brutal, colonial violence inflicted by European colonialists on non-European peoples. He was also speaking of the strength, flexibility, and subtlety of racial thought that led many non-European peoples to wear "white masks."

To be sure, these "white masks" were not just a simple, univalent guise. Appropriations of race science and racial thought were often mutually contradictory, strategic, subversive, cynical, and exclusionary all at once. As empires, at least the formal ones, gradually crumbled, their places were taken by new nations. Many of these nations were defined by a complex blend of ardently anticolonial politics, aspiring ambitions for international prestige, highly unequal societies produced by exploitative colonial regimes, powerful technocratic elites, and the lasting intellectual and institutional influence of colonialism. As Alison Bashford perceptively points out, "although historians sometimes wish and will 'anticolonialism' to meet current political benchmarks, in fact it forms part of the history to be explained." Anticolonial movements "had no necessary relation to the left or to indigenous presence, at all."[7] It was within this context that racial thought in general, and race science more specifically, flourished, often explicitly mobilized by postcolonial nationalisms.

Science or Pseudoscience?

It was long thought that race science was a nineteenth-century distortion of scientific rationality that reached its abhorrent and perverse apogee with Nazism, before finally being banished forever from the world of science. Several historians writing on nineteenth-century raciology thus referred to it as a "pseudoscience."[8] Such ascriptions of the label of "pseudoscience" have remained particularly powerful in South Asian histories of race science.[9] The conclusions, however, have not been borne out by recent historical studies.

The argument that race science was a "pseudoscience" is founded upon three interlocking assumptions. First, that there is a clear and universally acceptable line reliably dividing "science" from "pseudoscience." Latent in this assumption is that science properly so-called has a fundamental method or rationality, namely, the "scientific method," which pseudoscientists violate.[10] Yet, as Michael Gordin has pointed out, "No one in the history of the world has ever self-identified as a pseudoscientist." "Pseudoscience," Gordin continues, is "a term of abuse, an epithet attached to certain points of view to discredit those ideas." While the word is a "combative notion" and performs important political and rhetorical work in scientific disputes, it is also patently a word "without real content."[11] There is no clear line that demarcates "science" from "pseudoscience" or lays down clear, universal criteria for considering something pseudoscience.[12]

The related and implicit faith in a core scientific method or rationality that is violated by the pseudoscientist crumbles as soon as we recognize the semantic hollowness of the appellation "pseudoscience." Several historians of race science have explicitly described the proximity of race scientists to the mainstream science of their times. The pioneering historian of Latin American race science, Nancy Stepan, for instance, writes that, "though many of the scientists who studied race in the past were indeed guilty of bias in the collection and interpretation of their data, of failing to consider contrary evidence, and of making hasty or facile generalizations, few of them knowingly broke the accepted canons of scientific procedure of their day." Moreover, "scientists who gave scientific racism its credibility and respectability were often first-rate scientists struggling to understand what appeared to them to be deeply puzzling problems of biology and human society."[13] Likewise, Chloe Campbell, in her study of eugenics in colonial Kenya, writes that, "although such biologically based racial thought is now recognized by most as a profoundly mistaken dead end in intellectual history, when placed within its own historical context it was often not considered to be fraudulent or pseudo-scientific. In fact, it was considered a valid subject for respected scientists attempting to ascertain biological truths through accepted methods."[14]

Campbell's comment, in fact, simultaneously evokes the second assumption, which grounds the claim that race science is a pseudoscience: namely, that "biologically based racial thought is now recognized by most as a profoundly mistaken dead end in intellectual history." Deeply unfortunate as it may be, nothing could be further from the truth. There has been an enormous resurgence of scientific interest and investment in race in the twenty-first century. As Dorothy Roberts points out, "instead of hammering the last nail in the coffin of an obsolete system, the science that emerged from sequencing the human genome was shaped by a resurgence of interest in race-based genetic variation."[15] Race has thus been re-created upon a new, genetic foundation.

What is more, this re-creation itself was not entirely unexpected. Historians have increasingly found that the post–World War II demise of race science was grossly overstated. Lisa Gannett has argued that, rather than looking for a demise of race in scientific circles, the post–World War II transition is better seen as a replacement of one concept of race by another—a transition from a "typological concept" of race to a "populational concept" of race. This populational concept of race has subsequently fed into the human genome mapping projects that in turn have resuscitated and reenergized race for a twenty-first-century world.[16] Jenny Reardon is blunter. She asserts that "the history of race and science did not include a period of enlightenment in the middle of the twentieth century in which scientists pierced the ideological veil of race to find the category wanting for material reality."[17]

Reardon also demonstrates how the UNESCO Statements on Race of the 1950s, which are often held up as having finally sounded the death knell for race science, were themselves highly ambivalent documents. Their origin and the scientific networks that produced them were energized, in large measure, by the need to shore up America's standing on a Cold War stage where its lurid internal context of segregation and lynchings was seen to be compromising its claims to international moral leadership.[18] Ambiguity did not simply mar the birth of these manifestos. Intellectual ambivalence was written into the texts themselves. For one thing, "no consensus emerged on the fundamental issue of how rational scientific knowledge could be distinguished from the irrational forces of ideology."[19] Yet, the First UNESCO Statement's main objective was emphatically to elevate and authorize "scientific" deployments of race over and above what it viewed as "irrational" understandings of race.

The upshot of the First UNESCO Statement, far from being a dismissal of race science, was a call for its intensification. It did not call upon scientists to desist from using race, but rather recommended "that society should use scientists' understanding of race to guide their moral and social choices."[20] Notwithstanding this ambivalent stance, several scientists pushed back against even the minimal constraints that the First Statement had tried to impose, such as barring the racialized mapping of intellectual or emotional traits and the complete denial of typological races in favor of populational ones. Instead, these scientists pushed for and achieved revisions to the UNESCO Statement that essentially reopened the door to the *possible* existence of biologically definable, typological races and of race-specific emotional and intellectual traits. In the end, all that the final UNESCO Statements ended up doing was emphatically underlining the authority of scientists to conduct research into questions of race, while drawing a strong but under-defined line between "science" and "society." They simply stated that biological concepts of race "did not have any inherent fixed social meaning." This, as Reardon rightly points out, is "a very different claim from the claim that race is biologically meaningless, a claim social scientists and historians would later read into the UNESCO Statements on Race."[21]

Reardon's research also leads us to the third and final assumption that often underwrites the appellation of pseudoscience: that the concerned scientists were racists, and it was those prejudices that essentially shone through. Once again, this assumption is premised upon a fundamental misunderstanding of the key issues. As Reardon, and many others, have pointed out, many of the genomic scientists involved in the Genome Diversity Project thought of their work as "merely a scientific, humanistic, and anti-racist endeavor to understand the history and evolution of the human species."[22] These were not hollow self-delusions, either. Some of the foremost names associated with the project, such as Luigi Luca Cavalli-Sforza, had indeed publicly debated and opposed the Berkeley physicist William Shockley in the 1970s when he advocated the sterilization of African American women. Likewise, another researcher, Robert Cook-Deegan, had been involved with the organization Physicians for Human Rights, while yet another, Mary-Claire King, had used her genetic knowledge to help the Grandmothers of the Plaza de Mayo trying to locate their grandchildren who had been kidnapped during the Argentinian Dirty War.[23] None of these people were "racists."

More recently, scholars have noticed how race science was often appropriated by those who had previously been its subjects. We see similar trends in Iran, Israel, Egypt, Turkey, South Korea, and, I am sure, in many other young nation-states that were born or reborn in the mid-twentieth century.[24] What Nadia Abu El-Haj says of Jewish "genealogical scientists" studying the biological histories of Jewish communities could well be said of any of the race scientists who hailed from Iran, Egypt, Turkey, South Korea, or India. They did not approach their research as unalloyed "racists." Rather, for them race research was a form of "self study," that is, their endeavors were "biological research projects in which the subject and the object of research [were] represented as or taken to be the same," thus providing an "ethical and political alibi for those who fear the specters of race and its history of violence."[25]

This did not mean that hierarchies and power did not come into play. Of course, they did. Each of these "self study" projects classified and taxonomized the self in essentialized ways that were deeply political and frequently contested. As Elise Burton has recently pointed out, Israeli researchers who thought of their research as "self study" did not necessarily produce a science that was somehow more objective or less historically and politically contingent than those who sought to study the Other.[26]

The Feralness of Race

The difficulty of studying the history of race science is its tendency to constantly reappear in new disciplinary formations. In the nineteenth century alone, we can identify a welter of different disciplinary formations, such as comparative linguistics, ethnology, craniometry, medical jurisprudence, and comparative anatomy, where the question of race enjoyed a significant place. In the twentieth century, as "gentlemanly science" gradually gave way to more corporatized science and disciplinary boundaries came to be policed more rigorously, race still maintained the ability to be relevant across disciplines.[27] The early twentieth century witnessed researchers in sexology, tropical medicine, seroanthropology, eugenics, physical anthropology, zoology, and, increasingly, statistics explore issues of race. By the mid-century, even as some of these disciplines, such as seroanthropology, were beginning to wane, at least in certain parts of the world, other new disciplinary sites emerged. Human biology, population genetics, anthropobiology, anthropogeography, and such now became new homes for race science.[28] Later still would emerge genomics, biometrics, and other forms of algorithmic racialization.[29]

Each of these new formations mobilized new research tools, new methods, and new ways of defining race. Where craniometry had developed a whole range of tools, terms, and techniques for revealing racial identity from skull measurements, seroanthropology looked to the statistical analysis of human blood groups to determine race. Radiographers relied on x-rays to study the osteological basis of race. Genome researchers use wet-lab techniques for genetic determination together with advanced data-processing software to reveal racial groupings. The sheer plethora of techniques and methods, along with the diversity of social, cultural, and national contexts in which these disciplinary formations have flourished, makes any history of race science difficult to sufficiently delineate, let alone successfully pursue. The topic itself seems to elude analysis by its adamantine polymorphism.

One recent historian has argued that the only way to tackle the "indisciplined" polymorphism of race science is to conceptualize it as a set of historically evolving "transdisciplinary coalitions."[30] Such an approach, however, raises two problems, one conceptual and one practical. Practically, it still does not tell the historian how to bound the historical object they are interested in tracking. Conceptually, it reifies disciplinary boundaries across which "transdisciplinary coalitions" might be built.

Relocated outside Europe and its settler colonies, the feral notion of race renders itself nearly invisible by camouflaging within older and vernacular vocabularies of community and difference. Omnia El Shakry's fascinating discussion of the Egyptian appropriations of the race concept have demonstrated the two types of habitats in which race was able to find new sustenance. On the one hand, it was inserted into older terms, such as *thaqafa* (culture) and *hadara* (civilization), thereby transforming these older notions. On the other hand, it also inspired the development of new categorical terminologies, such as the notion of a "national personality" (*shukhsiyyat al-umma*) and its "ineffable quality of national character and identity" (*huwiyya*).[31] In South Asia, similarly, Luzia Savary has wonderfully described how race is able to camouflage itself in an older tradition of "progeniology," a branch of knowledge aimed at advising newly married couples about how to produce ideal children.[32]

To complicate matters further, "race" itself, as we have seen above, underwent a radical reformulation from a typological concept to a populational concept. Indeed, some scientists insist that with the mutation from "typologies" to "populations" there was also a transition from "essences" to "frequencies," and that this shift constitutes such a fundamental rupture that it is utterly unfair to see "population thinking" as a continuation of "race" properly so called.[33] A number of researchers, however, have pointed out that the shift did not constitute any clear break. Alexandra Widmer, for example, has demonstrated that typological conceptions of race continued to be used well past the alleged 1950s cutoff by famous researchers such as the Nobel laureate Daniel Carleton Gajdusek.[34] Looking beyond famous scientists and surveying a number of journals in the United States, Britain, and Israel, Snait B. Gissis found that while "race" had certainly been "reconstructed" at various junctures, there was no clear break, and the concept continued to be fairly widely used in scientific and medical literature throughout the second half of the twentieth century.[35] More recently, Veronika Lipphardt and Jörg Niewöhner have clarified that the carefully crafted public pronouncements of scientists on the issue of race do not align with everyday laboratory practice, where typological concepts of race continue to predominate for a variety of practical reasons.[36] Reardon has thus argued that it is misleading to equate race as a concept merely with the Nazi emphasis on pure races. While the Nazi conception of race was certainly undermined after World War II, that does not mean race per se disappeared. It was repeatedly redefined and re-created.[37] Going further, historians of race science in Latin America have pointed out that it is misleading to equate any avowal of "race mixture" with a disavowal of the race concept. The dominant national cultures of Latin America, some of which are premised upon claims of "racial mixture," articulate forms of racial thought where the notion of race and research into it are in fact intensified through explorations of "mixture."[38]

The daunting disciplinary prolixity of race science is thus matched by its conceptual elusiveness. Historians have mobilized a range of terminological strategies to delineate the powerful yet shifty histories of race science. Some, such as Susan Lindee and Warwick Anderson, have chosen to track the history backward from the present to ask "how humans became genetic."[39] Others, such as Gannett and Reardon, have tracked forward and tried to grasp the extent to which the older typological notions of race have lived on in new, population-based advocations of "human diversity," yet have largely avoided issues of disciplinarity.[40] Lindee and Ricardo Ventura Santos have followed a different path; cleaving closer to the seam of physical anthropology, they have adopted "biological anthropology" as the object of their investigations.[41] Writing about blood group research in Egypt, Omnia El Shakry has described "serology" as "an established field within physical anthropology concerned with the scientific determination of race."[42] Jenny Bangham, by contrast, has traced the history of research into "blood groups" in a space in-between transfusion medicine, physical anthropology, and population genetics.[43] Traveling along yet another historical and methodological path, Burton has chosen as her object "human heredity," while accommodating under it a range of disciplinary formations ranging from craniometry to seroanthropology.[44] Dissatisfied with the nomenclature offered by the actors' categories, some scholars have even delineated their own categorical objects. Abu El-Haj, for example, has outlined what she termed a "genealogical science" within which to accommodate three distinct explorations of race and heredity, whereas Anderson, in his pioneering study, has chosen to trace the history of the "cultivation of whiteness."[45]

Even this partial catalog of the diversity of categorical choices scholars have made will demonstrate both the challenge and the basic methodological fact that the objects of historical analysis, in some senses, are never pregiven. As Suman Seth writes of his choice to use the category "race science" in the eighteenth century, "there is no need to take understandings of race in the period in which race science became 'normal science' as entirely definitional of the concept."[46] Indeed, to do so is also ahistorical since it often obscures both actual actors' categories as well as historical continuities that might lurk beneath superficial terminological shifts. Historians have to choose which object to map and what boundaries to draw around it. These choices materialize particular histories by bringing into view specific connections between actors, concepts, and objects of the past. Making other choices would illuminate other connections. Instead of disclaiming these choices and rendering them seemingly automatic, therefore, I would prefer to make my choices explicit.

At the heart of this book is the category of "seroanthropology." Emerging from the putative intersection of empire and warfare during the Great War, it rapidly found a home in India and elsewhere. At its core was the assumption that the frequencies of various serological factors—initially blood groups but later a wider array of inheritable blood factors—varied by race. Notwithstanding some initial developments suggesting its impending emergence as a discipline, seroanthropology never really became a discipline properly so called. But neither was it entirely subordinated as a "mere technique" within other, better known disciplinary formations. Instead, starting in the 1920s, it remained a loose formation of techniques, personnel, and objectives that persisted until the mid-1960s in India. Conferences and publications continued to invoke the label "seroanthropology" throughout the roughly four and a half decades that this book covers. A major two-day conference organized by the Anthropological Survey of India in July 1970 to take stock of all the "bio-anthropological" research being done in India devoted one of its two largest sessions entirely to "seroanthropology."[47] Yet, it would have been easy to classify under other labels the researchers who presented at that session and the work they presented. "Bio-anthropology" and "physical anthropology" were the two labels under which the proceedings of the conference were published, though the title also mentioned "allied disciplines." "Genetics" might have been another viable option, since all the researchers mentioned genes, as also would have been "human heredity." Four out of the seven papers presented in the session dealt with medical conditions, such as the sickle cell trait, thalassemia, and glucose-6-phosphate dehydrogenase deficiency. Hence it is entirely plausible that "medical genetics" might also have captured some of the field of research. The point is that seroanthropology, by the summer of 1970, was both less than a discipline and more. It was a loose formation within which a set of research questions linking serological factors and the inheritable differences between human groups, which most contemporary Indian researchers still interchangeably called "race" and "population," were gathered up.

Those who practiced seroanthropology, by the 1950s and 1960s, were employed either at the Anthropological Survey of India (ASI) or in one of the growing number of university departments of anthropology. By the end of our period of study, other institutions, such as the Indian Statistical Institute, began to emerge as key players as well. The ASI alone published nearly eighty research papers on seroanthropology in the first two decades of its existence. Though their relations were sometimes marred by bitter institutional rivalries, particularly between the ASI and some of the older university departments, the seroanthropologists all knew each other and often had also worked together at some point or other. They also occasionally attended conferences, such as the one in 1970, and exchanged ideas. Most important, the city of Calcutta frequently provided a common physical and social context for seroanthropology. Both the oldest university department of anthropology and the ASI were based there, and several of the leading researchers, such as B. S. Guha, D. N. Majumdar, P. C. Mahalanobis, and S. S. Sarkar, had strong personal ties to the city. Consequently, despite lacking some of the marks of a full-fledged discipline, seroanthropology achieved a degree of relative coherence among its practitioners.

In tracking seroanthropology I eschew other neighboring formations, such as "dermatoglyphics" (the study of fingerprints) and "anthropological demography," which also featured at the 1970 conference. These boundaries were not watertight. Many seroanthropologists also occasionally published dermatoglyphic studies. But the methods pursued for the latter were distinct, which may well explain why they were separated into different sessions at the 1970 conference. Yet, it is important to note this exclusion, precisely because it demonstrates that seroanthropology alone did not exhaust the epistemic space covered by race science in mid-twentieth-century India.

Focusing on seronathropology permits me to grapple with the feralness of race and illuminate with particular clarity the continuities between pre–World War II, imperial race science and its postcolonial avatars. One of the papers presented at the 1970 conference, for instance, was simply called "Seroanthropology of the Indian Mongoloids." The paper included sections like "The Negrito Element" and wrote of "other Asiatic Mongoloid populations [such as] the Chinese, Japanese and Thais."[48] The author of the paper, P. N. Bhattacharjee, was a seroanthropologist with the ASI and had conducted other studies from the mid-1950s onward in which he regularly and unapologetically deployed racial categories.

The most powerful institutions in history are feral by nature. That is, they are able to run wild, live, and adapt to new and apparently inhospitable climates. Race and race science are no exceptions. Their power and persistence derive from their feral character. This in turn makes them difficult to track—to recognize and to trap by our more context-dependent and rigid analytic styles. To account for race and race science, thus, I contend that we need more supple definitions—definitions more attentive to both the words and the practices of our actors and less beholden to overarching typologies, global chronologies, and carefully calibrated self-presentations. As Seth points out, we need to pay attention to the "functioning of categorical divisions, without reifying them or regarding them as absolutes." This cannot be done merely by "blurring of extant categories," but rather requires historically mapping the "socially imbricated, tentative and complex coming-into-being of categories and binaries in the first place."[49]

Thus, race science or not-race science is not the right question. Nor indeed is it to ask if it was seroanthropology or genomics properly so called. Rather, what we should ask is how contingent, tentative categories emerged within inchoately formed disciplinary locations to shelter new, intensified, and re-created notions of race. In short, how did race survive as a feral formation even after its old haunts had been bulldozed? Only by acknowledging the feral nature of race and allowing our analytic categories the necessary flexibility can we illuminate the ways in which race has flourished in twentieth-century India, and indeed has done so in plain view and with the benediction of postcolonial Indian state science.

Failure to apprehend the persistent investment of a large section of Indian science with the troublesome and vexed issue of race has sustained a view in Indian political and intellectual circles that frequently positions science unproblematically as a progressive force promoting tolerance and social justice. Indian activists, social scientists, and humanists have often reposed their faith in "science" as the way out of divisive and hierarchical forms of human difference, such as casteism and communalism (which, in India, designates religious intolerance and prejudice).[50] Indeed, not just activists and scholars in general, but even Indian science studies scholars like Meera Nanda have argued that "modern science combines in it the power of disenchantment and universalism. It is time it was recognized, once again, as an ally of social justice, peace and advancement all around the world."[51] She looks back approvingly at the Nehruvian period as one that had cultivated the appropriate "scientific temper" in the "service of social justice and secularization."[52]

Clearly, neither the unabashed faith in science's alliance with "social justice, peace and advancement" nor the more particular nostalgia for Nehruvian science's unqualified commitment to social justice can be sustained if we take cognizance of the Indian history of race science. After all, this was not just science done during Nehru's tenure. This was science directly patronized by the Nehruvian state through institutions such as the ASI. Yet such views flourish not only among the ignorant, but also among scholars of science studies.

It is only by recognizing the feral nature of race, its ability to hide in plain sight, to constantly slip out of one disciplinary precinct and lurk in another, that we can begin to see why it has been invisible to scholars like Nanda.

An Agnotology of Seroanthropology

The invisibility of race in scholarship on Indian science is aided by the complete amnesia about seroanthropology. Despite its prominent presence in Indian research circles for at least half a century, from the 1920s through to the 1970s, the term has almost completely disappeared from public, scientific, and historical memory. It is important to account for this forgetting.

Noted historian of science and gender Londa Schiebinger has argued that historians of science need to balance their familiar interest in epistemology by simultaneously engaging in "agnotology." Adopting the conceptual tool from Robert N. Proctor, Schiebinger defines agnotology as "the study of culturally-induced ignorances." It "refocuses questions about 'how we know' to include questions about what we do *not* know, and why not."[53] Given that we are speaking here of fairly technical scientific work, rather than the knowledge about plants that Schiebinger explored, the public amnesia is fairly understandable. The scholarly and historical forgetting of seroanthropology, however, deserves an agnotology of its own.

Such an agnotology, I argue, is constituted by four interlocking scholarly aporias. The first of these aporias affects the historical scholarship on race in India. There is indeed a rich and diverse body of scholarly works that interrogate the histories of race in British India. This body of work has generally been focused on six principal sites, namely, Orientalist scholarship, state ethnography, colonial law, military, medicine, and more general social practice.

Historians like Thomas Trautmann arrived at the histories of race through their studies of Orientalist scholarship that commenced in the mid-eighteenth century. Trautmann's work has pointed out how race was initially articulated within the studies of comparative linguistics and framed by a "mosaic ethnology" that took the biblical Flood as an organizing principle.[54] Later, by the second half of the nineteenth century, these linguistic ideas of difference were supplanted by more clearly physical ideas about racial difference.[55]

With this shift, colonial knowledge production moved from an eighteenth-century preoccupation with ancient texts and philology to a post-Mutiny "Orientalist empiricism" that preferred ethnography over philology.[56] Mutations of colonial liberalism also informed this ethnographic curiosity and enabled the invention of "traditional societies."[57] Indeed, the colonial state in the second half of the nineteenth century has even been called an "ethnographic state."[58] Recently, Christopher Fuller has described the ethnographic knowledge produced during the so-called "high noon of empire" as "official anthropology."[59] This new ethnographic imperative produced the most blatantly racialized administrative frameworks by means of the anti-Thuggee initiatives and the various Criminal Tribes Acts and Regulations.[60] While it remains doubtful to what extent the actual assignment of criminality to particular groups was a discursive novelty, it is clear that the practical effects of such administrative practices have outlived empire.[61] The figure of Sir Herbert Hope Risley and his anthropometric researches, alongside his work as Census Commissioner in 1901, have also received considerable historical attention, resulting in Risley becoming something of an exemplar of colonial race science in scholarly works.[62]

Colonial law has been another fertile site for inquiries into the histories of race. Legal historians in general have explored explicitly racially discriminatory legal instruments as well as the various forms of racialized outcomes in colonial courtrooms.[63] Particularly revealing have been the so-called "boot and spleen" cases in which racialized understandings of "the Indian body" were used to acquit British culprits who had beaten Indians to death.[64] In partial contrast to these relatively straightforward accounts of racialized laws and interpretations, historian Mitra Sharafi has illuminated a much more complex and ambivalent range of operations of race within the colonial legal apparatus. She has demonstrated, for instance, that racial assumptions about "native mendacity" operated within a colonial legal machinery marked by tacit legal pluralism[65]—or, in a different context, that racial assumptions generated a degree of tacit legal toleration of abortion, even when it had been legally outlawed.[66]

Like the bureaucracy and the courtroom, the colonial army, too, has provided historians with illuminating histories of race. Particularly important here has been the so-called "martial races" theory, which promoted the military recruitment of certain groups of South Asians, such as Sikhs and Gurkhas, while barring others, such as Bengalis, from employment in the army.[67] Going beyond recruitment, recent historians have further demonstrated how racial assumptions marked every aspect of colonial military service, from wartime rations to the medical aid provided to soldiers.[68] Medico-racial logics prevalent within the colonial army also allowed a set of violent and racialized practices to develop in the civilian spaces within which colonial armies were located during peacetime. This was particularly the case with the notorious "lock hospital" system, which sought to cater to the sexual needs of underpaid, British subaltern soldiers by providing access to South Asian sex workers.[69]

In fact, the historiography of medicine, even beyond the focus on military medicine, has produced some excellent insights into the ways in which race and empire have mutually constituted themselves. Mark Harrison's work on changing ideas about the "seasoning" of temperate "constitutions" in "tropical climates" has provided a powerful potential explanation for why South Asia did not become a settler colony.[70] Verily, "tropicality" became a major intellectual tool in medical circles for racializing diseases and patients. Disease entities like "tropical diabetes" were putatively constructed through the racializing logic of tropicality.[71] Indeed, Seth has argued that the emergence of a notion of "tropical diseases" was closely entangled with the development of race medicine within a British imperial context marked by the slave trade.[72] Medical institutions, such as mental asylums, were also racially separated and frequently worked on highly racialized logics.[73] Moreover, both medical practice and medical knowledge came to be racially coded, as categories like the "Black doctor" or "Aryan medicine" appeared both self-evident and de rigueur.[74] Even anatomical dissections and physiological descriptions were pressed into demonstrating the obvious truth of race.[75]

Finally, several historians have explored how race shaped the many disparate spheres of colonial social interactions. Mrinalini Sinha's pioneering work showed, for instance, how the complementary myths of the "manly Englishman" and the "effeminate Bengali" were constitutive of the political culture of the Raj.[76] Sinha's intervention also highlighted the intersection of race and gender as co-constitutive axes of imperial power. Elsewhere, she has explicitly gone beyond significant political events and excavated instead the "set of class, gender and race relations" that enabled British imperialism by tracking imperial notions of "clubbability."[77] Satoshi Mizutani has similarly explored the racial anxieties of the "Domiciled British" community in colonial India. His work has been particularly important in highlighting the ways in which educational institutions and child-rearing practices were racialized.[78]

Notwithstanding the unquestionable richness of this scholarship, the vast majority of this work focuses on the period before the Great War and looks mainly at race as an idiom operating to reinforce the colonial divide. Relatively few works have looked at the interwar and post-Independence eras. In the few cases where they have looked beyond World War I, scholars have tended to focus principally on the place of race in various South Asian traditions of nationalism. Christophe Jaffrelot, for example, has explored the place of race in Hindu nationalist writings of the interwar period. Javed Majeed has delineated the place of race in the Muslim nationalism of Muhammad Iqbal.[79] Dagmar Hellmann-Rajanayagam has explored the place of race in the Dravidian Movement.[80] Shefali Chandra has explored the role of the intersecting and mutually reinforcing axes of race and gender in Hindu and moderate Indian nationalisms.[81] Very few scholars have looked for race beyond the textual ambit of nationalist ideologies. Harald Fischer-Tiné's investigations of the role of physical culture within an explicitly articulated racial framework among Hindu nationalists, and Sumathi Ramaswamy's accounts of the entanglements of archaeology and classroom geography with ideas of a Tamil race are rare exceptions.[82]

Compared to this robust and variegated historiography on race in colonial India, there is practically no scholarship on race in post-Independence India—despite the enormous scholarship on caste throughout the twentieth century. Indeed, several of the scholars who have written on caste have been explicitly concerned about denying the equation of caste and race.[83] Such denials emerged in India mainly in the context of the attempt by Dalit activists to argue at the World Conference against Racism, Racial Discrimination, Xenophobia and Related Intolerance organized by the United Nations in Durban, South Africa, in 2001. This is not the place to pursue the politics of such refusals to equate race and caste, but it is pertinent to point out that such positions were built on an inability to think of race beyond the typological, Nazi conception thereof.

Science has always been somewhat peripheral to histories of race in South Asia, with the majority of scholars labeling the scientific investments in race "pseudoscience" and concentrating mainly on ideological aspects of race.[84] Even this limited and peripheralized attention to science practically disappears once we move beyond the Great War. This allows historians of eugenics in South Asia to suggest that in South Asia, unlike elsewhere, eugenics was mainly a matter for amateur enthusiasts rather than for scientists.[85] Illuminating and insightful as they are, Fischer-Tiné's account of physical culture and Ramaswamy's account of archaeology also look primarily at how scientific notions were taken up and circulated beyond scientific circles. As Ramaswamy puts it, her account sits at the "intersection of professional archaeological interpretations of the remains, and the putative 'wild' and 'nonsensical' speculations of the public intellectuals."[86] Scientific research into race therefore remains invisible in this historiography.

A very different though complementary aporia emerges in the small body of contemporary ethnographies of race and science in India. These works have done a great deal to bring to light the new and complex technical and public work that is being done by geneticized notions of race. Particularly significant here is the work of anthropologist Yulia Egorova. Starting from the attempts by various Indo-Jewish communities to obtain genetic verification for their Jewishness, Egorova has increasingly explored the much wider issue of the kinds of authority and politics that are being articulated around genomic research in India.[87] Sociologist Subhadeepta Ray has similarly explored the entanglements of molecular and national life being constituted in Indian genetic laboratories.[88] In a more highly theorized vein, Kaushik Sunder Rajan has described the crucial importance of India within the tight international networks of "biocapital" through which the human genome came to be mapped and "postgenomic life" assembled.[89]

Once again, seroanthropology—or, indeed, the mid-twentieth-century fields of practice that assembled the institutional, intellectual, and human infrastructure for these late twentieth- and early twenty-first-century scientific developments—remains conspicuously absent from these studies. To be fair, these analyses are produced within fields such as anthropology and sociology, and not history. Yet, the absence of the longer genealogies of scientific engagement with genetically conceived ideas of human heredity willy-nilly contributes toward the forgetting of both seroanthropology in particular and race science in India more generally.

A third aporia can be diagnosed in the histories of Indian anthropology. Like the foregoing sociological and anthropological works on science, this is a small and still emergent field. Happily, it has already produced some excellent scholarship, among which an edited collection titled *Anthropology in the East* takes pride of place.[90] Yet, neither this capacious volume nor the few works that have been published on the subject affords much space to seroanthropology, or indeed physical anthropology more generally.[91] Indeed, writing of Irawati Karve, someone whom we shall meet repeatedly in this book, Nandini Sundar declares that, while Karve was in fact influenced by the "German physical anthropology tradition which attempted to provide a genetic basis for the existence of a variety of groups," in her case "fortunately, this was shorn of its racist implications."[92] And yet Karve remained throughout her life an active collaborator with L. D. Sanghvi, a leading geneticist who had actually participated in the drafting of the UNESCO Statement on Race that reformatted race in populational terms.

The final aporia that constitutes the erasure and invisibility of seroanthropology involves the historical accounts given by contemporary Indian human geneticists. Unquestionably the foremost institution that has provided the intellectual and material scaffolding for Indian researches into human genomic variations in recent times has been the Indian Genome Variation Consortium (IGVC). The Consortium began as the Indian Genome Variation Initiative in 2003 and was initially "tenured for 5 years." Its principal constituents were the six laboratories that make up the Council for Scientific and Industrial Research (CSIR), along with other institutions such as the Indian Statistical Institute, Kolkata (ISI) and the ASI. The IGVC originally sought to map genetic variations within the Indian population in order to explore differential susceptibilities to disease and responses to drugs.[93]

Central to this task was the initial conceptual separation of the whole Indian population into constituent "subpopulations." The project stated that "the vast majority of the people of India (≈80%) belong to the Hindu religious fold. Hindus are hierarchically arranged into four sociocultural clusters of groups (castes) and there are set rules governing marriage within the Hindu religious fold. About 8% of the population is constituted by tribals, who are ancestor worshippers and are largely endogamous. The remaining belongs to other religious groups, including Muslims, Christians, Buddhists, Jews, etc. Primarily, marriages occur within the religious groups. In addition, language and geographical location of habitat serve as barriers to free gene flow."[94] Practically every aspect of this statement is historically questionable. More pertinently, the assumptions about caste and religious endogamy that underwrite the statement were developed by mid-century seroanthropologists within a specific historical context. This is precisely the history we will explore in this book.

Similarly, the IGVC asserts that "all the four major morphological types—Caucasoid, Mongoloid, Australoid and Negrito are present in the Indian population."[95] (See figures I.1 and I.2.) Once again, notwithstanding the avoidance of the word "race," the names of the four groups,

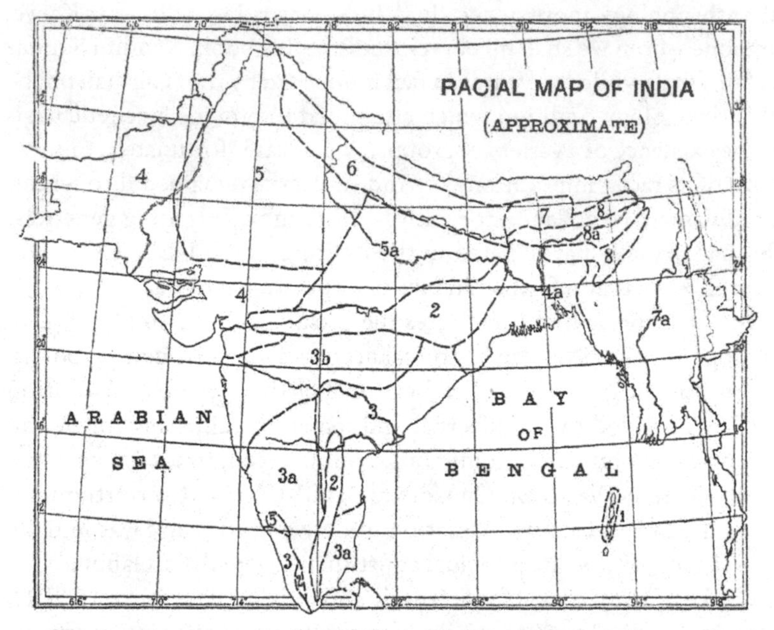

TEXT-FIG. 2.—Racial Map of India (approximate).

1. Negritos; **2.** Proto-Australoids and Negritos; **3.** 'Basic' Dolichos and Proto-Australoids; **3a.** 'Basic' Dolichos Alpo-Dinarics and Proto-Australoids; **3b.** 'Basic' Dolichos, Alpo-Dinarics and Proto-Australoids with small elements of Orientals and Proto-Nordics; **4.** Alpo-Dinarics and Orientals with a sprinkling of Proto-Nordics; **4a.** Alpo-Dinarics and 'Basic' Dolichos with small proportion of Indus and Proto-Nordic types; **5.** Proto-Nordics and Indus with some amount of Orientals; **5a.** Indus and Proto-Nordics with 'Basic' Dolichos as a small element; **6.** Orientals and Brachy Mongoloids; **7.** Brachycephalic Mongoloids; **7a.** Short Brachy Mongoloids; **8.** Dolichocephalic Mongoloids; **8a.** Dolicho-Mongoloids with a small proportion of 'Basic' Dolichos and Alpo-Dinarics.

Fig. I.1. B. S. Guha's Racial Map of India, 1937, along with a key showing the distribution of racial types. Source: Guha, *An Outline of the Racial Ethnology of India.*

affirmation of their presence in India, and even the regional concentrations outlined by the IGVC generally cleave close to the insights of earlier seroanthropologists.

The historical narrative that informs this project is highly telling of how the legatees of the earlier tradition of seroanthropological research recall their past. One of the scientists most prominently associated with the IGVC and who has often published synoptic histories of the research he and his peers are now engaged in is Partha P. Majumder, a professor of anthropology and human genetics at the Indian Statistical Institute. He usually begins his narratives from "about the turn of the [twentieth] century," omitting entirely the work done in the nineteenth century. This was when "physical anthropologists" began to "systematically" study the "physical characteristics of Indian peoples." The "overt goal of most of the studies conducted until about 1950," Majumder continues, "was typological—to classify the peoples of India into a number of 'racial' types." Thereafter, "over the course of the next few decades such typological studies increasingly fell out of favor, and physical anthropologists turned their attention to quantifying variability and studying relationships in specific regions such as Uttar Pradesh, Gujarat, Maharashtra and Bengal."[96]

While Majumder clearly deploys the familiar trope of suggesting that the shift away from typologies to populations was a significant rupture, he does not present this rupture as a sudden and complete break. Instead, he says only that the older studies gradually "fell out of favor" over the course of the "next few decades." Even more interesting are the exemplary works he cites in each group and their respective dates of publications. Among the typological studies, Majumder refers to the work of Risley (1908), B. S. Guha (1935), and S. S. Sarkar (1954). In the

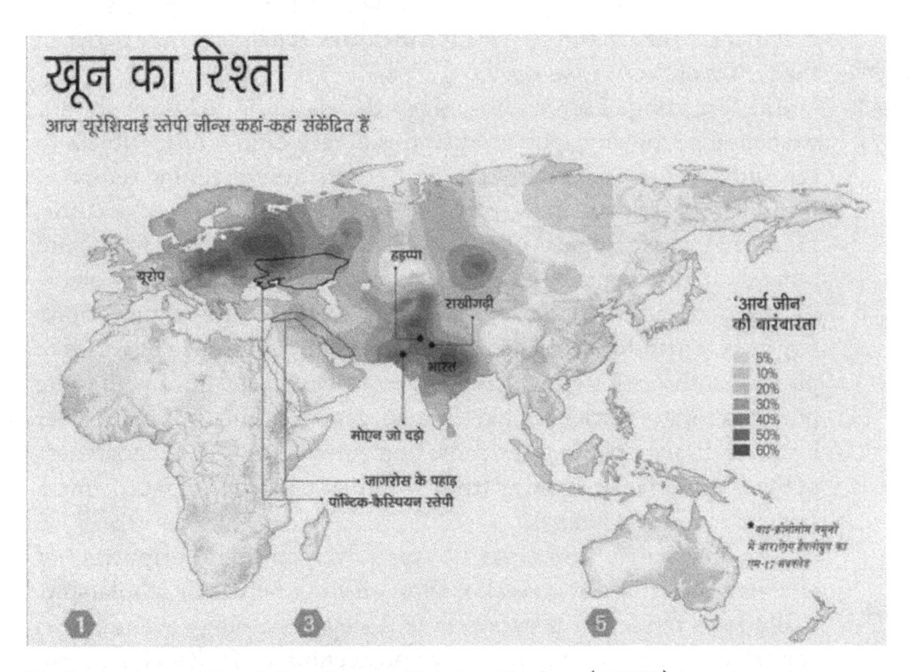

Fig. I.2. Projects such as Indian Genome Variation Database (IGVDb) have given rise to a popular discourse on genetics and race in contemporary India that is reminiscent of the colonial-era discourses. This map, published in 2017, shows the "Frequency of the Aryan Gene [*Arya jeen ki barambarta*]" and is captioned "The Ties of Blood [*khoon ka rishta*]." Source: Aaj Tak, https://www.aajtak.in/india-today-hindi/.

latter group of studies that quantified and studied relationships, Majumder cites D. N. Majumdar's work from 1949 and Irawati Karve's work from 1951. Even if we leave aside other difficulties with drawing a clear line between the earlier typological studies and the later populational ones, the chronological conundrums bear out Majumder's statement that the transition was gradual and lasted for decades.

Driven as such accounts are by a need to connect the oldest researches with the latest ones along a linear chronology, any intellectual formation that existed in the middle decades of this story can only be made sense of in terms of its resemblance to either the older, eclipsed studies or the then-emergent ones current today. Their independent existence, which might share features with both older and newer research paradigms without aligning with either, cannot be fully acknowledged. Historians of chemistry, for instance, have outlined how early modern "chymistry" was similar to and yet distinct from the "alchemy" that went before and the "chemistry" that came later.[97] Likewise, seroanthropology was an intellectual formation that was neither Risleyan anthropometry nor indeed contemporary genomics. But this autonomous existence cannot be assimilated within the linear, progressivist chronologies of practitioner-written histories. It falls precisely between the planks of the old and new.

The forgetting of seroanthropology thus is fabricated through four overlapping scholarly aporias. Histories of race cannot bear witness to seroanthropology's existence due to both its overwhelming temporal focus on the era before the Great War and its general neglect of scientific practice. Anthropological and sociological scholarship on genetics and human variation remains equally oblivious of seroanthropology owing to its general lack of interest in the longer past. Histories of anthropology miss seroanthropology as they generally peripheralize histories of physical anthropology, including anthropology's natal links with more purely laboratory sciences like human genetics. Finally, practitioner histories, though perhaps the most perceptive about the inchoateness of the alleged transition away from typological notions of race, remain trapped in linear timelines.

Recalling seroanthropology not only overcomes these aporias but also serves to connect several of these distinct bodies of scholarship. It allows us to see, for instance, how Risleyan raciology eventually is connected to current discussions of postgenomic life. Recalling seroanthropology repositions the history of anthropology itself by locating it within a genealogy that includes not just sociology, but also human genetics. These connections, along with the political and intellectual moorings that enabled them, can only be glimpsed once we bring seroanthropology out of the historical dark room.

The Double Gesture of Histories of Seroanthropology

Any history of seroanthropology that seeks to recall it from the historical oblivion to which it has been relegated must perform a double gesture, namely, to recover and to critique. I borrow the notion of history as a "double gesture" from the sociologist of science Amit Prasad. In seeking to understand why histories of successful bricolage and tinkering with technoscientific machines are constantly erased by Indian scientists in favor of newer and updated diffusionist narratives, Prasad invokes Jacques Derrida's notion of a "double gesture." Such a "double gesture" must overturn a traditional concept history, while simultaneously marking the interval in a way that forestalls its reappropriation. In the process, a "discursive clearing" can be created where new histories of "hierarchical and polycentric circulations of technoscience" can be written.[98] Thus, to prevent the reappropriation of any recovered histories into linear, teleological, progressivist, and diffusionist histories of science, we must also embed the recalled past in its own historical milieu. The work of such embedding is precisely what critique does. In so doing, it acknowledges the messy and hierarchic networks that interlace the global and the local, while also etching out the fractured nature of the "local" context itself.

British Indian soldiers were a very significant part of the first seroanthropological study, conducted in Greece during the Great War by a Polish scientist couple. It was this study that launched seroanthropology. Later, Indian scientists such as Sanghvi worked at the heart of the research networks that revamped human genetic research for the post–World War II world. Indeed, Sanghvi was the scientific secretary to the influential UN commission that was tasked to study the potential genetic impact of the atomic bomb. Through Sanghvi and his early mentor, V. R. Khanolkar, Indian notions of caste segregation, which suggested that there could be biological isolation of populations even when there were no geographic barriers to intermixture, became a key feature of much postwar genetic research. Indian subjects, data, concepts, and researchers thus formed a significant part of the story of mid-twentieth-century researches into race and human variation. Yet, most standard and critical histories of the subject almost entirely ignore the Indian presence. Reardon's excellent history of the UNESCO Statements on Race, for instance, mentions the presence of Indian scientists in the group, but says nothing more about them.[99] Gannett's work dwells on two scientists that Sanghvi worked closely with—Theodosius Dobzhansky and L. C. Dunn—but makes no mention of Sanghvi himself.[100] Only Lipphardt is awake to Sanghvi's and Khanolkar's place at the heart of genetic research networks. But she overlooks both their Indian networks and the centrality of notions of caste endogamy to their work.[101] This leaves Sanghvi and Khanolkar appearing as precocious outliers, rather than as a bridge that connected a vibrant Indian research scene, with its own investments in racialization, to a Euro-American one.

To recall this history is both tempting and risky. It can easily be recuperated, on the one hand, into a universalist history of science that ignores hierarchies and political ambiguities of mid-century genetic research. On the other hand, it can also, with equal ease, feed into stories of Indian nationalist chest-thumping. Neither of these alternatives is likely to illuminate the contradictory political and intellectual stakes of Indian seroanthropology with any degree of insight. It is to forestall such reappropriations that the double gesture is necessary.

Writing about antiracism and its checkered history, anthropologist Ghassan Hage has recently remarked on the bivalent sense of the word "recall." Drawing upon Bruno Latour's writings, Hage notes that the word "combines both recalling in the sense of remembering to ensure we build on past achievements, but also recalling in the way a company recalls a product when it realises it has some defect."[102] The double gesture of recall is similar. It simultaneously remembers and critiques.

Throughout this book I recall the forgotten history of Indian seroanthropology to attest to India's important place in the emergent networks of human biological research in the twentieth century, as well as to bear witness to the troubling political imaginaries and unequal social relations that enabled such scientific participation. While my focus, unlike much of the historical scholarship on race in India, is squarely on scientists, their ideas, and their practices, following Banu Subramaniam's felicitous metaphor, I simultaneously trace the "fibrils of science" that "seek support structures, and thus come to scale political and social scaffolding in different contexts."[103]

Given the remarkable lack of any significant archival culture to retain the private papers of scientists or major research projects of the past in India, writing the kind of "thigmotropic" histories of science that Subramanian calls for seems daunting at first. Yet, it is even more remarkable to see how many of the fibrils and tendrils can actually still be discerned from available published scientific works. Strategies of "reading against the grain," pioneered by historians of peasant insurgency to read colonial police reports, serve historians of science admirably in spotting trails of friction and refusal in the field from dry scientific publications, even when a more fulsome archive is missing.[104] It is mainly through such strategic readings of a wide and extensive corpus of published reports of seroanthropological researches that I recall, in its dual senses, the history of race science in mid-twentieth-century India. But I also supplement such a reading with a set of readings of race in more fictive literary archives. This latter approach allows me the space to reflect more directly on the elisions whose traces are no longer discernible in the published research.

Critical Fabulation and Plurigeneric Pasts

Writing histories of race and inequality immediately brings the historian face to face with the limits of the archive. Indeed, Fanon himself confessed that in writing *Black Skin, White Masks,* "I have not wished to be objective. Besides, that would be dishonest: It is not possible for me to be objective."[105] The limits of the archive are particularly palpable for historians of science in South Asia. Private papers and unpublished laboratory records are seldom preserved or accessible. Despite my best efforts, which included trying to trace living descendants of scientists and contacting the institutions they had worked in, I have not been able to trace any significant collection of private papers relating to several of the major figures who populate this book: Eileen Macfarlane, S. S. Sarkar, B. S. Guha, and L. D. Sanghvi. All these figures enjoyed significant international scientific reputation and thus were distinct from the many other, relatively minor and more obscure researchers who occasionally appear in this book. Yet, whatever fragments survive usually exist as parts of other people's correspondence or, more commonly, as documentation submitted to satisfy bureaucratic requirements. As Eram Alam has so ably demonstrated, the "documentary subjectivity" that emerges in response to such bureaucratic requirements is not only an artifact fundamentally shaped by power relations, but also a tool for the production of racialized identities.[106]

Being faced with an archive of the race scientist's carefully curated public presentations of their selves and their researches threatens to overwhelm the historian's critical apparatus to "overhear" other pasts.[107] Any attempt to tell the story of those who were objectified through the racializing discourse is doomed to feeling like an "untimely story told by a failed witness."[108] This is not merely a matter of archival gaps. As Ashis Nandy has pointed out, the discipline of history itself might be implicated in some of the most troubling and violent aspects of modernity. Indeed, Nandy sees both "science" and "history" as part of the same totalizing aspiration of modernity that is dismissive of alternative ways of being in the world.[109] Recent critiques of science's role in colonial violence have also often shared Nandy's mistrust of historicism and the discipline of history.[110] Feminist approaches to science and technology, studies (STS) have also pushed back against the narrative monoculture that narrowly historicist depictions of the development of science have engendered.[111]

These perceptive critiques have inspired a radical rejection of the objectivity claimed by the discipline of history. Instead, these scholars have increasingly turned toward myths, stories, and legends.[112] Nandy himself had explicitly advocated writing "alternative mythographies" rather than histories.[113] Yet, in the context of majority nationalist polities, such as those of South Asia, this is also a politically dangerous exercise that has been blamed for unwittingly reinforcing xenophobic, patriarchic, majoritarian nationalist mythologies.[114]

A somewhat different approach has been developed by historians of gender and Atlantic slavery, who have long struggled with such limits. Saidiya Hartman, writing about the enslaved women who died during the Middle Passage, has proposed "critical fabulation" as a strategy. Critical fabulation is a method that engages in "thinking with a twofold attention" that can encompass both the "positive objects and methods of history" and the "matters absent, entangled and unavailable by its methods."[115] It is, in effect, a mode of fabulation grounded in rigorous archival research. Instead of "recovering" lost voices or "redeeming" the dead, all that critical fabulation aims for is "paint[ing] as full a picture of the lives" of the objects of racial discourse and archival erasure as possible.[116] In the process, it also hopes to "illuminate the contested character of history, narrative, event, and fact, to topple the hierarchies of discourse, and to engulf authorized discourse in the clash of voices."[117] Such a strategy seems much more adaptable to South Asian history than the kind of strategic essentialism of mythography and indigeneity.

Yet, it also raises some further practical dilemmas. First, does the pitting of factual history against fictional story adequately capture the critical narrative choices available to us? South Asian historians have, for instance, described the wide array of complex vernacular forms of history-writing that have developed on the subcontinent since the nineteenth century.[118] Second, "fiction" itself is engendered in myriad distinct genres that make particular claims about objectivity, history, cosmology, and so on.[119] A science fiction novel, for instance, makes very different representational claims than does a social realist novel. But even these seemingly obvious generic distances are in themselves fluctuating historical artifacts. As I have shown elsewhere, at the turn of the twentieth century, and under the gloomy pall of colonial modernity, the demarcation between science fiction novels and theological writings became exceedingly porous.[120] This in turn raises the question of whether cosmological or theological writing is or is not to be categorized as "fiction." How should we deal with such science fiction/theology in the work of critical fabulation?

While these questions might be moot for the context Hartman writes about, for us, such questions are unavoidable. Writing about a past that bristles with a plurality of factual, fictional, and transfactual genres, not only do historians of South Asia have a wider array of narrative choices, but they are also confronted with a more dispersed archive. Where the traditional archives of the state or statist science end, there begins a more voluptuous vernacular archive of literary and nonliterary texts and traces. This does not mean that the latter archive simply supplements or fills in the gaps of the former. Rather, they sometimes seem utterly unrelated and disparate. This seeming disparateness in fact is an artifact of the larger processes of authorization, which extricates and elevates certain voices from within the "clash of voices."

To reconnect these seemingly disparate archives and their overdetermined genres, we need to repopulate a frame of social life that actively refuses the restrictive presumptions of disparateness between factuality and fiction, science and literature. Methodologically, this would mean rethinking how we historians provide "social context" for our critical histories. Rather than the more widely available view of the "social" as a pregiven moral community that conforms to common rules, we need to think more expansively of a possible "social" grouping that is united by much more rudimentary intersubjective interactions based in historical propinquity and common language competence. This is precisely the view of the "social" espoused by Gabriel Tarde, now resurrected by Bruno Latour and others, which had been marginalized due to the dominance of Émile Durkheim.[121] One way of extending the Tardian notion of "the social" as a rudimentary network of individuals held together by mutually imitative actions, such as speaking the same language, to historical determinations of social context is to look more expansively at the spatial and linguistic milieu of our historical actors.[122]

This in turn serves to reconnect the scientific and the literary voices that shared the same geographic, historical, and linguistic spaces. Literature, written in the same place, at the same time, and in the same language as science, immediately becomes an archive in the history of science. But this archive needs to be handled with care. Too many historical works disregard the moral, aesthetic, and epistemic distinctiveness of literary genres and mine literary works for morsels of usable information. Such usage, to me, is counterproductive. Rather than extending the methodological advances of critical fabulation, it flattens the literary archive to the same epistemic constraints as the state's documentary archive.

In order to optimize the potential offered by a networked notion of the social and the literary archive as an instrument of critical fabulation, I would argue that we need to draw not only upon the contents of the literary archive but also upon the generic possibilities offered by such archives. The critique of historicism that I recognize most in the writings of Nandy and others is not so much about historical methods as such, but rather about the historicist hubris that history alone has the exclusive authority to illuminate the past. Much of this criticism can be addressed by restoring the plurigeneric character of the past. Indeed, did Fanon himself not turn to poetry, theater, and epic poetry to develop his radical insights?[123] We never approach the past exclusively through "histories." Memories, anecdotes, myths, legends, novels, and much else mediate our relationship to the past. If historians recognize, acknowledge, and incorporate such generic diversity in their own writing, they will, I believe, be much better placed to practice critical fabulation.

To me, then, critical fabulation entails not the seamless blending of fact and fiction, but rather the careful braiding of multiple genres. Such generic choices themselves, I further hold, need to be grounded in genres that were available to the historical actors who produced the documentary archives. By neither denying the historically meaningful generic and archival distinctions between fact and fiction, or science and literature, nor allowing these calibrations to overwhelm our critical projects, I believe, we can ramp up the work of critical fabulation. Such a critical fabulation enacted through plurigeneric narratives about the past can both address the epistemic limits of standard disciplinary history and avoid the politically dangerous collapsing of all lines between "fact" and "fiction."

Once we align plurigenericism to critical fabulation, we find earlier models for similar multigeneric critical endeavors, though naturally with different ends in mind. Tarde, for instance, besides writing sociological works also penned a science fiction novel.[124] The novel form provided Tarde with an alternate narrative format for imagining the relationship of the past, archives and social forms, without having to adopt the generic and terminological constraints of established social analyses.[125] Yet, it is equally important that he wrote in both these genres, without collapsing their differences. Fanon, despite his genre-bending critical texts like *Black Skin, White Masks*, did much the same by keeping his plays, academic essays, and critical texts separate from each other. Closer to home, a clearer example of the past in multiple genres is available in the oeuvre of the pioneering Bengali historian, Rakhaldas Bandyopadhyay (also known as R. D. Banerji). Although he is mainly remembered today as the discoverer of the Indus Valley civilization, Bandyopadhyay also wrote three historical novels, all set in ancient India. The novels allowed him to narrativize usable pasts for a nation-in-the-making without surrendering the disciplinary rigors of archaeology and history. They acknowledged the limits of disciplinary histories and also cleared a space to transcend them by embracing the fictional genre.[126] Later, Amitav Ghosh famously turned to the novel in order to explore pasts he had already developed in a historical essay written out of fragmentary archives.[127] More recently and at a much smaller scale, Dipesh Chakrabarty has incorporated the ghost story genre into a monograph on the history of history-writing in India, while Rochona Majumdar has deployed the epistolary form to reframe the history of Bengali cinematic modernity.[128]

In my turn here, like Majumdar, I turn to the epistolary form, especially in the way it was often embedded within novels. My immediate inspiration, in terms of both the generic form and the "overheard" pasts I evoke, is a Bengali science fiction novel published in 1935. Written by Hemendrakumar Ray (1888–1963), one of the most successful authors of early to mid-twentieth-century Bengali genre fiction, the specific novel in question, *Amanushik Manush* [Inhuman Humans], is a particularly generative model for several reasons.[129] As literary scholar Bodhisattva Chattopadhyay points out, the novel is recognizable as a work of science fiction not simply because of what it calls science within its narrative but "more specifically because it is framed within a cluster of science fiction tales that allows us to identify it as part of a genre."[130] It traffics not only in multiple familiar science fiction plots, but also invokes a number of other authors and narratives, both European and Bengali, which help locate the text generically.

What particularly attracts me to the work is that Ray's novel shared many of the thematic concerns that occupied Bengali and Indian sero-anthropologists at the time. In Chattopadhyay's words, Ray "takes up the themes of degeneration, eugenics, supermen, primitivism, myth and history in one sweep as they relate to the Bengal context."[131] Yet, he subverts these themes through a self-reflexive discourse that confounds totalizing discourses of race, science, progress, and so forth. Chattopadhyay rightly points out how, distancing himself from some of his own earlier attempts to nationalize imperial narratives and genres, in *Amanushik Manush* Ray explicitly calls out the "racism of the Indian towards the African . . . the

Bengali towards other communities of India" by a "self-critical gesture revealing the silliness of such narratives of superiority."[132]

Although the novel is a bit of a generic mélange, one of the key elements that simultaneously drives the plot and creates a dissonance within it is the use of the epistolary format. The original narrator writes a letter communicating the narrative of a diary he has found in the pocket of a dead man. The diary writer's voice is often critical of race science, but it is ironically communicated through the letter-writer who seems to embrace casual racism. This multiplication of narrative voices, and the innate tension between the two, helps forestall any clear identification between the reader and the narrative. It also serves to convey in print something of the ephemerality of "overhearing," undisciplined orthogonal possibilities engendered in anonymized "words one has heard [*shona kotha*]."[133]

In *Brown Skins, White Coats* I draw on both the epistolary format and much of the putative narrative of Ray's novel. But I also amplify his strategy of using the letters to invoke other works. I draw almost entirely on Ray's own fairly large oeuvre, but I weave in other plots and stories into the broad plotline for *Amanushik Manush*. I embed these letters further within Ray's social world, referring to events and people from that world.

Ray's perception of his social world, and particularly the socially marginalized populations of the city, was particularly acute. Pseudonymously, he authored in the 1920s an ethnographic account of Calcutta's shadowy netherworld. The work was based on decades of deliberate, intimate socialization with his less privileged neighbors and was intended to humanize them.[134] His humanist eye provides a moral foil for the scientific objectification through seroanthropology that we encounter in the chapters of this book.

The simultaneous overlap of concerns, social worlds, languages, and objects of representation between Ray and the seroanthropologists (many of whom were also Bengalis or worked in Calcutta-based institutions like the ASI) draws our attention to two simple facts. The first is the existence of a Tardian "social" network that connected these alienated representational economies. The seroanthropologists almost certainly knew of Ray and possibly even read his works. Yet, there is no trace of him, his concerns, or his critiques of race science in their publications. Second, Ray's humanism, built upon a form of intimate observation that at times resembles anthropological fieldwork, might offer an alternative way of engaging and representing human difference. Indeed, in Ray's attempts we might catch sight of an early experiment in what Paul Gilroy calls a "post-anthropological" form of "planetary humanism."[135] I replay Ray's postanthropological experiments against the histories of seroanthropological race science that I document in an effort to pry open the critical, plurigeneric possibilities at the interstices of disparate and mutually alienated archives.

One further strategic move I make is to attach a name to the originally anonymous narrator of Ray's novel. At first this might seem to violate the aesthetic of overhearing. Yet, the act of naming itself, for me, is inspired by overhearing Ray's own social life. I choose to call the narrator Najrul Islam in my letters. My choice is informed by Ray's close friendship with the famous Bengali poet, Kazi Nazrul Islam (1899–1976).[136] Ray not only considered Islam as a younger brother, but also said of him that no one epitomized the spirit of the age in Bengal more than Islam.[137] The Najrul of my letters is not the historical Nazrul. But they certainly resemble each other in many personal and biographic features. The name also recalls a memorable friendship across the religious divide that split Bengal in 1947, and serves to remind the unwary reader that there is no necessary reason to assume that the unnamed narrator of Ray's original novel was a Hindu Bengali, as is so often done. My act of naming is thus intended to reopen a set of ephemeral, but very real, socialities of Ray's world that seem to have evaporated since.

Structure of the Book

The structure of this book may strike readers of historical monographs as being somewhat unusual. There are two parallel narratives that run simultaneously but apparently independently through the book. The first, a factual and historical one, is developed through seven independent chapters, making a set of recognizably historical case studies of seroanthropology. The second, the work of fabulation, is carried out betwixt and between, in what I have called "interchapters." The eight interchapters consist of letters, which are relatively short, interleaved between the chapters.

The chapters themselves are thematic. Chapter 1 recalls the development of seroanthropology and its arrival in British India. It is particularly attuned to the mutual transformations that are necessitated by the marriage of seroanthropological technique and caste society, and outlines the broad contours of the 'biometric nationalism' that emerges through such transformations. Chapter 2 moves from caste to religious communities. Focusing largely, though not exclusively, on seroanthropological studies of Indo-Jewish communities, the chapter tracks the cross-fertilizations between imaginations of caste and religion that eventually produced a geneticized notion of isolated Mendelian populations. The chapter also evokes the importance of exogeneity, as a foil to the articulations of genetic indigeneity. These first chapters thus delineate the basic conceptual terrain and the social field within which seroanthropology operated.

The next two chapters explore the expansive reach of seroanthropological research. Chapter 3 dives into attempts to racialize the sense of taste and explores the practical problems in defining taste objects and human phenotypes based on such objects. Connecting histories of race with sensory histories, this chapter describes the gradual, though inchoate, marginalization of sensory and affective stimuli in favor of genetic traits only fully knowable by researchers through experimentation and statistics. Chapter 4 follows a similar line, though in a more tragic vein. It maps the way the sickle cell trait, rather than sickle cell anemia specifically, was racialized in India. The chapter demonstrates that while most of the critical scholarship on race and sickle cell has been preoccupied with narratives of African origin, in India the sickle cell trait was mapped onto particular lower caste groups, which were also often considered among the most "primitive" components of the Indian population.

Chapters 5 and 6 take a different path, where the scientists are relatively decentered. Chapter 5 opens up the central material object of seroanthropological research, namely, blood. Following Annemarie Mol's exposition of the "body multiple," the chapter distinguishes among three types of blood with close attention being paid to the terms of their access, availability, and usage. In the process, this chapter, much more than the others, moves away from what scientists thought or wrote to what they did and how they did it. By taking up the use of caprine blood, sourced from Calcutta's slaughterhouses, for the production of key testing sera, the chapter also reveals the complex ways in which human biological truths came to be entangled with animal tissues. Chapter 6 leans even further away from the intellectual worlds of the seroanthropological scientists. While chapter 5 provides some broad outlines of the kind of frictions that marked the attempts of researchers to draw blood from their subjects, this chapter restores these subjects and their worldviews to narrative centrality. By exploring the cosmological, praxiological, and political motivations that led to particular groups or individuals refusing to give their blood to seroanthropologists, this chapter goes beyond the framework of resistance that continues to privilege the scientist's views and initiatives. The chapter also offers an alternative narrative of subcontinental human variation that is not premised on biological difference. Especially through the account of a Catholic theologian's refusal of genetic determination, the chapter also explores the historical dynamism of refusals that responded to developments in biogenetic truth-claims by updating and reformulating the grounds of refusal.

Finally, chapter 7 zooms out and back to the scientists again. This chapter tracks the shifting and programmatic investments in the future by four Indian practitioners of race science. Commencing with the early formulations by Sir B. N. Seal that influenced some of the stalwarts of the first generation of interwar Indian race scientists, the chapter moves through the futurities imagined by S. S. Sarkar, Irawati Karve, and L. D. Sanghvi. By recalling their distinctive ideas about the future, the chapter simultaneously outlines the speculative nature of seroanthropological work and fully explores the successive reformulations of the race concept in Indian scientific circles. By directly invoking the futurity of scientists, this chapter also reminds us that speculation is not merely a facet of literary fabulation. It is a much more pervasive feature of knowing and living in the world.

Finally, the conclusion returns to Fanon. It relocates what the chapters show us through the Indian history of seroanthropology within Fanon's paradigm of alienation. It also draws upon the Fanon-influenced scholarship of Paul Gilroy in order to propose a new chronology and geography for the emergence of what Gilroy called "nano-politics." The conclusion, therefore, is particularly committed to situating Indian seroanthropology within the tradition of post-Fanonian and anti-essentialist critical writing on race, liberation, and selfhood.

INTERCHAPTER

Letter 1

Najrul Islam
Kampala
5 November 1933

Dear Hemenbabu,
Please forgive me for this unsolicited letter. We have not met, but I have been an avid reader of your books for over a decade now. I was but a boy in 1919 when I read your first book. Since then, wherever I have traveled, I have left standing instructions with my bookseller in Calcutta to send me your books as soon as they were published. Indeed, I might even say that it was your novels that have, in a way, led me to my current predicament.

It was your books that inspired in me a love of action and adventure. I was nauseated by the sloppy sentimentality that generally passes for Bengali literature. Nay! I was angered by it. Is it a wonder that the world smirks at us Bengalis? What else can we expect if our boys grow up reading such maudlin rubbish? Your books inspired me to turn my back on such insipid nonsense and instead seek out adventure, to look danger in the eye and challenge destiny. I wished to do so not merely for my own pleasure, but also to show the world that we Bengalis are no whimpering cowards.

As a man of means and leisure, I was fortunate enough to be able to pursue my noble quest. Naturally, *shikar* was my sport of choice. I roamed from one forest to the next hunting down every big game known to man. And, in case you wonder, no, I do not hunt in the modern fashion from the safety of an aircraft. That would be a travesty. I do it the old-fashioned way. Rifle in hand, and on foot.

In a few years I had exhausted all the adventure I could find in our vast Indian forests. It is true that our Sunderbans are beloved by every major hunter in the world. But after a while, the tigers, crocodiles, and so forth cease to challenge one. It becomes too easy and degenerates to petty slaughter, rather than being the noble *shikar*. Bored, I tried to distract myself by reading. Something I don't much enjoy. I wrote to the agents of Thacker, Spink & Co. enquiring about books on *shikar* and enclosed a cheque. By return post, I received a book by Major William Robert Foran, *Kill: Or Be Killed*.

I tell you Hemenbabu, I had never read anything like it. I, who crave action, romance, and adventure and do not waste my time in reading, that very me, finished the book in one sitting. I realized that no *shikari* can really prove his mettle till he has hunted big game in Africa. In Foran's reminiscences I glimpsed the true nobility of *shikar*. Where beauty and fatal danger jostle together and where a man comes face to face with nature in all its majestic fury. Where else and how else might a man prove the true nobility of his character than to confront this beauty and this brute power face to face and prevail upon it? I decided I had to go to Africa.

You must be wondering why I am telling you all this. Please bear with me. I know you are a busy man, and I do not wish to waste your time. Indeed, I hate people who waste time. This brief background is necessary for the matter upon which I am writing to you.

Having arrived in Uganda and recruited an experienced group of native coolies, I decided that the best place to start my adventures would be in the forests around Lake Kivu. I had heard both of the natural beauty of the place and the prospect of hunting gorillas in the area. Even the most accomplished hunters agree that hunting a gorilla requires the greatest skill and courage. Yet, they are easier to track than some of the other, bigger African game. So naturally, I thought it would be best to start my sojourn with a gorilla hunt.

Unfortunately, the very day we reached the banks of Lake Kivu, I came down with a nasty bout of malaria. For a Bengali to come down with malaria in Africa seemed almost like bumping into an old friend in a faraway land! Be that as it may, the malaria laid me low and forced me to camp on the banks of Kivu till I recovered. It was here that two remarkable incidents occurred.

Both incidents have shaken my belief in humanity, progress, and man's destiny. I no longer know what to believe and why. I have lost my confidence and my sense of values. I feel utterly at sea and am no longer sure of what to do next. This is why I am writing to you. Having read your books, I feel that you, of all the writers, poets, and philosophers of our country, would be in a position to understand and guide me.

The first of these incidents occurred even as I was recovering from my bout of malarial fever. It started with my finding small items of food going missing from my tent every night, as I slept my fevered sleep. Enquiries amongst the coolies brought forth vigorous denials of their having pilfered the stuff. Eventually, along with the *sirdar* of the coolies, I set a trap to capture the food thief. After a couple of failed attempts, we did manage to briefly capture the thief.

But alas, the success did not last long. The thief bit my coolie *sirdar* viciously and called to her comrades. We were stunned to realize that the thief, stealing food in this impenetrable forest, was in fact a young teenage girl. But imagine our astonishment when, in reply to her cries, there emerged from the bamboo thickets an enormous posse of gorillas. The pack of animals behaved more like a well-formed human army of the sort I have seen on parade. These brute beasts, acting in concerted and self-disciplined fashion, rescued the thief from amidst us. Notwithstanding our numerical strength, our modern guns, and, above all, our much-vaunted human intelligence, we were left utterly defeated and vanquished.

My coolies later told me that some fifteen years ago the gorillas had abducted a one-year-old girl from a neighboring village. Since then, rumors have been rife in these parts that the gorillas have raised her as their daughter and that she thinks of the gorillas rather than the humans as her family. The people of the jungle-villages near Kivu speak of her as Tana, the Daughter of the Gorillas.

Can you believe this, Hemenbabu? A human child raised as a gorilla! A human child who loves and lives with these animals as her kinsmen and attacks humans with all the viciousness of an animal? Are these gorillas the animals I wished to hunt? How can they be so human? I am increasingly confused. What is Tana today? Is she a gorilla or a human?

The second incident followed only slightly later but has rattled me even more. Shaken by my experience with gorillas, I thought I would focus on hunting a lion. Indeed, there were rumors of a man-eating lion a little farther along, and I thought here the moral lines that defined my quest would be more neatly drawn. As I followed the lion's trail into the Ugandan forests, however, things became utterly bewildering. Members of our camp first refused to proceed farther on the pretext that a certain small range of mountains ahead was cursed and inhabited by Jujus. Naturally, I scolded them for their laziness and superstitions, and emphasized that there is no such thing as a Juju. Yet, events around us confounded my reasoning.

Initially several of the coolies reported seeing strange barrels rolling about by themselves after dark. These barrels seemed to have bright eyes peering out of them and prehensile limbs that could be stretched and contracted at will. I too was eventually confronted by these barrels one evening while taking a walk outside camp after sunset.

Even as my retainers refused to move farther and began to gradually desert my camp, I remained adamant that I would bag my lion and solve the mystery of the barrels. I will spare you the agonizing couple of days it took me to bring things to a conclusion. Suffice it to say, that I left no stone unturned in my bid to achieve my dual ends. Finally, I did succeed in shooting the lion. Though mortally wounded, however, it refused to die and withdrew into its lair in the very mountains that my coolies feared. I followed him alone to make sure that he was dead.

By the time I reached his lair, the lion was indeed dead as the proverbial dodo. But in his lair, I was surprised to find a possible key to the other mystery that haunted me. I found within the lair the remains of a man long dead. A victim of the carnivore, no doubt. Amongst his scattered belongings however, I found a diary.

The tragedy of the man's death, for me, was amplified manifold when I realized upon opening the diary that he had been a Bengali like you and me. His name was Amal Sen, though I have not been able to find any further information about him. More surprising still were the bizarre narrative recorded in his diary. He spoke of a mysterious and secretive civilization that has made its home in these hills.

The denizens of the grand secret cities there, it seems, are an advanced race. Their ancestors, again like us, had been Bengalis, and they continue to use our beloved mother tongue. But they have given up all else that their ancestors held dear. Ever since they were shipwrecked on the shores of Africa, this small group of Bengalis have used science and technology to completely reinvent themselves and their way of life.

With new and advanced scientific techniques that are beyond the wildest dreams of even Acharya Jagadish Bose, they have managed to completely re-create their bodies and minds. They no longer possess bodies like you and me. Instead, they are like jellyfish and live in barrels, somewhat like snails. Upon the need arising, however, they are instantly able to take on any physical form, including ours, in the blink of an eye. They can also extend or shrink their bodies as necessary. According to Mr. Sen's diary there are a million such changes they have wrought upon themselves. So much so that it is hard now to think of them as humans anymore!

I know most people will think I am losing my mind, but believe me, Hemenbabu, I am as much in possession of my sanity as you are. I can also vouch that Mr. Sen's descriptions are not figments of a crazed mind. I have myself seen these barrels and enough of their powers to know that even though I have not had the experiences that Mr. Sen did, his descriptions are more than plausible.

I will be happy to communicate the narrative to you in greater detail if you wish. Indeed, I would also be happy to send you the diary itself. But, Hemenbabu, I beseech you to please guide me. Confronted both by Tana, the Daughter of the Gorillas, and these barrel-shaped future-humans, I am no longer certain what is what. Who is a human, and what is nature? What is it that makes us human? Where is the line that divides the human from the animal? Is it right for scientists to play God and reshape our future being?

The doubt that assails me is crippling. I hate to be thus confounded and unable to act. If I do not find a way out of this moral morass, life itself will become a burden. I do sincerely hope you will be able to show me a way ahead.

Please do write back at your earliest.

I remain,
Yours truly,
Najrul Islam

CHAPTER 1

Seroanthropological Races

> This awareness is especially acute since the bourgeoisie is incapable of forming a class. Its organized distribution of wealth is not diversified into sectors, is not staggered, and does not nuance its priorities. This new caste is an insult and an outrage, especially since the immense majority, nine tenths of the population, continue to starve to death. The way this caste gets rich quickly, pitilessly and scandalously, is matched by a determined resurgence of the people and the promise of violent days ahead. This bourgeois caste, this branch of the nation that annexes the entire wealth of the country for its own gain, true to its nature, but nevertheless unexpectedly, casts pejorative aspersions about the other blacks or Arabs, which recall in more ways than one the racist doctrine of the former representatives of the colonial power.
>
> —Frantz Fanon, *The Wretched of the Earth*

The Great War changed the British empire forever. Two of these changes were to define the career of race science in twentieth-century South Asia: first, the emergence of the new discipline of seroanthropology; second, the rapid Indianization of the scientific services in South Asia. These two developments in the 1920s prepared the ground for the blossoming of biological researches into human difference in the subcontinent.

Seroanthropology owed its origin to a short, four-and-a-half-page article published in the *Lancet* in October 1919.[1] Titled *The Serological Differences between the Blood of Different Races* and authored by Ludwik and Hanna Hirszfeld—a Polish Jewish husband-and-wife team who had spent much of the just-concluded war in Greece supporting the Serbian nationalist militia—the paper relied on the then still fairly novel concept of human blood groups.[2] The Hirszfelds argued that the frequency with which each blood group appeared in any racialized "national type" was constant .

Discovered by Karl Landsteiner in 1901, blood groups were still a novel and insufficiently understood concept. The Rhesus factor—that is, the "positive" or "negative" groupings—was still unknown. Even the basic ABO terminology had not yet stabilized, and alternative classificatory systems, such as the MNP, the I, II, III, and so forth, were still in use by some researchers. This choice itself was important since different classificatory systems tracked different antigens in the blood and consequently grouped different sets of human beings together.[3] Based on this choice, however, the Hirszfelds created a new "biochemical race" concept that could be neatly represented, mathematically and visually, in a Biochemical Race Index (fig. 1.1).[4]

The Biochemical Race Index allowed the Hirszfelds to resurrect the nineteenth-century debates about human polygenesis. They declared

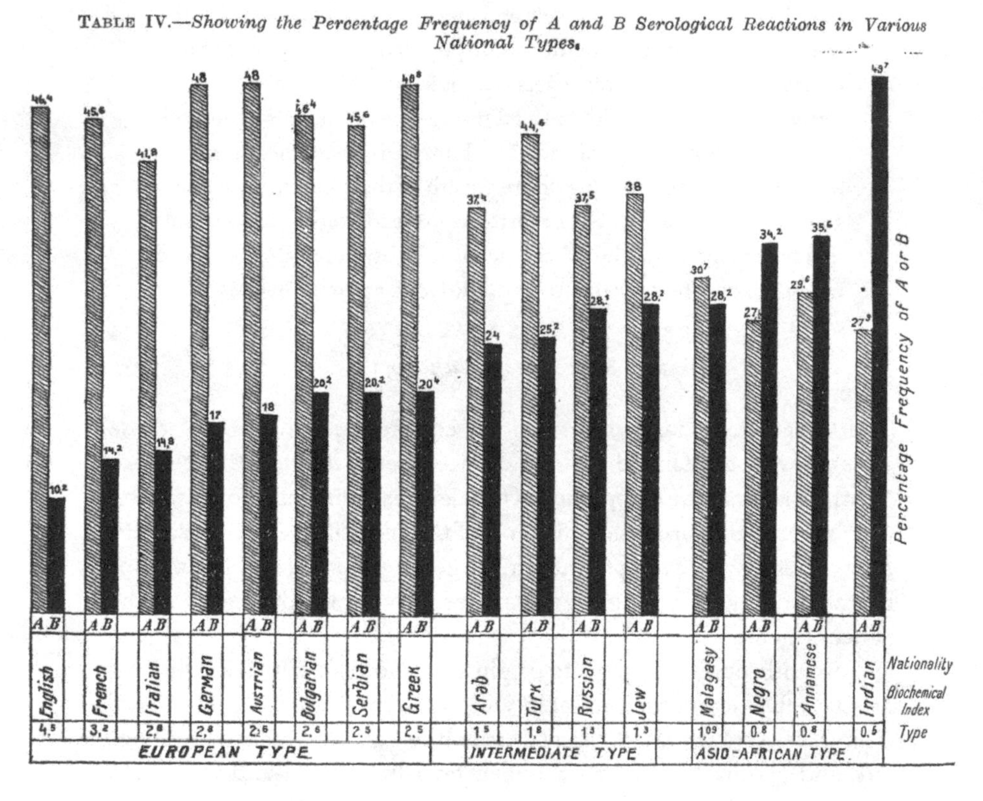

Fig. 1.1. Hirszfeld and Hirszfeld's Biochemical Race Index. Source: Hirschfeld and Hirschfeld, "Serological Differences between the Blood of Different Races," 678.

that "it is very difficult to imagine one single place of origin for the human race in view of our statistics, since it would then be inexplicable why A diminishes from west to east and south, while B increases."[5] From our perspective what is perhaps most remarkable is that the graphical representation of the Biochemical Race Index was bookended by the "English," on the one hand, and the "Indian," on the other. Indians had the highest percentage of blood group B. Next to them, but with lower percentages of blood group B, were such groups as the "Annamese," the "Negro," the "Malagassy," and so on.

The Hirszfelds' article would go on to whelp an entirely new discipline of science combining anthropology and serology. Within a decade "seroanthropological" studies would become widely popular and give a new lease of life to racial thinking. One study, focused exclusively on Britain, the United States, and France, calculated that at least 1,158 articles on seroanthropology were published in these three countries in the two decades between 1919 and 1939.[6] Notwithstanding occasional criticisms by researchers based in Europe or the United States, seroanthropological researches also quickly proliferated across much of the colonized world in the interwar decades. As Elise Burton notes, the "alluring simplicity" of seroanthropological assertions of racial affiliation led to the quick uptake of seroanthropology by both foreign and local researchers throughout the nations of the Middle East.[7] Similarly, Jaehwan Hyun describes how in the Far East, Japanese researchers led by the serologist Furuhata Tanemoto quickly established similar seroanthropological research programs in colonial Korea.[8] Researchers in British India were no less enthusiastic about seroanthropology.[9] Writing mainly about the Portuguese empire, Ricardo Roque argues that the "exponential growth" of seroanthropological literature in the interwar years was "fed by winds of nationalism and scientific racism."[10]

What further spurred Indian researches was the researchers' explicit disagreement with the way the Hirszfelds had described the racial identity of "Indians." Having drawn their samples from the British and French imperial troops stationed in Greece, they had included data on the blood groups of an undifferentiated group of "Indians." To Indian researchers, such homogenization of multiple different subcontinental communities was deeply problematic. Writing jointly in 1929, Major Reginald H. Malone of the Indian Medical Service and Dr. Mahendra Nath Lahiri, a Bengali physician, expressed their strong disagreement in an article in the *Indian Journal of Medical Research*. "Their theory is based on a complete misconception of India and Indians," declared the authors, adding that "it is curious that this lack of knowledge about the great diversity of races constituting the population of India has not been pointed out by writers who have supported or criticized the Hirschfelds' theory."[11]

The content of this pioneering study was as important as the structure of its authorship in anticipating what was to follow in the subsequent decades. The end of the Great War had brought about a rapid Indianization of the scientific services in India, thus creating the space for men such as Lahiri to obtain prestigious research positions within the imperial state.[12] The Hirszfelds themselves had commented on meeting "several highly educated and cultivated Hindu [*sic*] physicians" in Greece.[13] Recognizing this aspect of interwar science in British India is crucial to understanding how and why interwar trends remained so powerful in postcolonial India, despite the watershed of decolonization.

Malone and Lahiri collaborated on several research projects, including working side by side on the high-profile Bacteriophage Inquiry Committee (1927–39). Misunderstandings about the nature of Indian participation in interwar scientific research have led some historians to misconstrue the nature of Lahiri's involvement in these projects. William Summers, for instance, has erroneously assumed that Lahiri's being an "Indian physician" meant that he was "essentially ineligible for the Indian Medical Service."[14] In fact, there were several South Asians in the IMS, not to mention the subordinate medical services, by the late 1920s. Indeed, the Hirszfelds met some of them during the Great War as well. Rather than indicating some racialized exteriority, Lahiri's nonmembership in the IMS signaled the emergence of a new pathway for biological and even medical research in dedicated research institutes that were no longer principally linked to clinical medicine.

Both Malone and Lahiri, notwithstanding their medical degrees, were principally research scientists in a way that was rare for earlier physicians in British India. The novelty of their career paths becomes clear by even a cursory perusal of their biographies. In 1923 Malone officiated as the assistant director of the Pasteur Institute at Rangoon. In 1925 he became the acting assistant director of the Bacteriological Laboratory in Bombay. The following year, he became the acting assistant director of the Central Research Institute at Kasauli. Likewise, Lahiri after earning his MB (bachelor of medicine) degree from the University of Calcutta, was employed in the Public Health Laboratory in Patna in 1929. Subsequently, he earned a DTM (doctor of tropical medicine) degree from Liverpool University and joined the Medical Research Department of the Government of India, where he rose to the position of officiating assistant director of the Central Research Institute at Kasauli, the same position Malone had held earlier. In 1950, he was appointed an associate professor of microbiology at the All India Institute of Hygiene and Public Health in Calcutta. Six years later he was made a full professor at the same institution. In their careers, therefore, we witness a new type of physician who was principally employed in research institutes. Their collaborative research, including the seroanthropological work, needs to be located within this context.

Race was no longer a matter that concerned only the ethnographer-bureaucrat or the clinical physician, as it often had before the war. Instead, it became an object of inquiry within new research institutions and was pursued by research scientists. This trend would grow and evolve further after Independence in 1947.

This is another reason to take note of the character of the collaboration between Malone and Lahiri. There is evidence to suggest that Lahiri, though he was the junior scholar and the second author on the article, was in fact the principal actor in the seroanthropological work. Whereas Malone's publications were all on microbiological research and he published nothing else on blood groups, Lahiri published another single-authored article in the same year and in the same journal as their co-authored piece, on the medico-legal applications of blood groups.[15] Lahiri's interest in the subject also underlines the way South Asian researchers, and not just British officers, took up seroanthropology early on. Moreover, it was these Indian researchers who, in their later leadership and professorial roles, institutionalized race science in postcolonial India.

Many of the features we notice in Malone and Lahiri's collaboration were also to be seen in another prescient and influential study on race done in the 1920s. A chance meeting during the 1921 Nagpur session of the Indian Science Congress had brought together Nelson Annandale, a talented Scottish zoologist who was then the director of the Zoological Survey of India, and a young Cambridge dropout who had been studying for a physics degree, Prasanta Chandra Mahalanobis. Though Mahalanobis had left Cambridge without a degree when the Great War broke out, he had taught himself the then-novel tools of statistical analysis.[16]

At the time, Annandale had been studying mixed-race Anglo-Indians in Calcutta using older anthropometric methods. Mahalanobis, eager to demonstrate his new statistical tools, offered to statistically analyze Annandale's measurements. This resulted in another jointly authored paper in which, using his new tools, Mahalanobis claimed to demonstrate that the mixed-race population was mostly derived from lower-caste Indians, rather than upper-caste ones.[17]

While Annandale's untimely and sudden death closed off further collaborations between the two, Mahalanobis continued to be deeply interested in the statistical possibilities for determining issues of racial origin. Indeed, in 1925 Mahalanobis was elected president of the newly formed Anthropology Section of the Indian Science Congress. For a man with no background in anthropology whatsoever, this was a remarkable ascent. He pursued this interest along two lines. First, he reanalyzed and validated the anthropometric datasets collected by Sir H. H. Risley in the late nineteenth century.[18] Second, he personally mentored and used his own enviable social connections to enable some of the most significant seroanthropological surveys of the following decades.[19]

Major seroanthropological researchers, such as the Cambridge-educated Dhirendra Nath Majumdar, owed their ability to mount major seroanthropological surveys and eventually to build highly successful academic careers in large measure to the early, enthusiastic support of Mahalanobis. The latter's involvement also injected new statistical tools into seroanthropological research. While the uptake of such tools remained patchy beyond those projects which were directly connected to Mahalanobis and his circle, in the long run they improved sampling techniques and error calculations. Above all, however, Mahalanobis's interest in the subject meant that by the end of the 1960s human genetic research had found a new and influential home in the Indian Statistical Institute, the flagship institution built by Mahalanobis. As Sayori Ghoshal has recently written about Majumdar and Mahalanobis, they "produced modern, scientific knowledge around racial difference, racial origins and mixture of blood lineages of Indian communities, even as they contested certain European race science notions, such as the desirability of racial purity or the scientificity of skin color as evidence of race."[20]

Thus, in these collaborations we find two important origins: in Malone and Lahiri's collaboration we have the beginnings of seroanthropology proper, and in Annandale and Mahalanobis's collaboration we notice the origins of powerful national traditions of race science that would eventually be crucial in creating both the intellectual and institutional basis for the furtherance of seroanthropological inquiries.

From Head Count to Blood Count

Race science remained a vibrant and fast evolving pursuit through the 1930s, but few South Asian researchers at the time published seroanthropological studies. Most continued to depend principally upon updated versions of osteology and craniometry in this research and those young South Asian researchers who did take up seroanthropo-

Fig. 1.2. Eileen Macfarlane. Source: Macfarlane, *Eileen Macfarlane: Lecturer, Traveller, Scientist* (flyer, 1946).

logical work only began publishing their results in the next decade. This, however, did not mean that seroanthropological inquiries were forgotten in the 1930s. In fact, a remarkable number of publications and research projects date from this decade. The drive and energy behind the vast majority of these projects came from a single remarkable woman who lived in British India for a little less than a decade. Her name was Eileen Macfarlane (fig. 1.2). A genetic scientist by training, Macfarlane built upon Malone and Lahiri's critique of the Hirszfelds, and undertook more detailed studies of serological differences within British India.

Born Eileen Jesse Whitehead in London in 1899, she moved to Ann Arbor and joined the PhD program in botany at the University of Michigan in the early 1920s.[21] She obtained her PhD in 1928 for her work on the genetics of wild roses of America. She published extensively in the following years and on the basis of these publications obtained a second doctoral degree, a DSc, from the University of London in 1934. In 1933 she traveled to India as a visiting professor in genetics at the Maharaja's College of Science in the princely state of Travancore. The reasons behind her seeking out this position seem unclear, but a combination of professional and personal circumstances might have pushed her to do so. Whatever might have driven her to Travancore, once there she became fascinated by the human diversity in the region and commenced some of the earliest studies in seroanthropology. In 1934 she married a British employee of the Burma Shell Company then stationed in India, thereby also acquiring the surname Macfarlane. She was also able to subsequently find some funding from the University of Michigan for an ambitious project titled "Blood Groups and Types in Castes and Tribes from Malabar and Madras to the Borders of Tibet." While in India, Macfarlane was also active in promoting genetics in Indian academic circles. After a failed effort to organize an International Genetics Congress in India, she guest-edited a special issue of the official journal of the Indian Academy of Sciences, *Current Science*, on the subject in the hope of stimulating interest among Indian scholars.

Three aspects of Macfarlane's work in the 1930s are worth underlining here. First, she frequently collaborated with young Indian scholars. Her first study of blood groups in the princely state of Cochin, for instance, was done in collaboration with P. Narayan Menon,[22] and her work on the blood groups of the Bagdis in Bengal was done in collaboration with Sasanka Sekhar Sarkar.[23] Her final and most comprehensive work on blood groups in India, published in 1941, was also co-authored with Sarkar.[24] Sarkar would later become a professor at the University of Calcutta and one of the most influential scholars of race in South Asia. Second, Macfarlane managed to publish an enormous amount in the eight years (1933–41) that she spent in India. Though, mostly in keeping with the trend at the time, her sample sizes were relatively small, her work covered many different groups across a wide swathe of southern and eastern India, thus creating a baseline of work for future researchers. Finally, Macfarlane was one of the few researchers to argue that the lower castes were biologically much closer to each other than to the upper castes. Eschewing the ethnographic logic that studied particular groups as discrete entities, she clearly drew upon the political language of the decade to argue that the lower castes (Dalits) could all be biologically grouped under the category "Depressed Classes" rather than be separated into a cornucopia of regionally specific groups.

This final point is worth belaboring. Macfarlane repeatedly invoked the "apparent homogeneity of blood group mixture in the Depressed Classes right across the Deccan and into Bengal" precisely at a time when leaders such as Dr. B. R. Ambedkar were seeking to establish a common, pan-Indian Dalit identity.[25] It is also significant that many nationalists and imperialists refused to admit such a shared pan-subcontinental identity for Dalits. Even liberal-minded "official anthropologists," such as E. A. H. Blunt, thought that such a pan-Indian identity for "Depressed Classes," though it might be politically expedient, could not be anthropologically justified.[26] As late as in 1946 many still believed that a leader such as Dr. Ambedkar would have no political support in Bengal, precisely because the specific caste he belonged to did not exist in Bengal.[27] Macfarlane's claim of a shared biological identity cutting across regions and uniting the "Depressed Classes" of the Deccan with those of Bengal is therefore particularly poignant. Even more telling is Macfarlane's use in her article of the phrase "Adi Hindu," or "Original Hindus," to designate these Depressed Classes.[28] The term emerged as part of a concerted effort made by lower-caste intellectuals, such as Swami Achyutanand, since the 1920s. Adi Hindus adopted organizational tactics from Christian missionary and Hindu reformist organizations in order to advocate for a wider, transregional identity of lower castes and untouchables.[29] The Adi Hindus also rejected any affiliation with caste Hindu society and claimed that "Dalits were the autochthonous inhabitants of the subcontinent who had been conquered by invasion and subsequently subjugated to Brahmanic rule."[30]

In advocating for this deeper Adi Hindu/Depressed Class identity, Macfarlane clearly went beyond the researches of Malone and Lahiri. The latter pair had mostly used older categories suggested by H. H. Risley, namely, Turko-Iranian, Indo-Aryan, and Dravidian, to designate the diversity of the subcontinental population.[31] Macfarlane, in contrast, rejected and criticized such "lumping." She studied each group discretely, but then pointed out the similarities between them as a way of positing a deeper, shared identity.[32]

Even more significant for the long-term development of race science in South Asia was Macfarlane's robust repudiation of any link between blood group A and Aryanism. Ever since the Hirszfelds, blood group A had been associated predominantly with Europeans and group B with Asians, and particularly Indians. In the South Asian context, where there had developed since the nineteenth century a hoary tradition of assimilating the upper castes, and particularly upper castes from the northern regions, with Europeans into a category called "Aryans," this meant that the blood group B became a marker of lower caste status. Malone and Lahiri emphasized this position: "The Dravidian type has the highest frequency and the Turko-Iranian the lowest of B; the frequency of B decreases as we travel up the Ganges valley to the North-West Frontier . . . These findings are quite in accordance with what might be expected from the views expressed by Risley with regard to the origin and relationship of these types."[33] This statement was all the more remarkable because the researchers also noted that tracking the distribution of group A in their data seemed to contradict this. But they remained content to bracket off this contradiction by merely saying that "no explanation of the discrepancy can be offered at present."[34] They ended their article somewhat coyly by proposing that this contradiction might mean that groups A and B had both originated somewhere in "India or in some other portion of the great Indo-African continent."[35] Macfarlane, by contrast, embraced this possibility wholeheartedly and developed a hypothesis about both the principal blood groups having developed in different parts of the subcontinent and among the various "tribal" peoples. She argued that "there is in general an increase in the frequencies of genes A and B from the south northward and in group B and AB from east to west across Central India. It is suggested that there may have been two original racial stocks one resembling the Paniyan or Maler, with little of B, and the other resembling the Oraons, with little of A and plenty of B."[36] Both "Aryans" and upper-caste Hindus thus became incidental to the historical origins of the blood groups. This position naturally further reduced the biological importance of any group, such as the upper-caste Hindus who claimed to have entered South Asia from beyond,[37] thereby underscoring the Adi Hindu position in favor of a biological autonomy and autochthony of the Depressed Classes.

To fully understand the import of Macfarlane's studies, and especially her position on caste, it is crucial to situate her work within the larger social and political context of 1930s South Asia. Following the Great War, both "politicised caste and 'casteism,' as well as religious communalism, were strengthened during the inter-war period."[38] Whereas the last prewar official census, taken in 1911, had focused on studying religious differences and reduced the emphasis on caste, the postwar dyarchy reforms that expanded the electorate, together with the emergence of large-scale mass politics around M. K. "Mahatma" Gandhi, inserted caste back into the vision of the colonial state's official anthropology. Indeed, as Fuller argues, although categories like "caste" were unquestionably reified by the official anthropology of the colonial state, the officials progressively "lost control" of these census categories after 1909 as they were politicized by an increasingly involved public.[39]

The politicization of census categories became particularly striking at a series of Roundtable Conferences organized between 1930 and 1932 to discuss the devolution of political power. Several groups, including Muslims, Depressed Classes, Anglo-Indians, Indian Christians, and domiciled Europeans, began to demand separate electorates. The idea of separate electorates had emerged in municipal elections since the late nineteenth century and was essentially a form of electoral democracy where constituencies were organized "according to types and classes rather than areas and numbers."[40] Originally intended to defuse inter-community conflict and provide a kind of proportional representation, the separate electorates over time significantly hardened social fractures.

More important, the existence of separate electorates now raised difficult questions about counting. Who was to be counted in which category? Were "Depressed Classes" Hindus or not? Was there any basis for grouping a number of socially disadvantaged groups in different provinces under a single label? Many of those who could be classified as "Depressed Classes" often had religious practices that resisted easy categorization as "Hindu," "Muslim," or even "Christian." How then were these people to be counted?

From the point of view of the Indian National Congress political party, the constituency of the Depressed Classes was the most challenging. Accepting a separate social and political identity for the numerous Dalits (lower castes) would substantially damage the claim of the Congress to be the main representative of the largest number of colonized South Asians. Hence, the Congress did everything in its power to convince the Dalits to remain within a common electorate. Ambedkar, the leader of the Dalits, eventually struck a deal with Gandhi in 1932. Known as the Poona Pact, this agreement provided that the Dalits would give up their demand for separate electorates in favor of reserved constituencies. In this latter system, while the candidate had to be from the Depressed Classes, the voters included everyone living in the same territory.

In the long run this system would have dramatic effects. Dwaipayan Sen has shown how this system operated in the 1940s to keep Dalit politicians who were not members of the Indian National Congress out of crucial elected bodies, even though they enjoyed large support bases among the Dalits. This in turn led to the Congress being able to significantly control the process of decolonization and the shape of the postcolonial Indian state.[41]

The 1930s, apart from other things, were crucial in working out the relationships among Dalits, upper castes, and the state. As Ian Duncan points out, "in the 1930s, the implications and consequences of counting and classifying the population acquired an additional and more immediate political significance in the context of the impending constitutional reforms. As a consequence, challenges to official taxonomy became urgent and better organized."[42] While even the simplest classificatory and enumerative schemes had always brought forth myriad difficulties for the so-called "ethnographic state," the new stakes and the better organization of the challenges made it clear that the officials no longer controlled the census categories.

There is a rich and ever-expanding literature on social enumeration in South Asia. One of the main points of contention concerns whether precolonial states counted populations or not.[43] This debate has been related to the debate over whether the colonial state significantly transformed, or even invented, certain social identities. This preoccupation with "did they count" or "didn't they," however, has obscured any significant interest in *how they counted.*

The literature has also failed to take note of the ways in which head counts, or "human inventories," to use Norbert Peabody's phrase, gradually gave way to blood counts. Building on earlier forms of anthropometric counting, such as craniometry, seroanthropology developed robust traditions of blood group counting as a way of both determining and consolidating social identities. Apart from other things, such blood counts gave a new sense of objective physicality to group identities by grounding them in legible and enumerable bodily difference. Yet, unlike the earlier inventories, blood counting did not aspire to count every member of any given group. Indeed, the sample sizes were mostly remarkably small, even for groups whose membership was thought to run into the hundreds of thousands. It is therefore both surprising and regrettable that the entire historical debate on whether earlier, "fuzzier" notions of group identity were transformed through colonial counting practices into more sharply defined, authoritative, and administrative categories should have entirely neglected blood counting (fig. 1.3).

It might well be argued that blood count developed rather late in the career of colonialism and had limited direct administrative impact. This objection can be answered by highlighting two simple facts. First, as I have argued above, the decade of the 1930s was particularly important for constituting the later, postcolonial senses of identity, representation, and participatory politics. Second, as Fuller has recently pointed out, colonial enumeration and administrative policy had always been weakly articulated.[44] It would be incorrect, therefore, to neglect seroanthropological counting simply because of its seemingly late entry onto the colonial stage.

The mutation of colonial enumeration from head counts to blood counts must be analyzed independently of its uptake within administrative circles. The general political climate of the times, which helped crystallize new anxieties over the validity of earlier classificatory schemes that informed human inventories, fed directly into the growth of new technocratic enterprises such as seroanthropology. The increasingly

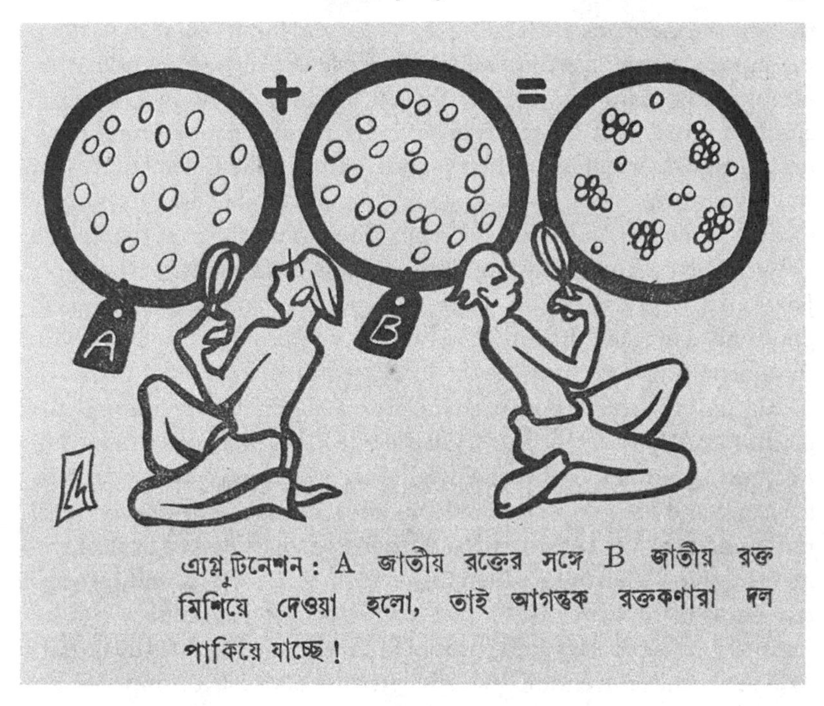

Fig. 1.3. A popular visual representation of blood groups and how they are determined. Source: Chattopadhyay, *Payer Nokh theke Mathar Chul* [From Toe Nails to the Hair on the Head], 23 (illustration by Debabrata Mukhopadhyay).

urgent need to work out issues of proportional political representation in governance, as well as the overriding importance of the question of relatedness between various castes, helped to consolidate seroanthropological researches into human difference in South Asia.

Macfarlane's researches provided a new technocratic possibility for enumeration of castes and their mutual relatedness, one that entirely skirted the increasingly unreliable markers of social and discernible physical difference. But remarkably, in her work this technocratic possibility was inextricably bound up with political positions that resonated with the radical Dalit politics of contemporary leaders like Ambedkar and Achyutanand.

Though her published research only hints at her political views, Macfarlane was more candid about her politics in her personal correspondence. In a letter to her longtime friend and supporter, J. B. S. Haldane, the geneticist and communist,[45] she wrote of her sympathies for the Depressed Classes, stating that "the Depressed Classes who have any education, and many who have not, realize that the Caste Hindus are their tyrants rather than anyone else. They were treated by the Hindus the way the Germans are treating the Poles and for milenia [*sic*]." Writing in 1942, when anti-German sentiments were high, the association of upper-caste Hindus with the Nazis was telling. Turning to colonial rule, she agreed that "British rule has been an exploitation," but immediately added that "it is a mild revenge by fate on the Caste Hindus for their tyranny that they gave a spiritual sanction to." She was also highly critical of Gandhi, writing that his positions were "those of the British Labour Party of 1910." Among the members of Congress, it was only Jawaharlal Nehru whom she seemed to admire, saying that he was "a good and able man," though he tended to "compromise too much with sentimentality."[46]

Macfarlane's researches are a crucial and enigmatic expression of the contradictory pulls of the 1930s. On the one hand, there was a new technocratic ambition to reinvent the faltering old regimes of human classification and provide a more reliable, objective basis for proportional representation. On the other hand, there were strong political movements for social justice and empowerment being articulated through strident Dalit movements. Empire itself was losing legitimacy, but Indian nationalism led by upper-caste Hindu elites in turn was beginning to look like the reinvention of an ancient form of ritualized imperialism.

Patronizing Biometric Nationalisms

Macfarlane left India aboard a Free Dutch cargo ship in the fall of 1941, headed for New York. She never returned and never succeeded in obtaining a stable research position in the United States that would allow her to continue her seroanthropological work. Her departure, in many ways, marked a new phase in the history of race science in soon-to-be-decolonized India.

This new phase, commencing in the 1940s, came to be defined overwhelmingly by various forms of biometric nationalism, that is, the use of biometric measurements to explore questions of nationalism.[47] Unlike the "biometric state" whose colonial and postcolonial emergence has been so ably mapped by Keith Breckenridge, the biometric nationalism that I document here was not as clearly tied to population registries and actual administration.[48] Instead, it sought to provide technocratic and ideological answers to difficult questions of political belonging at a time when multiple potential nations were being nourished on the dying colonial apparatus in South Asia. As Frantz Fanon insisted, "the anticolonial struggle is not automatically written from a nationalist perspective." The "fight for democracy against man's oppression . . . arrive[s], sometimes laboriously, at a demand for nationhood." This contingent, laborious process is marked throughout by the alienation of the "elites" and the "masses."[49] Technocratic frameworks such as biometric nationalism sustain the elite's alienation while also promising a way to neatly demarcate the national body. Fanon is incisive in identifying the "racism" that comes to mark postcolonial state nationalisms.[50] Indeed, it might not be entirely accidental that Fanon repeatedly chose to use the word "caste" to designate the racialized outlook of the postcolonial national bourgeoisie.[51]

This new epoch was marked by the emergence of a smorgasbord of new institutional structures for patronizing race science. While Macfarlane had mainly cobbled together the resources for her studies from a range of academic institutions that were not exclusively devoted to racial research, from the 1940s new institutional structures of patronage developed that were emphatically invested in researches into human variations. These institutions can be grouped into at least four main categories: private think tanks, the mammoth Anthropological Survey of India (ASI), a growing array of university departments of anthropology devoted overwhelmingly to biological studies of human variation, and finally, a small but influential number of medical research bodies interested in race.

The private think tanks financing race research were a new phenomenon in the 1940s and did not long outlive the decade. They were, however, significant in financing a number of large studies in that decade and in clearly establishing the polymorphous political stakes of such research in the closing years of colonial rule. In 1942, for instance, Atul Krishna Sur published *Bangalir Nritattwik Parichay* [The Anthropological Identity of the Bengalis], which sought to determine the precise racial identity of the Bengalis. The book was published by the Hindu Mahasabha, a right-wing Hindu political party. In 1947, Dhirendranath (D. N.) Majumdar and Irawati Karve coauthored a book titled *Racial Problems in Asia*, which was published by the Indian Council of World Affairs (ICWA). The ICWA was a think tank founded in 1943 by Sir Tej Bahadur Sapru, a liberal nationalist politician and a relative of the Nehrus. In 1950, Majumdar alone published another work, titled *Race Realities in Cultural Gujarat*. Majumdar had carried out his research for this project since 1941, and it was supported and published by the Gujarat Research Society (GRS).[52] Each of these organizations had its own specific political interests in patronizing such research. The GRS, for instance, was part of the Maha Gujarat movement seeking to bifurcate the then-unified Bombay State into separate Gujarati- and Marathi-speaking states. The movement was particularly challenged by the difficulties in determining whether inhabitants of the forested Dang region were closer to the Gujaratis or to the Marathis and hoped that Majumdar's research might illuminate this. Likewise, Tejbahadur Sapru had his own nationalist political vision and at the founding of the ICWA had reiterated the need to investigate the suitability of Western democratic institutions for Asia. He was also the one who had mediated between Ambedkar and Gandhi to create the 1932 Poona Pact, a momentous electoral agreement that assimilated Dalits into a Hindu-dominated, mainstream nationalist constituency. The Hindu Mahasabha in early 1940s Bengal was agitating for the division of the province on religious lines, and here again the vexed question was whether the Dalits would affiliate with the Muslim political bloc or the Hindu one.[53] The patronage of these bodies therefore clearly implicated seroanthropological research in the rough and tumble of the new political processes seeking to redraw national political boundaries.

In the foregoing decades seroanthropological research had mainly been sponsored by academic institutions. Malone and Lahiri's research was supported by the Bacteriological Laboratory at Kasauli; Annandale and Mahalanobis's was done under the aegis of the Zoological Survey of India; and Macfarlane's research was financed by the University of Michigan and the Royal Society. None of these institutions was openly political. By contrast, entities like the Hindu Mahasabha, the ICWA, and the GRS were all eminently political outfits with clear political reasons for supporting research. In them we see a clear blossoming of biometrical nationalism whereby technocrats hoped to resolve complex issues of political affiliation and national belonging through racial research.[54]

Whereas these private and fairly overtly political sources of patronage proved to be short-lived, the Anthropological Survey of India, which emerged in the same decade, had a much longer and sustained impact on Indian race research. In time, by the end of the 1980s, the ASI would become the largest single institutional employer of anthropologists anywhere in the world, with a staff strength of 750 anthropologists.[55] The ASI was a somewhat curious and hybrid institution hovering somewhere between a government ministry and a university department. It sought to align academic research and statecraft under a single umbrella. The structural origins of "Surveys" lay in the late eighteenth century with the Trigonometric Survey of India.[56] Since that time, a number of other Surveys had emerged, gathering pace from the late nineteenth century. These included the Botanical Survey of India (BSI), the Archaeological Survey of India (ArSI), and the Zoological Survey of India (ZSI).

It was Mahalanobis's collaborator Annandale, director of the ZSI, who first proposed an Ethnographical Survey in 1916. Similar proposals were made by R. B. Seymour Sewell, Annandale's successor at the helm of the ZSI in 1927 and, later, by the famous missionary-turned-anthropologist, Verrier Elwin, in the early 1940s. The colonial government, however, remained unwilling to act on these proposals until near the end of colonial rule. Only then, and acting on the initiative of the eminent anthropologist B. S. Guha, did it sanction the formation of the ASI.[57]

Born in Shillong in 1894, Biraja Sankar Guha undertook graduate studies at Harvard and in 1924 became the first South Asian to obtain a PhD in anthropology. His dissertation was titled "The Racial Basis of the Caste System in India."[58] Upon his return from Harvard, in 1926, he initially joined the University of Calcutta as a lecturer in physical anthropology. The following year, however, Guha left the university and joined the ZSI as an anthropologist in its Anthropology Section. His professional rise from then on was nothing short of meteoric. In 1928, only four years out of graduate school, he was elected president of the Anthropology Section of the Indian Science Congress. In 1931 he was deputed to serve on the Census Commission. Five years later, in 1936, he was instrumental in founding the Indian Anthropological Institute in Calcutta. In 1938, he was elected president of the Anthropology Section of a joint meeting of the British Association for the Advancement of Science and the Indian Science Congress. Later that same year, he was elected vice president of the Physical Anthropology and Human Biology Section at the International Congress of Anthropology in Copenhagen. By the end of the 1930s, therefore, Guha was one of the foremost names in Indian anthropology. Later, in 1945, upon learning of plans by the government to invest more in the discipline as part of its postwar reconstruction plans, Guha, with the support of the then director of the ZSI, R. B. Seymour Sewell, submitted a plan for the creation of an independent new Survey for anthropological research. The proposal moved quickly, and the Anthropology Section of the ZSI was severed from the parent body on 1 December 1945, a little more than a year before the formal decolonization of British India. Guha was initially appointed as officer on special duty to head the newly created entity and later took over as the first director of the ASI.[59]

Guha remained at the helm of the ASI until his retirement in July 1954. Even after retirement, however, he continued to try to control the ASI and was instrumental in mobilizing the staff against his successor, Dr. N. Datta Majumder, eventually managing to get the latter removed from the position in 1958.[60] Guha's true influence upon the organization, therefore, far outlived his formal tenure as director.

From its inception, and owing no doubt to Guha's own academic interests, the ASI was overwhelmingly focused on research in race science. One of the first people to be employed at the ASI was Macfarlane's erstwhile collaborator, S. S. Sarkar, who in turn introduced seroanthropological research at the ASI. Sarkar had also been picked to be Guha's deputy and potential successor, but the two men subsequently fell out, and Sarkar was fired from the ASI.[61] Notwithstanding Sarkar's departure, the ASI remained deeply invested in seroanthropological researches throughout the 1950s and even into the 1960s. In the nineteen years between 1948 and 1967, the ASI published a whopping total of seventy-eight research papers on these topics. This was by far the largest cache of publications among all of its output. The next largest category, anthropometry, lagged behind at fifty-six papers in the same period.[62] Together, however, seroanthropology and anthropometry, both of which were interested in questions of race, by far dominated the ASI's research publications. Much of the credit for this robust seroanthropological publication profile goes to a Swiss anthropologist from the University of Zurich, Dr. E. C. Büchi, who was employed at the ASI during the early 1950s. It was Büchi who, after the departure of Sarkar, maintained the seroanthropological work and trained a number of younger Indian scholars in the field. These latter included many of the researchers we will meet in later chapters, such as Dr. S. R. Das, Dr. R. S. Negi, Dr. S. R. Gupta, and Mr. Narendra Kumar.[63]

The ASI bears strong testimony to the depth of the new postcolonial state's direct investment in race science. Guha's original proposal had imagined an organization that would serve two main purposes. First, it would be a "scientific institution for the most advanced research in anthropology comparable to the best in Europe and America." Second, it would be a "training centre where administrators who require to deal with tribal populations could get a systematic training in laboratory methods and the principles of science along with advanced students of the discipline."[64] He further hoped that through this union of science and statecraft, "knowledge regarding various races and tribes" could be produced and disseminated with a specific view toward the "unifying and gradual welding of variegated and diverse cultural components into a unified whole." One of the foremost ways of pursuing this end was through "physical measurements accompanied by the investigation of different blood groups and their percentages in each."[65]

Guha's vision was an intensification of biometric nationalism as well as its mutation into direct statecraft (fig. 1.4). While its political goals were comparable to the research sponsored at the time by organizations such as the GRS or the ICWA, its location within rather than outside

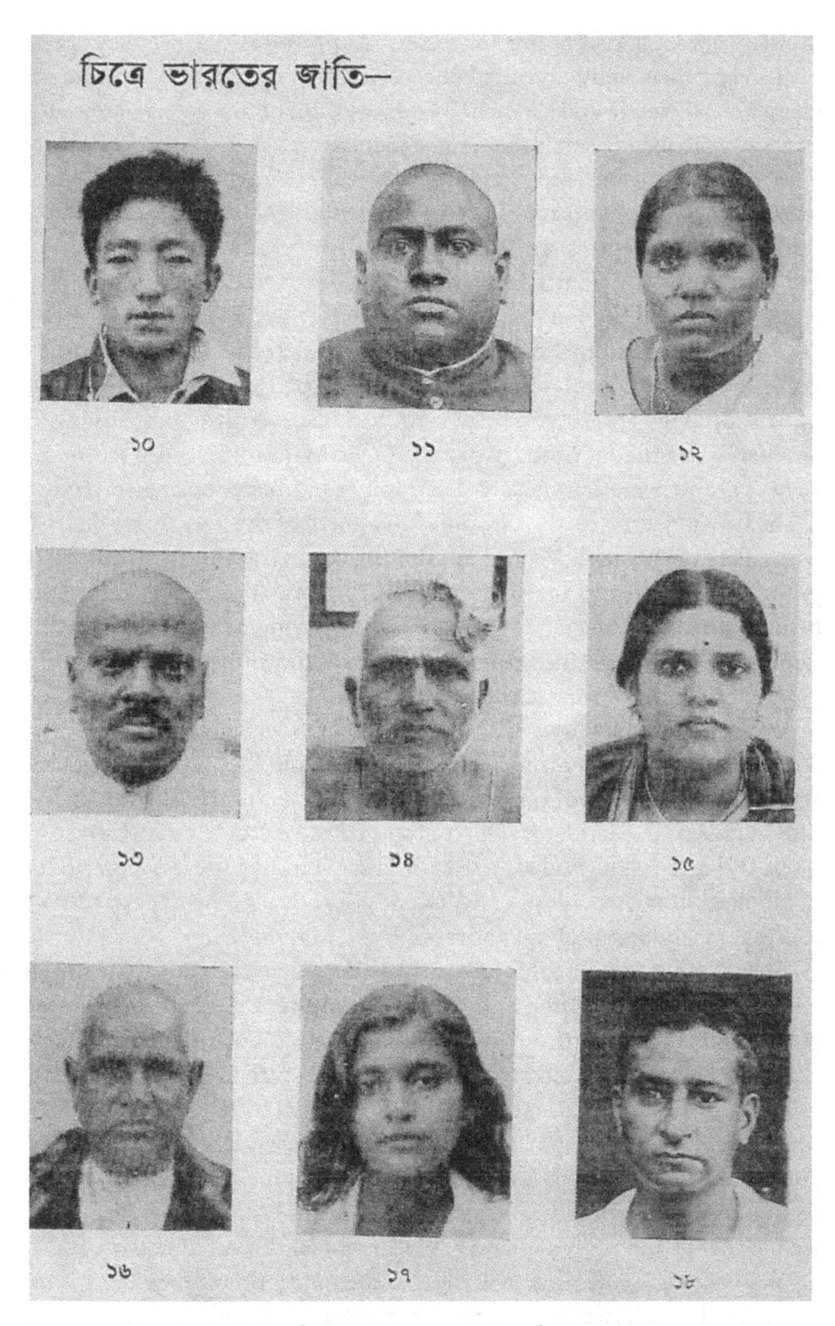

Fig. 1.4. *Chitre Bharater Jati* [India's Races in Pictures]. Source: Guha, *Bharater Jati Porichoy* [India's Racial Identity].

the state gave it access to greater resources, power, and legitimacy. The Nehruvian era is rightly recalled as a period that saw a strong bond develop between state and science.[66] Guha's vision of biometric nationalism was ensconced within that larger picture.

A third institutional location for race science were the various university departments of anthropology and closely allied disciplines. Though the first department of anthropology in British India was founded at the University of Calcutta in 1921, the first doctoral degree in the subject was awarded only in 1949. The recipient of this Doctor of Science (DSc) degree was none other than Sasanka Sekhar Sarkar, Macfarlane's young assistant and collaborator.[67] Indeed, in the 1920s the department of anthropology at Calcutta University had been so obscure that when another early student of the program, D. N. Majumdar, chose to enter it, his parents were immensely disappointed at his choosing to study a "shady" subject.[68] It was through the efforts of these early students, particularly Sarkar and Majumdar, that by the 1940s and 1950s anthropology in general and racial anthropology more specifically, far from being a "shady" subject, had emerged as a thriving new discipline at a number of new departments, such as those at the Universities of Delhi (1947), Lucknow (1950), and Utkal (1958).

By 1965, a total of fifty-eight teaching departments in the country offered PhD degrees in anthropology.[69] At the time there were a total of seventy-six professors of various ranks teaching in these departments.[70] The Lucknow department, which emerged as one of the most prestigious and active, is a good example for understanding the larger growth of the discipline through the mid-twentieth century. The department's success was the result of the efforts of D. N. Majumdar.

Born in Patna in 1903, Majumdar was initially educated at Calcutta University. After obtaining one of the first master's degrees in anthropology from the Calcutta department in 1924, he joined the Department of Economics and Sociology at Lucknow as a lecturer in "primitive economics." In fact, it was the then-head of the department, Radhakamal Mukherjee, who "brought him in when he sensed the talents of this young man."[71] This in itself is suggestive of Majumdar's intellectual direction. As Alison Bashford has recently shown, Radhakamal Mukherjee pioneered an influential and explicitly racialized discourse on anticolonial ecology grounded in ideas about tropical physiology.[72]

At the time, however, Majumdar himself had mainly been interested in the ethnography of India's so-called "tribal" populations. In 1933 he went to Cambridge for his doctoral work and studied physical anthropology with G. M. Morant. It was during this time that Majumdar was introduced to seroanthropology by Reginald Ruggles Gates, the botanist and eugenicist based at King's College, London, as well as studying serology at the Galton Laboratory. Upon his return to India, he was invited by the census commissioner in 1941 to undertake a seroanthropological survey of the various communities of the United Provinces, one of the largest provinces in British India. In 1942 he was invited to teach a course on "Races and Cultures of India" at the Indian Civil Service Training Center at Dehra Dun. Five years later, in 1947, he founded the *Eastern Anthropologist,* a new academic journal published from Lucknow University. This journal did much to highlight Lucknow's position as a powerhouse of anthropological research. By 1950 Majumdar had become a full professor in the Economics and Sociology Department, but that very year he convinced the university to create an independent department of anthropology. He also established several international collaborations, especially with Cornell University, thus connecting the fledgling department to an international network of anthropologists. In 1953 he was elected a Fellow of the American Association of Physical Anthropologists. By 1955 he had been appointed a member of the Standing Committee on Research Programs at the all-powerful Planning Commission of India, which gave him a prominent place in determining the country's research priorities. In 1958, his interests in human variations also led him to be invited to become a member of the Indian Council for Medical Research (ICMR). When he died suddenly of a cerebral hemorrhage in May 1960, he was also the dean of the Faculty of Arts at his university.[73]

Though few anthropology departments achieved the prominence of the Lucknow department, the sheer number of new departments created new career opportunities for those interested in human diversity, opened up publication opportunities, and created the need for new research topics. By the late 1960s hundreds of students were enrolled every year in master's courses in anthropology and a smaller handful of students were pursuing doctoral research. In 1965, for instance, there were 258 students studying for master's degrees in the subject and another 50 pursuing PhDs.[74]

Finally, as Majumdar's own invitation to the ICMR demonstrates, medicine provided yet another productive site for both institutional patronage and intellectual rationale for pursuing race science. Medical genetics is often treated separately from seroanthropological work, but as we shall see throughout this book, the wall dividing these two areas of race science remained extremely porous throughout our period. Not only were the researches done in these areas of research mutually referenced, but the actual researchers were themselves frequently socially and institutionally connected. Majumdar's role in the highest national body to coordinate medical research demonstrates this porosity.

The man whose career best illustrates the rise and development of the medically oriented strand of seroanthropological work in India is Labhshankar Dalichand Sanghvi. Born in Morbi, Gujarat, in 1920, Sanghvi had come to Bombay as a young boy for his education. He obtained a bachelor's degree in mathematics and physics in 1941, followed by a master's in pure mathematics in 1943. In 1944 he obtained a research position to assist Prof. V. R. Khanolkar in the Blood Bank at the Tata Memorial Cancer Hospital, Bombay. This was his first introduction to serology. He quickly became interested in human genetics and biological variation, and published a number of papers on the topic. In 1949 he won a fellowship from the World Health Organization to study for two years with L. C. Strong at the Yale School of Medicine and L. C. Dunn at Columbia University. A subsequent grant from the Wenner-Gren Foundation allowed him to continue to study with Dunn and L. S. Penrose at the Galton Laboratories in London. He eventually obtained a PhD in 1953, having worked under the legendary geneticist, Theodosius Dobzhansky.[75]

At the time Sanghvi finished his PhD, the discipline of genetics was undergoing a transformation. Leading geneticists, such as Dobzhansky, were in the process of repositioning genetics away from its prewar associations with eugenics and toward new medical applications. In the midst of the nuclear fears evoked by the atom bomb, these scientists argued that genetics could illuminate the true, long-term impact of radiation on human health. The institution that was tasked to coordinate this study was the United Nations Scientific Committee on the Effects of Atomic Radiation (UNSCEAR).[76] Sanghvi headed the Scientific Committee of this key Cold War–era international scientific body.

Later, Sanghvi would also serve as the official geneticist at the head offices of the World Health Organization at Geneva, before returning to India and joining the Tata Memorial Hospital. There, with financial support from the Dorabji Tata Trust and the Rockefeller Foundation, he established a new section on genetics. By 1965, this section had grown into a much larger, Human Variation Department at the Indian Cancer Research Center (later, Cancer Research Institute [CRI]).[77]

Sanghvi did much to promote not just national, but also international research into Indian populations and their biology. These researches, along with the institutions he helped establish, such as the CRI, created a new set of professional opportunities and motivations for studying racial differences, much as the ASI and the various new university departments had done. Yet, it is crucial to remember that throughout his career Sanghvi remained closely connected, both socially and intellectually, to those in more standard anthropology departments. As we shall see, he engaged in a long-term collaboration with the famous Berlin-trained anthropologist Irawati Karve, at the University of Pune. He was also one of the small handful of friends and collaborators to contribute a paper to the memorial volume published upon the sudden death of S. S. Sarkar in 1969.

Taken together, what we see from the 1940s onward is a new mosaic of research institutions, career pathways, and intellectual motivations for studying human difference in India. These sustained race science throughout our period and laid the foundation for much of the contemporary genomic researches into human variation. Quite literally, and from the perspective of this book, it was the troika of new university departments, ASI, and medical research centers such as the CRI that provided the "white coats" through which race was articulated in postcolonial India.

Seroanthropological Races

André Béteille, writing in the journal *Daedalus* in the spring of 1967, argued that "race" was a "social category." In support of his contention he pointed out that "studies by P. C. Mahalanobis, C. R. Rao, D. N. Majumdar, S. S. Sarkar and others have shown how complex the racial pattern might be even in a limited region or sector of society."[78] Later in the same article, he insisted that "clear physical types do not exist in India in the shape of concrete groups as they do in the United States and South Africa. They are, rather, constructs that enable anthropologists to order their data."[79] He added that "physical anthropologists are now gradually coming to discard the term race itself and to use in its place such neutral terms as *population* or *breeding unit*."[80] He singled out Sarkar for having demonstrated that "even tribal groups—generally assumed to be homogenous—show a great deal of internal variation when examined anthropometrically and serologically."[81]

Béteille's reading is common in social-science circles. Ever since the mid-twentieth century social scientists have developed a view that scientists have generally discarded, and indeed disproved, the idea of "race" by describing admixture. Yet the scientists they have cited have seldom gone along with such a reading. Speaking only a few months after Béteille's views appeared in print, Sarkar—whom Béteille cited in his support—clarified the issue of race at length in a lecture.

Sarkar began by acknowledging that race was an "incendiary term," but insisted the trouble was mainly about the "addition of an ism to it." He noted further that several "modern authorities," including UNESCO, had advocated discarding the term and replacing it with "ethnic group." Others had suggested alternative terms, such as "geno-group" and "cline." Yet, Sarkar was emphatically not in favor of these changes. He insisted that to a biologist, "race is a taxonomic category" and absolutely essential to the work of classification. "It is often said that species and races are arbitrary categories. This opinion however, is false." Though there might be "border-line cases" between species and races, an "expert taxonomist," in Sarkar's view, would always be able to distinguish the two. Even more important, with regard to Béteille's views, Sarkar clarified that the mere fact that two races are "sometimes imperfectly differentiated" did not undermine their identity as distinct races.[82]

What was needed, Sarkar thought, was a new definition of "race," and one that was based on "biological and genetic criteria." The one that best fit this requirement, he felt, was a definition offered by Dobzhansky: "Races are defined as populations differing in the incidence of certain genes, but actually exchanging or potentially able to exchange genes across whatever boundaries (usually geographic) separate them." Race, therefore, was a "genetically open system," while species was a "genetically closed system."[83]

Dobzhansky himself, in an exchange with Frank B. Livingstone in 1962, had explicitly defended both the term "race" and its objectivity. Like Sarkar, he had denied the contention that proof of admixture somehow undermined the concept of race itself. He argued that "races can be discrete only under exceptional circumstances," while also asserting that "race differences are [an] objectively ascertainable biological phenomenon." He posited that only the actual nomenclature of races was conventional, not the reality of their existence or difference.[84]

The key difference between Béteille's view of race and Dobzhansky's and Sarkar's understandings of it was that Béteille imagined race to exclusively be a closed, essentialized, and ahistorical group. By contrast, Dobzhansky and Sarkar both held that race was an "open system," and hence races emerged and disappeared historically. One did not have to be stuck with the same racial groupings throughout time. As Dobzhansky put it, "variability precedes race and serves as the raw material for its formation."[85] Sarkar explained in greater detail that races were transformed by three processes. First, and perhaps most significant, was geographic isolation. Second were environmental adaptations to the habitat in which the group was isolated. Third, "man's intellectual discrimination has led to various forms of mating systems, which have led to many changes." It was through the combined effect of these processes that "races are being formed," explained Sarkar, using the present tense and insisting that "race formation is therefore a process."[86]

These rival formulations, both advanced at the very end of the period covered in this book, demonstrate that race remained an explicitly formulated conceptual category for researchers like Sarkar. The social-scientific readings of seroanthropological studies were in fact misunderstandings premised on an older formulation of the race concept. Seroanthropological races, in the late 1960s, were no longer imagined as closed, ahistorical, and stable units. Rather, they were understood as dynamic, historical entities.

This historical element in turn was constituted by environmentally adaptive changes as well as changes to "mating systems" brought about by "man's intellectual discrimination." Though Sarkar himself did not fully explicate what he meant by this latter formulation, other seroanthropologists did explain this at great length. Vijender Bhalla, at the time a young anthropologist based at Punjab University in Chandigarh, explained that the "sex-mating-pattern of the caste system" had been responsible for the "stratification of genes" in the Indian population. He argued, therefore, that "caste was an evolutionary force in genetic change."[87] Pithily summing up this position, Bhalla explained that "the institution of caste system set up an intricate but well-knit framework of social barriers which in turn splitted up [*sic*] society into a large number of social isolates, i.e., the endogamous mating groups constituted by a number of exogamous familial lines." Furthermore, "evidently this resulted in the isolation of the groups so formed providing thereby enough latitude to the law of genetic drift to lead them into different directions of development."[88]

The notion of "endogamy," which was the cornerstone for most of the seroanthropological sorting of castes, had been developed in the mid-nineteenth century by John Ferguson McLennan, a Scottish ethnologist. It was rapidly picked up by British Indian writers, such as Sir H. H. Risley, and subsequently by a host of Indian intellectuals, including both radicals and conservatives. However, McLennan's own formulation had been built partially on his reading of South Asian social and historical data. As Durba Mitra reminds us, "McLennan utilized Hindu 'social custom' as a key empirical referent for a comparative universalist theory."[89] Far from being a one-way dissemination of European social categories to British India, therefore, the notion of endogamy was itself a product of the history of colonialism. It is also worth noting here that the colonial knowledge of Hindu customs that McLennan drew upon was itself a product of unequal collaboration between Brahmin pandits and British Orientalists.[90] To cite but one example, John Muir, whose work McLennan drew upon, was one of the key "forerunners" of what Michael Dodson has described as "constructive Orientalism," that is, an educational project that sought to "engraft" Western knowledge onto South Asian languages and social hierarchies.[91] Constructive Orientalism was crucially reliant upon Brahmin pandits, and one of its direct consequences was the reinforcement of the social authority of Brahmins and high Sanskrit knowledge as the exclusive bases of an "Indian civilization."[92] The notions about caste and endogamy that underwrote seroanthropological researches were, therefore, not just deeply indebted to colonial knowledge but also produced by earlier, unequal collaborations between upper-caste Hindu elites and British Orientalists.

Though Indian seroanthropologists made little explicit effort to systematically draw upon precolonial traditions of textual knowledge, their general efforts to reconcile seroanthropological truths with "prior knowledge" about caste identities are remarkably similar to the projects of early Egyptian seroanthropologists like ʿAbbas Mustafa ʿAmmar.[93] This idea that caste was a major factor that had led to reproductive isolation through endogamy and hence to a new process of race formation was almost universal among seroanthropologists. The last step of the process, however, was seldom fully articulated since many anthropologists explicitly hoped that this differential evolution could be changed so as to eventually form a kind of unified, national race. As Guha put it baldly in a lecture in 1958: "in our age-long history various races have sought India's hospitable shelter . . . If, however, we are to weld ourselves into a well-equilibrated nation, the existing group conflicts have to be resolved and different elements have to be integrated with a common outlook and purpose . . . a process of Indianization on lines similar to the concept of Americanization of the U. S. A."[94] Although Guha did mention maintaining some diversity within a unified national body, it was clear that the main emphasis was on "welding" the diverse "races" into an "Indian nation." Sarkar was even more explicit about the biological aspects of this welding: "the reasons behind race riots are the lack of the actual exchange of genes."[95]

Whereas Guha and Sarkar noted the importance of caste as an impediment to the formation of a uniform national body, Sanghvi was more taken by the research possibilities that the caste system offered. In 1966 Sanghvi published a programmatic piece titled "Inbreeding in India" in the *Eugenics Quarterly* in which he pitched India's caste society as a valuable object of research for geneticists. "No feature of the Indian caste society is more resistant to change than its institution of marriage," he declared. He acknowledged that much in India was rapidly changing, and even the actual marriage rituals, but "regulations governing marriage, however, have been little affected in their essential biological aspects."[96] Instead of simply seeing these caste groupings as a negative social form to be overcome, Sanghvi insisted that genetic studies of these groups could illuminate some of the foremost problems in human genetics.

Sanghvi's pitch was aimed at intervening in a long-running debate among geneticists called the Classical-Balance debate.[97] At its heart was the notion of a "genetic load." In the postwar world, the specter of radiation-induced mutations had come to loom large in scientific and public consciousness. The work of Hermann Muller, who was awarded the Nobel Prize in 1946, and the Atomic Bomb Casualty Commission from 1947, had promoted research in this area and thereby also rescued genetic research from the taint of Nazism by giving it a laudable, humanitarian goal. Muller, in a famous article in 1950, pointed out that deleterious mutations might remain hidden or submerged in a population until an individual inherited two of the mutant genes. Hence, it was crucial to try to detect the "genetic load" of submerged mutations.[98]

As Diane Paul points out, "genetic load" is not a neutral expression.[99] Dobzhansky and others challenged Muller's formulation. In 1957 Dobzhansky detailed his objections, pointing out that Muller assumed not only that mutations were necessarily bad but also that homozygotes—those who were genetically homogeneous, that is, had two identical alleles of each gene—were necessarily the fittest. Instead, Dobzhansky argued that "supervital homozygotes" were "narrow environmental specialists," whereas heterozygotes with one potentially lethal mutation might actually be better adapted to survive in certain environments.[100] Soraya de Chadarevian points out that this debate was crucial to population thinking's , surviving beyond the dramatic postwar transformations in the field of genetics.[101]

The debate effectively boiled down to a disagreement over basic values. For Dobzhansky, genetic variation was a "supreme value," whereas Muller sought a genetically homogeneous, though interestingly socialist, society. As Paul writes, "what Muller and Dobzhansky expressed was not so much scientific positions as world-views."[102]

Most insights on genetic mutations at the time were built on experiments with fruit flies, and scaling these up to the complexity of human societies proved unreliable. As Dobzhansky quipped, "men are not overgrown *Drosophila*."[103] This led geneticists to try to find ways of calculating the "genetic load" and its impact in actual human groups. In 1956, Newton Morton, James Crow, and Hermann Muller proposed a possible way to measure what they called "the total hidden mutational damage carried by a population" by studying the offspring of consanguineous marriages.[104] Sanghvi, who was both a student of Dobzhansky and the scientific secretary of UNSCEAR, entered the debate in 1963 by publishing a detailed critique of Crow's position in the *American Journal of Human Genetics*.[105] One of his key criticisms was that Crow's modeling was based on a "very static concept of environment" and one ought to study the problem "over a span of hundreds of generations."[106]

This was where Sanghvi argued that the Indian caste system provided excellent experimental possibilities by allowing discrete populations to be studied through hundreds of generations in real-life environments. In his 1966 article he explicitly referred to the work of UNSCEAR and the studies of Japanese children affected by atomic radiation. He insisted, however, that all the extant studies of the offspring of consanguineous marriages were "of very unequal scope and reliability for a quantitative generalization."[107] He repeated this claim several times in the following years, including in a memorial essay for Sarkar, after the latter's untimely death.[108]

The new seroanthropological conception of race, therefore, was doubly rooted. On the one hand, it was based in the Nehruvian rhetoric of unity in diversity. On the other, it was based in international networks of geneticists interested in notions of "genetic load" and atomic radiation. These two did not always pull in the same direction, but where they did agree was on a new, dynamic, historicist notion of race and on the role of caste as a key mechanism for race formation. Far from the displacement of race by population, what we see in Indian seroanthropological research is a mutation and intensification of race, through its embrace of admixture and historicism in place of the earlier notions of stable, typological races.

INTERCHAPTER

Letter 2

Hemendrakumar Ray
21, Pathuriaghata Bylane
Calcutta, India
6 December 1933

Dear Najrul,
You mention that you were still a boy when my first novel was published. So, I am sure you are much younger than I and hope you will not mind my addressing you by your first name.

A man in my position receives numerous letters almost on a daily basis from various fans. Most of them, to be perfectly honest, are tedious. Your letter was not amongst those. The singular experiences you have had and the moral crisis you write so movingly about both made your letter stand out. I must thank you for writing to me.

I am not sure that I can guide you out of your moral dilemma, but I will say this much: your situation is not unique. I think you, better than most of my writer friends, have articulated the spirit of our times. With every passing year there emerge new fossil finds that illuminate the many steps by which apes evolved into humans. Like lost family photographs of long-dead ancestors they bring us face to face with the simple fact that sometimes we might resemble those whose very existence we had largely forgotten. In fact, I have lately been working on a short novel around this theme. I think I will call it Man's First Adventure. *In it I want to imagine the first contact between the Cro-Magnons and the Neanderthals. Alongside the usual rivalry and war over territory and food, I want to explore the possibility of romance between the two peoples. After all, as we have repeatedly seen in India, no matter how rigidly the boundaries of caste, religion, or race might be drawn, love always blossoms. People cross the boundaries they are forbidden to cross.* Children are born between *parents of different castes, religions, and races. We know this. We see this every day around us. Anyway, let us see how the "adventure" develops! I will certainly send you a copy once I finish it. You are a man of adventure, after all, and perhaps you will like it.*

What I am struggling to understand is what these new discoveries might really mean for our evolutionary history. How might acknowledging the great adventure of such forbidden love change our view of our evolutionary past? Love between those we think of as human ancestors and those we think of as subhumans, after all, raises exactly the sort of questions you are raising.

Then there is the question of upbringing. So many of my readers urge me to write for children in a way that will instill in them good values, make them into good human beings. It is clear that they, and I, deeply believe in the role of education to shape character. But then, seldom do we reflect just how deep the influence of this education runs. Does it perhaps shape the very core of what makes us human?

You mention Tana, the Daughter of the Gorillas. A few years back another shikari *friend of mine who had also traveled to Africa to hunt gorillas sent me information of another young female who had similarly been brought up by gorillas. Her name was Numa. I have written a short story about her called* Jungle-dwelling Curse. *I wonder if Numa and Tana might be the same child? One can hardly rely on illiterate village people to get names straight!*

Whether they are the same child or not, the stories of these children force us to rethink not only our ideas about humanity but also how we think of men and women. The sheer physical strength and boldness of these girls seems to make us question how much of the notorious weakness of our women is the result of how we bring them up rather than of some inner, physical incapacity.

What was most singular about your letter, however, was not the story of Tana, but the other story. Tana, like Numa, compels us to look back, both at our evolutionary childhood and our real childhood. In a sense, they are the same as the evolutionary cousins whose existence is now suspected by many men of science. I read in the Statesman *some time back about François de Loys's ape in South America. Hardly a year passes without some report of the Yetis in the Himalayas. In fact, I firmly believe that the fossil evidence that Prof. de Terre and his Yale North India Expedition have uncovered will eventually show us that the Yetis are some evolutionary offshoot that has managed to hang on in the mountain fastness. More recently, my adventurer friends, Bimal and Kumar, have even told me that they have obtained reliable information that there might be a similar colony of surviving Neanderthals in the jungles of Kenya. All these accounts tell us that our evolutionary family tree is a banyan tree rather than a pine tree. There are branches of forgotten ancestors, missing cousins, and who knows what else! But the crucial thing is that all these stories compel us to glance back in evolutionary time: send us back to the time of that ancient monarch Mandhata.*

What your new account does is entirely different. It forces us to look ahead. To peer into the future.

We are modern citizens of a modern world, and with that has come a reticence to peer into the future. We leave that to the grubby astrologers one finds sitting on the footpaths outside the Indian Museum. Inside the Museum science tells us the story of our past, outside the Museum the half-literate star-gazer tells us our future. Isn't it ironic? Little wonder we have ended up calling Museums Jadughar. *A House of Magic, indeed!*

But perhaps the irony itself is hollow. Perhaps astrology too is a science, and science too will learn to predict. Indeed, isn't our Mr. Mahalanobis saying that he will soon be able to predict floods before they actually happen? I am still not sure whether he can actually do it, but Gurudeb Rabindranath seems to have great faith in him, so perhaps he will.

You, on the other hand, have already got a glimpse into the future. I must confess to a tinge of brotherly jealousy at your fortune in this regard. I cannot wait to learn more about these barrel-dwelling future humans you and the late Mr. Sen seem to have stumbled upon. It is a fascinating thought that science might re-create our bodies and minds so that we end up looking like snails. I must confess, my dear Najrul, of all the things I might have imagined in my novels about the future, I would not have thought of making human beings resemble the slimy denizens of our paddy fields! What an intriguing possibility.

Do, please, tell me more.

But when you say humans have played God in creating this future race, I must beg to differ with you. God is not just a biologist. Surely, there is more to Him than that. For many years now I have attended the séances that are frequently organized at the home of my illustrious neighbors, the Tagores. The things I have seen there have convinced me more than ever that there is much more to human existence than our bodies. I do not know whether we can call it a soul, or spirit, or something else, but it certainly is more than our bodies. It is that domain which lies beyond our bodies and over which God, or perhaps gods—and goddesses and saints and prophets—hold their dominion.

Since you like my stories, let me tell you one that may convince you of what I am saying. It was told to me by a very close friend who is also a doctor. My friend was also a close friend of the famous spiritualist, Anantababu. One day, as the two were engaged in one of their never-ending discussions about life, mortality, science and much more, they heard a commotion outside Anantababu's house. Upon walking out, they discovered that Anantababu's neighbor, Surenbabu, had fallen from the second-floor balcony onto the footpath.

As Surenbabu lay unconscious, my friend called in Dr. P. Ghosh, the leading specialist in cerebral injuries of this kind. Dr. Ghosh arrived immediately, but said there was nothing that could be done. Surenbabu would most likely expire within the hour. To make matters worse, Surenbabu's wife and the rest of the family were away on pilgrimage to Haridwar. Especially, with his son being absent, there was no one to perform the funerary rituals.

It was then that Anantababu volunteered to intervene. He said he could mesmerize Surenbabu so that even after he technically "died," his body would remain in a state of sleep as his soul would remain attached to the body.

As you can imagine, my doctor friend and Dr. Ghosh both scoffed at the outlandish and, some would say, unscientific idea and instead sent an urgent telegram to the family in Haridwar urging them to hurry back as soon as possible. But even if they left immediately upon receiving the telegram, it would still take them several days to reach Calcutta. The body had to be somehow preserved till they got back, so no one objected to Anantababu trying his technique, even though none believed it would work.

To everyone's surprise, next morning, when Dr. Ghosh and my friend returned to Surenbabu's house, they noticed that although the body was cold, had no pulse or heartbeat, and was stiff, it showed no signs of decay. What is more, when Anantababu called out to Surenbabu in a deep, soft voice, a reply seemed to emerge from the body. My friend and Dr. Ghosh were stunned! They did not know whether to trust their senses or the science they had learnt.

I will not bore you with the details of the incident. I have published a short report of it under the title Story of the Imprisoned Soul. *But suffice it to say that the body remained in the same state till the family arrived. It continued to respond, though with increasing difficulties, to Anantababu's calls and eventually begged to be released from its mortal coil. The final surprise came as soon as Anantababu eventually released the soul. There was a sudden sound—a loud bang—and, before anyone knew what was happening, upon the bed, where Surenbabu's body had lain for days, there lay a putrefying corpse whose disgusting stench made it difficult for anyone to remain in the room.*

How it all happened I cannot tell. Neither could my doctor friend. But he did vouch for the truth of what he witnessed, just as you vouch for the truth of your barrel-dwelling humans. Anantababu himself referred both to spiritualist insights and the strange powers of yogis. I have never been much of a philosopher and would not be able to explain any of this. But this much I have come to realize: man is not just his body and, like the Bard, I too might say today, "there are more things in heaven and earth, Najrul, than are dreamt of in your philosophy."

I shall look forward to hearing from you.

Affectionately,
Hemenda

CHAPTER 2

Mendelizing Religion

> Whenever you hear anyone abuse the Jews, pay attention, because he is talking about you.
>
> —An Antillean philosophy professor, quoted by Frantz Fanon, *Black Skin, White Masks*

"The serological attack on the race problem" was the heading of a key subsection in a lengthy research paper published by Eileen Macfarlane in 1937.[1] As we saw in the last chapter, Macfarlane dominated Indian seroanthropology throughout much of the 1930s, and studies of the "race problem" using seroanthropological techniques were not at all unusual. Yet the 1937 paper was different in one key respect. Whereas other studies were focused on castes and tribes, the 1937 study was focused on a religious group, namely, the Jews of Cochin.

In a sense this was not surprising. After all, "the Jew's body" was central to the construction of biological race science almost from the beginning. Nineteenth-century inquiries into the distinctiveness of the Jewish foot or the Jewish nose had already constituted the Jewish body as both a coherent object of scientific investigation and a precursor to the studies of Jewish genes and genomes that were to follow in the next century.[2] As Steven Weitzman points out, even as the earlier "race science was being discredited after World War II, the Jews were surfacing as a subject of genetic research."[3] Indeed, this singling out of the Jewish body is one of the most indisputable points of similarity between prewar race science and postwar genetics research.

Little of this research focused on South Asian Jews. Barring occasional fleeting references, these groups remained largely peripheral to the early discussions of Jewish biology. It was only after the formation of the state of Israel, and framed largely by its own national anxieties, that the *Edot ha-Mizrach,* or "oriental Jews," became a subject of intense biological studies in the 1950s and 1960s.[4] Studies such as Macfarlane's were therefore both out of place and out of time within the usual traditions of racialized studies of Jewish populations.

Indeed, British Indian Jewish communities had not featured in the earliest seroanthropological inquiries into racial difference on the subcontinent in the 1920s. Malone and Lahiri's study, for instance, had not included any religiously designated groups.[5] The researchers stuck to the usual colonial ethnographic groupings of "tribes and castes." After Macfarlane's 1937 study of the Cochin Jews, however, there gradually emerged a small but strong tradition of works where seroanthropological techniques were deployed to study particular religious groups.

By 1952, Bentley Glass, an American researcher based at Johns Hopkins University's Rh Typing Laboratory, along with lab members Milton S. Sacks, Elsa F. Jahn, and Charles Hess, proposed a new experimental object in human genetic research: the "religious isolate."[6] Explaining what he had in mind, Glass wrote that he thought of a "religious isolate" as one of the "smaller endogamous religious groups," such as the Amish or the Dunkers, who, in Glass's words, might be considered to be a "microgeographic isolate." The great advantage of studying such "religious isolates" was that while geographic isolation—such as on an island—meant the geneticist could never be sure of the factors that produced the genetic difference, in the case of religious isolation, the individuals still lived under geographic and environmental conditions identical to those of their neighbors and hence provided a better basis for comparison.[7]

Glass was clearly not interested in these communities for themselves, but thought they were good experimental objects that might help resolve one of the key debates in human genetics at the time. As he mentioned in the very first sentence of his paper proposing the study of religious isolates: "by general recognition race is now increasingly defined in terms of differing gene frequencies."[8] What was not clear was whether those differences arose through "genetic drift"—the accidental loss of particular alleles in small breeding groups—or through actual selection. Since most geneticists agreed that such questions probably originally arose in small populations and that the questions were best studied in small, "isolated" populations, Glass thought "religious isolates" would be ideal to explore such questions.

As historian Veronika Lipphardt has pointed out, "browsing through textbooks and journal discussions of human geneticists in the 1950s, one can [see that] endogamous groups, more often termed 'isolates,' made up a crucial part of theoretical and methodological considerations, and since 1949, a growing number of empirical studies on 'isolates' had appeared."[9] Yet, as Alexandra Widmer notes, "populations do not exist unproblematically in nature, easily enumerated by scientists."[10] The "isolate," just as much as its offshoot the "religious isolate," thus had to be constructed through particular modes of research design, sampling protocols, and citation networks. Within this larger history of delineating "isolates," researchers had usually turned to geographic rather than religious isolation. Island communities, therefore, became ideal research subjects.[11] Others turned to small, cutoff montane communities.[12] Still others looked at communities living in the midst of harsh deserts. These were the communities that geneticists easily imagined as being insular "genetic universes," sufficient unto themselves.[13] Lipphardt points out the "Darwinian fashion" in which "scientists turned to supposedly isolated populations . . . namely, to island inhabitants, to people separated by geographic barriers such as mountains or rivers, to people that could be studied abroad, in remote places, in colonial contexts."[14]

The significance of constituting a notion of "religious" rather than "geographic" isolation within this larger tradition of geneticists thinking about isolation has not been sufficiently appreciated in much of the extant historiography, which tends to group the two types of isolation together. Glass and his co-authors, however, were clear that this was an innovation, and they were equally explicit about their inspirations for it. They introduced their notion of a "religious isolate" by writing that "a significant study of this type has [already] been made by [L. D.] Sanghvi and [V. R.] Khanolkar."[15] To be sure, Sanghvi and Khanolkar's 1949 study that Glass and his colleagues mentioned was not really about religious groups. (It concerned six specific castes met with in Bombay.) But Glass allegorized caste to religion to argue that both were distinct forms of "microgeographic isolation" distinct from the sort of geographically or environmentally isolated communities that geneticists usually studied. Glass's citation of Sanghvi and Khanolkar, even though they were not really speaking of religious groups, is also remarkable because there were in fact earlier seroanthropological studies of religious differences in the Middle East, such as Ernest Altounyan's very early seroanthropological study of differences among Arabs, Armenians, and Jews in Aleppo, or a later study by Karima A. Ibrahim of the blood groups of Muslim and Coptic Egyptians.[16] That Glass ignored these Middle Eastern studies, while citing the Indian study, suggests a greater global visibility of Indian seroanthropology, at least relative to Middle Eastern seroanthropology.

Whatever might have informed Glass's selective citation of Indian research, it is worth noting that his formulations themselves were in turn quickly picked up in India. By the early 1960s, Glass's work on "religious isolates" was being cited by eminent human geneticists, such as K. R. Dronamraju and P. C. Dutta.[17]

The geneticization of religious difference in the mid-twentieth century was emphatically not a simple story of the conceptual dissemination of a "Western" model to India. Instead, the "religious isolate" emerged as an experimental object through a series of crisscrossing citational networks and overlapping research protocols that increasingly tied Indian race scientists into particular global tracks, and in which caste and religion could become geneticized allegories of each other. Sanghvi and Khanolkar in 1949 worked within an emergent Indian tradition of seroanthropology birthed and nourished by researchers like Macfarlane in the 1930s. By the end of the 1940s, however, men like Sanghvi were much more plugged into American geneticist networks. His own time spent at Columbia University and his later involvement in influential projects with the World Health Organization (WHO) meant that his work was increasingly read and cited by American researchers such as Bentley Glass. Younger researchers, such as Dronamraju, in the 1960s again encountered conceptual frameworks through Glass and others that had already emerged from earlier dialogues with Sanghvi's work.

Even as a race-marked and geneticized idea of religious difference was thus stabilized in India and beyond, roughly between 1935 and 1966, the key terms of this framework often remained underdefined. The word "race" continued to be used by researchers throughout the period, and yet its meanings were not always consistent. Likewise, "religion" was operationalized in highly selective, specific, and reductionist ways within these research projects. Interrogating these usages not only illuminates the process by which religion was reconstituted as a genetic object in both India and beyond, but also sheds light on the way seroanthropologists "made up people."[18]

What Is a Race?

The Hirszfelds had inaugurated seroanthropology by proposing a "biochemical race index." As its name suggests, they thought of race as essentially a biochemical object. Indeed, they wrote that "if we inject into dogs the blood of other dogs it is in many cases possible to produce antibodies. By means of these antibodies we have been able to show that there are in dogs two antigen types. These antigen types, which we recognise by means of the iso-antibodies, we may designate biochemical races."[19] According to their index, the Hirszfelds identified sixteen races in the population they studied: English, French, Italians, Germans, Austrians, Serbs, Greeks, Bulgarians, Arabs, Turks, Russians, Jews, Malagasies, Negroes (Senegalese), Annamese, and, last of all, Indians.[20]

The rapid uptake of seroanthropological research meant that more data and more groups accumulated quickly. The sixteen races of the biochemical race index seemed to require endless expansion. In 1925, Reuben Ottenberg, an American researcher, proposed a solution to this proliferation of categories. According to Ottenburg the various racial groups could be regrouped by the proximity of their racial index value into six more general "types." These types were European, Intermediate, Hunan, Indo-Manchurian, Afro-South-Asiatic, and Pacific American.

The Indian pioneers Malone and Lahiri, when they adopted seroanthropology, explicitly rejected both the Hirszfelds' and Ottenberg's classifications. They argued that neither of these had recognized the racial variety within British India. Instead, they fell back on Risley's schema. According to the Census of India in 1901, over which Risley had presided, the British Indian population could be divided into seven "types": Mongoloid, Dravidian, Indo-Aryan, Turko-Iranian, Mongolo-Dravidian, Aryo-Dravidian, and Scytho-Dravidian. As the hyphenated names suggest, in Risley's schema these types arose from the admixture of older, more primordial races.[21] Malone and Lahiri, inspired undoubtedly by the boldness shown by seroanthropologists like the Hirszfelds in rejecting the earlier, more rigid racial classifications, stated that though they had adopted Risley's "types," they would in fact "speak of these 'types' as 'races.'"[22]

In these early interventions we already begin to see a newer, more dynamic idea of 'race'. Unlike earlier authors like Risley who tried to hold on to an idea of races being primordial and anterior to the historically mixed groups actually available to them, these later researchers seemed to be inching towards a more dynamic idea of races. In this latter view, races were groupings that could arise out of older groupings. Malone and Lahiri, for instance, mentioned how the Turko-Iranian was a "mixed race" of recent vintage, while the Dravidian was the oldest race and indigenous to the subcontinent.[23]

Macfarlane, when she undertook her study of Cochin Jews, had to confront the definition of race directly. The Jewish community in Cochin had long been splintered into several factions. There were the white, black, and brown Jewish communities, allegedly designated according to their complexion and descent, that maintained separate syn-

Fig. 2.1. Photograph comparing "White" and "Brown" Jews taken by Macfarlane. Source: Macfarlane, "The Racial Affinities of the Jews of Cochin," pl. 1.

agogues and did not intermarry (fig. 2.1). The most numerous of these were the Black Jews, but they in turn were divided into the *Meyookasim* and the *she-enam Meyookasim*. The factionalism and rivalries between these communities meant that each community often questioned the Jewishness of the others. These challenges occasionally led to legal disputes as well. One such dispute had led to a petition to the Chief Rabbi of Cochin in 1881.

Macfarlane found that the Chief Rabbi's opinion was important for two reasons. First, the rabbi himself had used the word "race" in the opinion. Second, while affirming that the *Meyookasim* were "equal in racial purity to any of the Jews throughout the world," he simultaneously averred that if the non-*Meyookasim* performed the ritual bath called *tabila*, they too would become "equal to our Israelite brethren."[24] What was more, the rabbi based the latter opinion on an earlier opinion on a similar dispute in Cochin issued by a sixteenth-century Alexandrian rabbi, David Ibn Abi Zimra.[25] Based on these observations, Macfarlane observed that "the term 'race' as used by the Chief Rabbi signifies 'people' or cultural group."[26]

While noting the Chief Rabbi's understanding of race, Macfarlane herself maintained a distinctly biological notion of race. She stated that the Jews "must be recognized, like the Aryans, as a cultural group of mixed racial strain, they are not a race in the biological sense."[27] This idea of "race in a biological sense" becomes a little clearer in her concluding comments. Macfarlane wrote that "the White Jews have preserved a Near Eastern and European Semitic strain and show no indications of admixture with Malayalis . . . The Black Jews are the descendants of mixed Semitic and native Malayali ancestors . . . Judaism is a culture and Jews come from many races."[28]

In 1938 Macfarlane guest-edited a special issue of the journal *Current Science* published from Bangalore by the Indian Institute of Science. The issue was on genetics and carried articles by several major international researchers, such as the future Nobel Laureate Hermann Muller, American eugenicist Charles Davenport, Danish biotechnology pioneer Øjvind Winge, Japanese geneticist Hitoshi Kihara, and others. The issue also featured a frontispiece with Gregor Johann Mendel's photograph. In the introduction to this important volume, Macfarlane wrote that "genetics has already shown that all men are cross-breeds of various degrees and has thus removed the main prop of the antiquated theory of discrete races among us. Just as Darwin, Wallace, Weismann and other naturalists of the last century destroyed the theory of special creation of species, so modern geneticists have shown that both race and nation are dynamic, continuously mutable aggregations."[29]

This dynamic, but ultimately still clearly biological, view of race remained dominant among Indian seroanthropologists till the end of our period of study. In the influential 1949 study by Sanghvi and Khanolkar that inspired Glass, for example, after critiquing anthropologists such as B. S. Guha for having grouped various Marathi-speaking castes in a single "Marathi race," the authors insisted that "for any investigation on the distribution of genetical characters of the people of India, the ultimate racial units of importance are the endogamous groups."[30] These endogamous groups were dynamic in the sense that they could change over time through migration, intermixture, isolation, and other factors.

Sanghvi's PhD work at Columbia University had been supervised by Leslie Clarence Dunn, a man who did much to transform the notion of race, and Sanghvi's formulations were close to Dunn's. As Lipphardt writes of Dunn, "what he wanted to prove by investigating endogamous groups was that they were biologically distinct not because of some inexorable racial fate; instead, they had become biologically distinct in a relatively short time because they had been separated from other humans by social and cultural means."[31]

Dunn's and Sanghvi's views were part of a larger shift that took place from the 1930s and 1940s. With the rise of the new evolutionary synthesis as well as the emergence of population genetics, the older, more rigid and essentialized views of race began to mutate into more dynamic, statistically driven notions of human difference coded as "popula-

Fig. 2.2. Satyavati Sirsat. Undated photograph. Source: Bhisey, "Satyavati M. Sirsat (1925–2010)."

tions." Yet, the population concept, far from spelling the end of the race concept, continued to biologically reify race.[32] As Lipphardt and Jörg Niewöhner succinctly put it, "in the laboratories, however, away from the relative glare of public social and ethical scrutiny, race continue[d] to be used for rather pragmatic reasons of data availability, comparability, and marketing chances. The difference between politically correct, purified self-description and everyday practice is significant."[33]

The post-1947 studies of religious difference in India evince this disjuncture between a superficial, rhetorical avoidance of "race" coupled with a practical intensification of research into racialized populations. One of the most ambitious of these researches was Satyavati Sirsat's project. Sirsat was a Karachi-born daughter of Gujarati Theosophists educated at a prominent Theosophical school, and would become one of the pioneering Indian cancer scientists (fig. 2.2).[34] The seroanthropological project in which she explored religious differences was in fact her PhD dissertation. Submitted in 1951 and supervised by V. R. Khanolkar, it compared the blood groups of six different communities in and around Bombay. Four of these were religious groups: the Bene Israel, the Baghdadi Jews, the Parsees, and the Iranis. The remaining two were caste groups, namely, Vadnagar Brahmins and Desasth Brahmins.[35] Comparing the Marathas with the Bene Israel Jews, Sirsat wrote that "the impression of apparent homogeneity of the two groups as seen from the OAB [blood] groups is removed by extreme diversity in two other genetic traits."[36] Clearly, what is proposed is a more reliable basis for biological differentiation of the two groups, rather than the rejection of any such biological differences.

Another contemporary researcher interested in the biology of religious difference was P. N. Bhattacharjee. Employed at the Anthropological Survey of India, Bhattacharjee published two studies devoted to the seroanthropological mapping of religious differences. One of these compared "Muslims" and a local Brahmin group in West Bengal, while another compared "Muslims" with another local Brahmin group in Kashmir. In both studies he assiduously avoided any description of the two groups as "races" or even "populations," using the latter word only in the most general sense. Instead, he described the groups he was comparing as "communities" and "religious isolates."[37] As Elise Burton points out in her study of human genetics in the Middle East, however, "while terms such as 'population' and 'isolate' may have seemed more precise and objective than 'race,' the processes by which geneticists reified these group categories were no less politically contingent than earlier racial classification systems."[38] Bhattacharjee does not explain, for instance, why "Muslims," despite the long-known denominational, class, and caste differences among them, should have been considered a single "isolate."

Bhattacharjee also published a number of other studies that were almost identical in their designs apart from the fact that they studied groups labeled as "tribal" rather than as "religious isolates." In these latter studies, he seemed far more explicit in his use of racial categories. In a 1968 study of Ladakhis, for example, Bhattacharjee wrote that "from the serological data as a whole, it may be concluded that the Ladakhis have both the Caucasoid and Mongoloid strains towards their racial composition . . . Ladakhis show distinct Mongoloid characters such as high cheek bones and oblique slit eyes but contain fundamentally the racial characteristic of their neighbors, the Purigi and the Machnopa, variants of the Oriental race which settled very early in the Western Himalayas."[39] The following year, in another study, Bhattacharjee gave a more general sense of his particular take on race science when he approvingly cited Sir Arthur Keith to say that "nearly all who have sought to explain the differentiation of the population of India into racial types have sought the solution of the problem outside the Peninsula. They have never attempted to ascertain how far India has bred her own races."

Based on these comments, Bhattacharjee said that more studies of further genetic markers were "desirable to come to stable conclusions regarding this racial problem."[40]

In both Sirsat and Bhattacharjee we notice a steady increase in the number of genetic markers and the number of different groups being studied. Unlike former researchers, these postcolonial researchers seemed to actively avoid the word "race" as well as any firm conclusions. Some of Bhattacharjee's publications in particular looked like raw data rather than research articles with any argument. His comparative paper on Muslims and Rarhi Brahmins of Bengal is a good instance. There was no conclusion and no footnotes, only a raw listing of data and methods used. Especially when compared to his unabashed invocation of the "problem of race" in his other studies done around the same time, it seems there was some sensitivity about avoiding the explicit framing of religion as race. Yet, following Lipphardt and Niewöhner, beneath the carefully controlled presentation of these studies, their framework, protocols, and raison-d'être were all derived from a tradition of race science.

We see in general in this period a proliferation of racial categories that emerged with the demise of more rigid, fixed classificatory systems. In their place, at least potentially, any group could now become racialized. These ever proliferating races were not imagined as eternal entities, but they were most certainly understood as reified biological groups that emerged from genetic isolation.

Mendelian Similes

Any systematic attempt to examine the seroanthropological explorations of religious difference in India must begin by confronting the apparent categorical chaos that seems to run riot in these studies. The chaos seems perceptible on at least three levels. First, individual studies seem to compare apples and oranges. Thus, Bhattacharjee compares, for instance, Rarhi Brahmins, a fairly specific local Bengali *jati* group, with Muslims, an undifferentiated religious group.[41] Usha Deka Mahapatra and Priyabala Das likewise studied Brahmins—a *varna* group, Kalitas—a local Assamese *jati* group, and an unspecified group of Muslims.[42] Studies such as Inderjit Singh Bansal's comparison of unspecified Buddhists and Muslims in Ladakh, which was at least consistent in comparing two religious groups, were extremely rare.[43]

Second, once one tries to compare different studies done by different researchers at different times and in different parts of India, there is another level of categorical heterogeneity. With little pressure to develop consistent categories across studies, various studies often ended up using identical or similar terms without defining these terms or clarifying their usage. When Bhattacharjee, for instance, speaks of "Muslims" as an unspecified group in two of his separate studies (in West Bengal and in Kashmir), does he wish to suggest that members of these two groups would automatically belong together if they were to meet at a common location? Or are we supposed to implicitly understand the geographic location of the study to be a qualifier? But in that case, how did he ensure that none of the Bengali Muslims had Kashmiri ancestors or the other way around? This is especially relevant since Kashmiri adventurers are known to have served in armies of the precolonial nawabs of Bengal. Even more confusing in some ways perhaps is the question of how Bhattacharjee's "Muslims" in the Kashmir study related to Bansal's "Muslims" in the Ladakh study. Ladakh, after all, is usually seen to be a region within Kashmir.

Finally, a third level of confusion arose because different researchers described the boundaries of endogamy in overlapping groups differently. In 1940 when S. D. S. Greval and S. N. Chandra studied the blood group frequencies for various communities living in Calcutta, they used "Mohammedans" (Muslims) as a single, undifferentiated community. Bhattacharjee, as we have already seen, did this as well. Yet, other researchers refused to treat "Muslims" as a single, endogamous unit. Malone and Lahiri, for instance, had not made much of the common religious background in the groups they had tested, and they separated out Muslim groups into other endogamous units like "Pathans" and "Hazaras." Likewise, D. N. Majumdar in his 1943 study of "Muslim blood groups" in Lucknow distinguished between Sunnis and Shias.[44] Later still, in a seroanthropological study in the Dewas district of Madhya Pradesh, Narendra Kumar and Alok Kumar Ghosh studied a group called "Nayta Muslims." According to Kumar and Ghosh, the "Nayta Muslims" were "an agricultural people and they do not intermarry with other Muslim sects."[45]

Even while confining ourselves to the various studies of "Muslim blood," we can see how much categorical heterogeneity and apparent chaos reigned in the seroanthropological studies of religious difference. If we wish to compare these with the ways in which other religious groups, such as Jews and Zoroastrians, for example, were categorized, there would be even graver complications in finding a consistent classificatory logic. Santiago José Molina has found similar categorical heterogeneity in mid-twentieth-century Latin American genetics. Molina argues that this heterogeneity resulted from an "agnosticism towards the ontological status of the racial and ethnic categories" that itself grew out of the shift toward populationist thinking.[46] Though population thinking did not really undermine race thinking, it did undermine the ontological status of fixed classificatory schemes. As we have noticed above, this led to a constant proliferation of racial categories.

Molina also points out that the "semantic ambiguity of the term 'population'" allowed researchers from different backgrounds to collaborate around it. It was this fluidity of the term that also allowed mid-twentieth-century researchers to build large research and citational networks. The networks that connected Dunn, Sanghvi, Glass, and Bhattacharjee were eventually held together by the plasticity of "population" as a category. "Instead of formalizing nomenclature or eliminating 'race' from scientific vocabulary," writes Molina, "in order to mollify disciplinary differences experts leveraged the semantic ambiguity of 'population' to loosen ontological commitments to the reality of categories."[47]

There was one major difference between the Indian studies of religious difference and the Latin American studies. Molina points out that in the latter context, the categorical heterogeneity was eventually structured in large measure by a dichotomy between "primitive" and "industrialized" populations.[48] In the Indian context, while the notion of the "primitive" was certainly a redolent and powerful one, its longer history in colonial ethnology delimited its application to groups other than those that were classified according to religious labels.[49] Groups such as the Black Cochin Jews or Bengali Muslims or the Nayta Muslims of Dewas might have been classified as "agricultural" or "peasant," but there was too hoary a tradition of marking them off from groups such as the "Santals" or the "Paniyans." Indeed, when Macfarlane found that the blood group frequencies of the White Cochin Jews seemed to match those of the Paniyans, she wrote that "the White Jews and the Paniyans have nothing in common anthropographically but a high frequency of agglutinogen A."[50] As a result, she ignored the similarity in her analysis.

Given such a traditional delimitation of the category of the "primitive" to certain specific "tribes," the categorical heterogeneity in religious categories needed to be stabilized by other grids. This alternative grid proved to be the notion of "Mendelian populations." It was both the pliability and the possibility of having nested groups that made the "Mendelian populations" framework such a valuable resource for Indian researchers.

Indian researchers uniformly drew their notion of a "Mendelian population" from a formulation by geneticist Theodosius Dobzhansky in 1950. Sirsat, for instance, while studying the Bene Israel, baldly stated that "Jews in all parts of the world largely conform to Dobzhansky's definition of a Mendelian population."[51] G. N. Vyas and his colleagues were even more forthright, writing that what made "genetical variations in India highly complex and unique" was precisely that it was a "peculiarly heterogeneous community" made up of a number of "Mendelian populations."[52] Sanghvi, himself a former student of Dobzhansky, also repeatedly emphasized that "the people of India are broken up into a large number of endogamous groups (or Mendelian populations . . .) whose members are forbidden by social law to marry outside their own group."[53]

Dobzhansky had defined a Mendelian population as a "reproductive community of sexual and cross-fertilizing individuals which share in a common gene pool." These "reproductive communities" were, in Dobzhansky's words, "compound" in nature. That is, they could be nested within larger Mendelian populations. Hence the species was the largest Mendelian community, but species was then "differentiated into complexes of subordinate Mendelian populations, which may be referred to as subspecies, races, or local populations. Each of these subordinate gene pools may, like the gene pool of the species, be uniquely characterized in terms of frequencies of gene alleles and chromosome variant."[54] Gannett rightly describes the individuality of Dobzhansky's "Mendelian populations" as "the individuality of nested Russian dolls."[55] Finally, Dobzhansky specified that, "in addition to the geographic races, man has evolved national, linguistic, religious, economic, and other cultural isolates."[56]

Using the notion of a "Mendelian population," one could therefore get around two crucial problems. On the one hand, they could compare groups whose basis was very different from each other. Thus, a linguistic group could be compared with a religious group, a caste, a class, a group of island-dwellers, or any other group that could be argued as being somehow or other reproductively isolated. On the other hand, they could compare groups of very different sizes, because the concept jettisoned questions of scale. A larger group, even if it was internally heterogeneous, could be compared with a much smaller, more tightly knit, and more homogeneous group. As Gannett points out, from Dobzhansky's perspective, "no classification will include all existing Mendelian populations; racial differences are of different orders of magnitude, ranging from differences between neighboring villages to differences among continents." And yet, this seething variability of scale did not undercut the naturalness of the groups thus identified.[57]

While the scalar equalizing in Dobzhansky's program has been noted by scholars like Gannett, what is relatively underappreciated in the extant historiography is the capacity of Dobzhansky's Mendelian populations to render groups of very different kinds comparable. In effect, the notion of a "Mendelian population" was a fertile simile-producing machine. It was a highly productive tool for constantly rendering one human collectivity similar to another, no matter how different the two were. Castes, religions, classes, villages, and continents are extremely different kinds of collectivities. Yet, all of them can be compared by geneticists if they are tagged as "Mendelian populations." All that was needed to make them comparable was that they were somehow believed to be reproductively isolated. But reproductive "isolation," as Alexandra Widmer points out, does not just exist in nature. They have to be perceived and iteratively constructed as such. In Molina's words, they have to be iteratively made into "populations of cognition"; that is to say, "the iterative and pragmatic construction of the population as a boundary object through categorical work is an example of the intricate coordination and compromise required to produce knowledge through populations of cognition."[58]

The iterative production of religious communities within the networks of seroanthropological and genetic research happened primarily by rendering them as similes of "castes." The fact that so many of the comparative studies—those of Sirsat, Bhattacharjee, and Kumar and Ghosh, to name but a few—actually ended up comparing religious and caste groups overdetermined this simile. The "caste" groupings themselves, however, especially their understanding as perfectly endogamous entities with married, reproducing couples at their core, were a fiction of colonial ethnology.

Scholars have long pointed out that the very notion of "caste" is a colonially reconstructed category that stitched together distinct notions of *varna* (ranks in a fourfold, ideal typical classification with textual basis) and *jati* (localized groupings often with little textual basis that regulate various aspects of sociality); moreover, the notion of caste endogamy has been demonstrated to be historically problematic. Dalit feminist scholars, for instance, have drawn attention to the existence of forms of sexual labor that violated caste boundaries.[59] This was particularly conspicuous in the existence of an array of different forms of ritualized sexual labor.[60] There were also extramarital forms of sexual contact that operated between ritual and convention.[61] Finally, there were forms of reproductive sex, even in certain courtly and elite contexts, that did not follow the rules of caste endogamy or marriage.[62] As Durba Mitra has recently pointed out, colonial ethnology and upper-caste, national social science combined to naturalize caste endogamy and, through it, often violent forms of control of female sexuality. Instead of accepting this fiction, Mitra argues, we need to see endogamy as "an unstable and incomplete act of caste enforcement, a form of imposition based on the ontological life of caste domination."[63]

This is particularly important in our discussion because the existence of reproduction outside of caste endogamy undermines the very basis on which Mendelian populations were constructed. Contrary to Irawati Karve's position, the existence of "concubinage, housemaids and prostitution" was not simply a "biological exception" that could be bracketed off to preserve endogamy as the dominant social form of conjugation.[64] From a genetic perspective, such "biological exceptions" cannot simply be brushed aside—especially given the allegedly prolonged and extensive distribution of such forms of nonendogamous reproduction.

Strikingly even the high Sanskrit juridical texts, frequently presented as the fount of caste endogamy, accepted a wide array of forms of social reproduction and lineage survival that did not involve caste endogamy. The *Dharmasastra* had traditionally recognized twelve different types of sons who could all, with certain stipulations, inherit from their father and maintain the lineage.[65] These include the *Gudhaja,* a son born of the wife by an unknown man; the *Kshetraja,* a son born of a wife by levirate; the *Dattaka,* an adopted son; the *Kritaka,* a son bought from his (biological) parents; the *Sahodha,* a son born of a wife already pregnant at the time of marriage; the *Swayamdattaka,* a son who has offered himself for adoption; and the *Krittima,* a made-up son. None of these forms of reproduction sustains the assumptions that inform genetic constructions of Mendelian populations through endogamy.

It is important to note, as well, that the fiction of caste endogamy had already been challenged in 1917 by the eminent Dalit leader, jurist, and intellectual, Dr. B. R. Ambedkar. Though he dubbed it the "essence of caste," he argued that far from being an ancient custom, it was most likely a later imposition on an earlier practice of exogamy and then too remained incomplete in its application.[66] In fact, an earlier generation of Indian scientists interested in the history of caste, such as the eminent chemist Sir P. C. Ray, had also found that the rigors and even the practice of endogamy had varied in different historical epochs.[67]

Yet the myth of perfect caste endogamy remained powerful and productive for Indian geneticists. Some of Sanghvi's junior colleagues, for instance, stated that

> the Indian population offers an excellent opportunity to study the blood groups and other genetic characters in different endogamous groups existing in its peculiar social system. In Gujarat State in Western India, several well-defined endogamous groups are available. The

> members of these groups are endogamous to the extent of marriage within the caste but not between the blood relatives. Such a caste system in Gujarat and all over India is a well evolved system. It has been deliberately built up by transcendental seers and is unrelated to the linguistic or geographical distribution. The division of four Varnas—namely Brahman, Kshatriya, Vaishya and Shudra—is not arbitrary but is based on sound principles of heredity. The social system of *Chaturvarnas* (Four Classes) had been in existence in the Vedic Society for several millennia [Bhagwad Gita, chap. 4, v. 13; Taitariya Upanishad]. These four classes were eventually divided and sub-divided into smaller classes now known as castes, mainly by the process of intermarriage.[68]

Interestingly, the study compared one Jain group (one of very few studies to include Jains) with two Hindu caste groups. It was the fiction of perfect and quasi-eternal endogamy, not to mention a confusion of *varna* and *jati*, that allowed the researchers to iteratively produce the three groups they studied as genetically isolated Mendelian populations.

Snapshot Biohistories

Dunn and Dobzhansky pioneered the New Evolutionary Synthesis in the 1930s and 1940s. They promoted a new understanding of evolution as a "dynamic process driven by migrations, isolation, environments, mutation and random shift."[69] One remarkable outcome of this synthesis was the significant foreshortening of the time needed for evolutionary change. As Lipphardt puts it, the "time span of a few hundred years of isolation seemed enough to bring about a novel combination of allele frequencies."[70]

Such a shorter time span brought evolutionary research out of deep time and located it alongside historical time, familial genealogical time, and a host of other genres of temporal ordering. Moreover, as Lipphardt points out, the replacement of "pure races" by "isolates" compelled geneticists to directly reckon with narratives of reproductive isolation of the communities they studied. This meant that by the 1950s geneticists increasingly relied on several types of "nonscientific knowledges." "Accounts of isolation and endogamy from all kinds of sources, provided by linguists, ethnographers, historians, sociologists and others, as well as myths and claims of collective identity, turned out to be essential for studies of human genetic diversity."[71] Writing about the genetic fieldwork of Victor McKusick in the 1960s among the Old Order Amish in Pennsylvania, Susan Lindee points out that gossip, feelings, social consensus, and, above all, family Bibles (with their handwritten notes of births, marriages, and deaths) became key sources, alongside more recognizably "scientific" artifacts such as blood tests and x-rays.[72]

This is where Indian research traditions seem to proceed along a very different route. The diversity of sources and knowledges that entered seroanthropological research actually shrank during the period between 1930 and 1970. In Macfarlane's researches in the 1930s, we see a wide variety of sources and knowledges being used. Her study of Cochin Jews includes copperplate land grants, juridical verdicts by early modern and modern rabbis, family histories, community memories, and Dutch colonial records. She even cited some modern histories of Cochin, such as Francis Day's *Land of the Permauls, Or, Cochin: Its Past and Present*, though the works were slightly dated by Macfarlane's time.[73] By contrast, a little over a decade after Macfarlane, when Sirsat did her research, she relied only on some community memories and an extract from a single Parsee chronicle, the *Kissah-i-Sanjan*, and two published histories of the Bene Israel community written by members of the community.[74] For the other communities she included in her study, namely, the Baghdadi Jews, Iranis, Vadnagar Brahmins, and Desasth Brahmins, there were no concrete sources at all, merely a general claim that they were all perfectly endogamous communities.

Even Sirsat's more limited engagement with other records and knowledges often appears plentiful in comparison to the studies that followed in the 1960s. P. N. Bhattacharjee's studies of Kashmiri and Bengali Muslims, for instance, did not include a single archival source, mnemonic account, or social-scientific analysis in the bibliography.[75] Similarly, Narendra Kumar and A. K. Ghosh's study on Nayta Muslims and others in Dewas included two sources of nongenetic or statistical information. These were a colonial ethnological volume, R. V. Russell and Hiralal's *Castes and Tribes of the Central Provinces, Vol. 4*, and a volume of the 1931 Census of India.[76] Perhaps more egregiously, N. P. Parikh and his colleagues, in studying the Oswal Jains and others, included two nongenetic sources of information, namely, the *Taittiriya Upanishad* and Diwan Chaman Lal's *Hindu America*. The former is a religious work dating from before the Common Era, but its textual history, like most works of such vintage, is extremely convoluted and its relevance to the particular groups studied by Parikh's team is not at all clear. Chaman Lal's work, which Parikh cites approvingly as authoritative, was a kind of mid-century "weird history" that claimed that Indians had originally "colonized" and settled the Americas and that the Aztecs, the Mayas, and the Incas were all descended from the original Indian "colonizers."[77]

The work of Parikh and colleagues, with its reliance almost exclusively on works lacking the social-scientific scaffolding that renders such knowledge reliable in academic circles, raises the question of how these authors chose the extremely limited number of sources and knowledges they relied on. While they too, like many of their Euro-American counterparts that Lipphardt discusses, relied on nongenetic information to frame their chosen "isolates," it is unclear how they chose what sources and knowledges to rely on. It appears that from the high point of the 1930s, when Macfarlane made a clear effort to be rigorous in her usage of these materials, there was a progressive deskilling of geneticists in the use of nongenetic sources and knowledges. As a result, this part of the research was left entirely to the whims, fancies, and idiosyncrasies of individual researchers. That the study by Parikh and colleagues, with its extremely problematic grounds for recognizing the groups studied as "isolates," was still published in a leading international journal, *Human Heredity*, suggests that this part of the research protocols was not scrutinized with sufficient rigor.

This unskilled and cavalier approach to the nongenetic information upon which geneticists relied to constitute their religious and other isolates also overdetermined the kinds of narratives about the past that were produced in these studies. Given that the studies were enabled by a specific understanding of the past that allowed for the groups to be considered isolated in the first place and that they then produced a narrative about the groups' past, it is best to think of these studies as themselves being a genre of historical narrative. Indeed, Lipphardt calls them "biohistories."

Describing "biohistories," Lipphardt and Niewöhner explain that, in them, each piece of "genetic information" is "wrapped within a biological story." It is this story that explains how this variety emerged. These stories are evolutionary stories constructed using a small number of discrete concepts, such as "mutation," "selection," and "drift."[78] Nadia Abu El-Haj describes this reductionist emphasis on a small number of explanatory "events" in genetic histories in terms of what some geneticists call the "principle of parsimony."[79] Moreover, biohistorical narratives "are not limited to the domain of science." They are adopted and deployed by nations, families, ethnic groups, and others in the construction of their own identities and pasts.[80] What remains unexamined in such biohistorical reductionism is the fundamental question: is it at all "reasonable to assume history is best understood through a minimalist, parsimonious logic"? What if it takes more and not fewer events to understand a community's complex past?[81]

This generic compulsion to use a small number of discrete "events" to constitute the narrative already lends these historical narratives the aesthetic of a snapshot. The cavalier and instrumental use of nongenetic sources about the past affirms an explicit orientation toward "snapshot historiography." Writing about "snapshot historiography," Ahmed Ragab explains that snapshots have the "ability to remove context and reject connections, [and to] produce an easily movable, focused and concentrated image" of a community.[82] As a form of historiography, what is important, Ragab points out, is that "snapshotting" elevates the contemporary interests of the person taking the snapshot as the only meaning-making process that organizes and connects the snapshot with other similar snapshots.[83] In other words, denuded of all complexity and any attempt to locate the "events" leading to the alleged isolation in their own context, the geneticist is allowed to freely cherry-pick and organize a set of snapshots about the past so long as it authorizes their final conclusions. There is no need to consult a wide array of sources and archives in any systematic fashion, nor indeed is there a need to engage with the authority of nongenetic forms of knowledge.

Combining Lipphardt's notion of biohistories with Ragab's notion of snapshotting, we can see that what the Indian seroanthropologists produced were essentially "snapshot biohistories." Consider, for instance, Sirsat's accounts of the communities she studied. For the Bene Israel, she found that the community's seroanthropological markers differed from those that had already been reported of "Ashkenazim," "Sephardim," and "Oriental" Jews, while they matched closely with those of the "Marathas," a historically heterogeneous group that geneticists saw as a homogeneous general population in and around Bombay city.[84] She explained this by an initial migration into India, and an initial period of intermarriage, followed by endogamy—snapshotting elements of the Bene Israel's own accounts of their past as provided in the two published accounts by Haim Samuel Kehimkar and Moses Ezekiel.

Though the "exact date" of their arrival was not known, Sirsat wrote that "it is believed" that the ancestors of Bene Israel left Galilee sometime after 175 BCE during the persecution of the Jews by Antiochus IV Epiphanes. They traveled via the Red Sea and arrived at the Konkani port of Cheul. Their ships were wrecked near the Kanheri islands before Cheul, and only seven men and seven women survived the wreck and reached the port. Sirsat does sound a note of skepticism about the number seven, which she thinks might have been used because it was a "sacred number." In any case, she writes that the expatriates quickly adopted local language, customs, and dress, and became virtually indistinguishable from their neighbors. Finally in 1000 CE, a visiting rabbi, David Rahabi, "reinstated the active observance of the Jewish faith."[85]

As anthropologist Yulia Egorova points out, "there is no evidence for [the shipwreck story] of any kind."[86] Even more interestingly, she finds that other local communities have very similar origin stories. The Chitpavan Brahmins, who are also from the Konkan region, have an almost identical origin story, including the insistence on seven couples as the original number of expatriates. By the early nineteenth century, when others encountered the Bene Israel and the Christian missionaries started writing about them, they already existed locally as a caste of oil pressers who did not press oil on Saturdays and were known as the *Shaniwar Teli* (Saturday oil pressers). Other South Asian Jews tended to disregard their claims to a Judaic identity. Separate rabbinical courts in Baghdad and Jerusalem in 1914 had forbidden marriages between Bene Israel and other Jews. Baghdadi Jews in Bombay also refused to recognize them as Jews. After much of the community moved to Israel in the 1950s and 1960s, this precipitated a major legal conflict in Israel with the Sephardi Chief Rabbi, Yitzhak Nissim, forbidding intermarriage with the Bene Israel unless they were able to prove both their Jewish ancestry and the absence of intermarriage with non-Jews over several past generations. Though a large mobilization by Bene Israel organizations, protests, and intervention from the Knesset defused the situation in the 1960s, skepticism about the Jewishness of the Bene Israel did not die out. In 1997 the Chief Rabbi of the Israeli city of Petah Tikvah once again officially forbade his employees from validating new marriages with Bene Israel.[87] While the controversies in Israel were naturally driven by local issues and interests, their repeated eruption underlines that the Jewishness of the Bene Israelis, leave alone their migration history or endogamy, was far from the transparent truth that Sirsat presented it to be.

Naturally, in South Asia such disputes, mostly involved the immediate Jewish neighbors of the Bene Israel, the Baghdadi Jews. In these often long and bitter disputes, the Bene Israel often weaponized exactly the kind of fictionalized picture of perfect caste endogamy that has proven so attractive for geneticists. They argued that because of the caste system and its intricate endogamous hierarchies, they could not possibly have intermarried and therefore must be of pure Jewish descent. They responded to the similarities of their origin stories with those of the Chitpavan Brahmins, an extremely elite group in the Konkani social order, by claiming that the Chitpavans themselves had in fact been Jewish.[88] There have also been divisions within the community, with a split between *gora* (fair) and *kala* (dark), which seem to mimic the divisions in the Cochin Jewry. Though, once again as with the Cochin Jews, this did not often reflect actual complexions, it does attest to internal social differentiations.[89]

Two things are patently clear from this discussion. First, at the time when Sirsat was conducting her research, in the early 1950s, neither the Jewishness of the Bene Israel nor the shipwreck story was universally accepted. Other Jews and some non-Jewish locals were skeptical. The protestations of endogamy, too, met with similar skepticism. Nor indeed were these suspicions of intermarriage directed exclusively at the early years of the Bene Israel's historical memory. As is evidenced from Rabbi Nissim's 1997 demand for proof of halakhic marriage in the immediately past generations shows, at least in his mind the intermarriage issue was not a thing merely of deep history; it was a possibility in recent generations as well. Second, as is clear from both the Bene Israel's own responses and the ways in which others responded to them, their pasts and even the imagination of those pasts were closely and inextricably intertwined with the histories of other local groups like the Baghdadis, the Chitpavans, and the Agris (a caste whose practices allegedly some Bene Israel mimicked).

Sirsat's snapshotting of history jettisoned both the complexities of this past and the contested, relational nature of the claims she presented as factual. She did not give any details about how these ongoing contestations might have influenced other aspects of her research protocols, especially the choice of subjects to include. Most important, perhaps, she gives us no evidence of how and why she chose to unquestioningly adopt one specific origin narrative in a context where there clearly were multiple, competing narratives about the community's past.

Bhattacharjee's snapshotting was even more acute in its decontextualized interpretations. In his West Bengal study, for instance, he drew all his samples of both Rarhi Brahmins and Muslims from the Chandernagore subdivision of the Hughli district. Given his utter avoidance of any nongenetic archive or knowledge, he did not offer any explanations as to why these two groups should be treated as genetically isolated units. Muslims in Hughli and Bengal more generally, as everywhere else, are divided in numerous ways. Their connections to their non-Muslim neighbors are also often blurred, perhaps much more so than in other parts of South Asia.[90]

According to the district gazetteer for Hughli district issued by the government of West Bengal in 1972, while "an overwhelming majority of the Muslims of the district are Sunnis of the Hanafi sect [,] Shiahs [*sic*] are found in almost all areas of Muslim concentration but they are most numerous in and around Hooghly-Chinsurah."[91] This latter area,

Hughli-Chinsurah, was part of the Chandernagore subdivision where Bhattacharjee collected his samples. Moreover, most of the well-to-do and professional classes among the Muslims in the district tended to be Shias. The gazetteer also informs us that while the relationship between the Shias and the Sunnis was amicable, they did not usually intermarry.[92] This is in keeping with historian Rafiuddin Ahmed's comments that the vast majority of the rural population in Bengal was divided into "mutually exclusive social groups—whose membership was defined by ascribed factors of heredity–similar in many respects to the Hindu *jatis*." Besides this, there was also a major division between the *ashraf* ("honorable") and the *atrap* ("lowly"), who were divided by "an almost impassable barrier."[93] The *ashraf* were usually divided into four descent categories, namely, Syed, Shaikh, Mughal, and Pathan, that operationally resembled the fourfold, ideal-typical *varna* classifications whose relation to everyday divisions was aspirational, contested, and historically changeable. Indeed, elite Hindu converts were assimilated into the fourfold division with reference to their claimed *varna* ranks, with Brahmins becoming Syeds, Kayasthas becoming Shaikhs, and so on.[94]

Not only was there no intermarriage between many different groups of Muslims, there were also conversions into Islam from other castes, including Brahmins. Furthermore, there was the still more complicated issue of there being forms of social community where the line between Hindu and Muslim seemed largely blurred. Dirom Gray Crawford's 1902 history of the Hughli district mentioned the existence of several popular *Pir* (Islamic saints) cults whose devotees were both Hindus and Muslims, and whose rituals often combined elements of both religions. Satyanarayan/Satyapir was likely the foremost of these figures. He was represented either by a dagger resting upon a rectangular board and covered by a cloth, or a small mound of earth smeared with vermilion, and his worship required both a Brahmin and a Muslim priest co-officiating. Other, similar figures included Saichand Pir, Almon Sahib, and Shayamba Pir. Each of these figures commanded extensive and cross-denominational worship.[95] How this might have impacted marriage patterns or extramarital sexual relationships within these congregations remains largely unexplored.

Again, pointing out this complexity and overlap is not merely to expose how Bhattacharjee obviously misunderstood the nature of the communities he studied. Rather, the point is to draw attention to how these snapshotted, decontextualized images of communities and their pasts were precisely what enabled, reinforced, and authorized the kind of fictions about perfectly siloed, endogamous Mendelian populations that seroanthropologists mobilized. By ignoring social and historical complexity, systematically eschewing reliance on nongenetic knowledges, and working with frequently misleadingly simplified images of actual communities, geneticists simultaneously consumed and produced snapshot biohistories.

Exogeneity

Ellis David Zackay became a citizen of the United States of America sometime in the early 1920s. With his emigration, the once grand house of Zackay came finally to an end in Cochin.[96] The Zackays were one of two families that claimed to be descended from the kings of Shingli, the legendary ancient Jewish kingdom in Kodungallur (previously Cranganore).[97] Shingli's history and even its existence remain covered in the mists of time. Yet lore about it was rife among Cochin Jews and their neighbors. Most community disputes, origin stories, and so on all went back to the days of Shingli. Macfarlane argued that the kingdom had begun to collapse under pressure from the Portuguese in the early sixteenth century and that the last Jews left for Cochin around 1565.[98]

In Cochin, the Zackays had belonged to so-called White Jewish communities, and indeed, some of their ancestors were referred to in early modern Portuguese records as *brancos* (whites). Yet, as J. B. Segal has pointed out, not all the members of the royal house of Shingli were referred to as *brancos,* and it was clearly not a reference to their actual complexions.[99] Indeed, when the then thirty-two-year-old Ellis David Zackay landed in Quebec on 16 June 1919 having traveled from Liverpool by the *S.S. Tunisian,* his complexion was described as "dark," his hair as "dark," and his eyes as "brown."[100]

What makes Zackay's successful claim to whiteness even more remarkable is the situation of Jews in America in the 1920s. As Eric Goldstein points out, the racial logic of the Progressive Era had mainly distinguished "white" and "black," and tended to group the (mostly European) Jews as "white." However, things became more complicated after the Great War. There emerged vocal and powerful, if often contradictory and inconsistent, camps of "antisemites" and "philosemites."[101] In this milieu, while American Jews generally continued to be considered "white", their "whiteness" was much more fragile than earlier. The larger numbers of American Jews in urban, distinctly modern occupations also meant that much of American ambivalence toward postwar modernity itself often became attached to Jews, and "many White Americans cast the Jewish 'race' as an infiltrating force."[102] In a more biographic and impressionistic mode, Karen Brodkin states that even in the 1940s, American Jews were not usually considered "white folks."[103]

Most of the people Goldstein or Brodkin write about, however, were the children and grandchildren of European Jews, not people of Eastern origin, like Zackay.

Whereas American Jews managed to hang on to a fragile and vulnerable "white" identity in America of the 1920s, the decade saw a distinct rejection of the claims of Indians to be similarly included. Indeed, the decade of the 1920s was one of the most difficult times for Indian immigrants to the United States. While since 1894 a small number of Indians had indeed managed to obtain US citizenship by arguing that they were "white" or "Caucasian" according to the prevailing racial frameworks of the day, all that changed in 1922. In that year, Bhagat Singh Thind, a Sikh veteran of the Great War, was initially granted citizenship and then saw it revoked by the US Supreme Court on appeal. Thind had argued that he was "a high caste Indian and having no intermixture of Dravidian, or other alien blood, and coming from the Punjab, one of the most northwestern provinces of India, the original home of the Aryan conquerors."[104] Jennifer Snow has shown how the judgment against Thind's citizenship unyoked the previously tight linkage of "whiteness" and the category "Caucasian" by creating a discourse on "white civilization" from which the racialized figure of the "Hindoo" was excluded.[105] Doug Coulson finds the rejection of Thind's claims following from the "highly racist and xenophobic era of the 1920s." He adds that racist views in the 1920s enjoyed more "importance and respectability" than at any other point since the Civil War.[106]

Zackay was not unlike Thind in many ways. Born in March 1887, he was five years older than Thind. Like Thind, he had been in the United States before the outbreak of the Great War. Both had arrived in the same year, 1913, and in the same state, California. Zackay had traveled there from Hong Kong. In Hong Kong he had been employed as a merchant clerk at David Sassoon & Co. The Sassoons were Bombay Jews, and Zackay's employment with them might well have been arranged through Indo-Jewish networks. In any case, when the war broke out, Zackay joined the US Army in New York and served in the war with the British Expeditionary Force. He ended the war with a medal, an injured left arm, and an English wife, Edith, whom he married in Croydon.

Besides the biographic similarities, Zackay's claim to "whiteness" was also similar to Thind's. It depended on a claim that, notwithstanding nearly two millennia of residence on the subcontinent, the Zackays had not married into any local groups, thereby maintaining their exotic whiteness through perfect endogamy. Yet, where Thind failed, Zackay succeeded.

More important, Zackay's success was transmitted back to Cochin, where Macfarlane heard of it. White Jewish interlocutors in Cochin, citing the Zackay instance, said they were "regarded as part of the 'White Race' by the United States Government and one of them obtained American citizenship since the World War."[107] Despite Macfarlane's more critical approach to such oral material than would be evinced by those coming after her, she seems in the end to have accepted these claims. Her conclusions about the Cochin Jews posited that the White Jews descended on their maternal side from the Jewish lords of Shingli, while on the paternal side they were descended from Jewish immigrants who had arrived from Arabia, North Africa, and Europe some 450 years ago. By contrast, she thought the Black Jews were in fact descended from "mixed Semitic and Malayali ancestors."[108]

Recent scholarship on race and genetics has done much to interrogate and expose the ways in which indigeneity has been geneticized.[109] It has pointed particularly to how indigeneity has been instrumentalized in a variety of ways that have often been at odds with the life, livelihood, and dignity of the indigenous groups that have been the focus of such researches. In South Asia, and more particularly in these studies of religious difference in South Asia, however, the predominant figure is one of *exogeneity* rather than *indigeneity*.

Exogeneity has received nowhere near the critical attention that indigeneity has marshaled. In part, this might have to do with the distinctive political stakes of majoritarian, postcolonial democracies like India as opposed to settler colonial countries, such as the United States and Australia. In India, the politics of exclusion from rights and public life is usually articulated around being "outsiders" rather than "indigenes." This is not to deny the marginalization or exclusion of communities framed as "indigenous." It is merely to point out that the exclusion is usually not framed in terms of their indigeneity. Indeed, majoritarian Hindu nationalists since the mid-twentieth century have usually framed their own identities as "indigenes." Moreover, of late, through vigorous outreach campaigns, the Hindu right has often successfully mobilized the marginalized "tribal" communities against Muslims precisely along the lines of the insider/outsider dichotomy.[110]

By contrast, Jews, Christians, and overwhelmingly Muslims are marked off as being exogenes. As historian William Gould perceptively points out, "the foreigner in 'Hindu' rhetoric was positioned specifically and unambiguously in relation to the subcontinent itself. This demonization of 'foreigners' in India also had a relevance, sometimes explicitly stated, for earlier invaders of the subcontinent—the various Muslim communities who, despite differing ethnic origin and levels of assimilation into indigenous cultures, were easily characterized as 'invaders' by virtue of their religion."[111] This demonization of outsiders was a key formula by which the Hindu nationalists obtained "a means of national consolidation, national history, and mythology, in the sense of cultural purity."[112]

Scholars of diasporic Hindu nationalism have emphasized how "new forms of affiliation and expressing solidarity" are becoming possible today that no longer "seek nor assume identification with the nation-state."[113] Exogeneity, prima facie, seems to fit into this model. But whereas Hindu diasporic nationalism has deployed the idiom of religion as a way of eschewing the politics of race in the United States, what I am describing as exogeneity was clearly based upon an intensification of race as religion. Moreover, this politics was not peculiar to the US milieu in which Zackay landed. It had roots in British Indian ideas about the racial character of small religious minorities and continued to flourish in early postcolonial India.

The exogeneity of groups like Jews, Armenians, Parsees, and other similar groups that became popular research objects for race scientists in India was mirrored in the ways "Indian" communities were rendered exogenous by race scientists working in South Africa, Malaysia, and Singapore. We will hear more about these in chapter 4.

It is not surprising that, of all the seroanthropological studies conducted in the period between the 1920s and 1970, not a single one chose to essentialize the "Hindu" as a single category. Indeed, studies of Jains or Sikhs, too, remained extremely rare in the period under consideration. It was instead minoritized groups, such as Jews, Parsees, and Muslims, all seen to have come from the "outside," who were repeatedly studied.

Ellis David Zackay's fate, however, reveals another dimension of exogeneity. Exogeneity could also open up certain specific pathways for moving out of India. As race science continues to remain a protean component of immigration regimes across the world, it is important to attend to the ways in which claims to being exogenous enable certain specific forms of interstate mobility and political citizenship. In the Indo-Jewish context, recent anthropological work has shown how the right to perform *Aliyah*, that is, to settle in Israel, has increasingly become entangled with the possibilities of proving Jewishness through genetic testing.[114] It has opened up a new pathway for Indo-Jewish communities such as the Bene Menashe and the Bene Ephraim whose Jewishness, like the earlier cases of Bene Israel and the Black Jews of Cochin, is under suspicion by other Jewish groups.[115] What I want to underscore here is that these genetic tests are all premised on subcontinental exogeneity.

Whereas articulations of genetic indigeneity have often been aligned with political claims to occupy the land upon which a community resides, sometimes even to the exclusion of other communities,[116] exogeneity is frequently linked with mobility and claims to a land away from that upon which the community resides. Such mobility, expectedly, is not exclusively driven by the claims of exogeneity alone. Economic rationalities undoubtedly inform the eventual decision to move or not move, but claims to exogeneity can both reinforce and enable such mobility.

Zackay's move to the United States was not driven merely by his inherited claim to exogeneity. He was an economic migrant and in fact followed a familiar pathway for migrating out of South Asia, often via Southeast Asia, and to the Americas that had gradually developed since the late nineteenth century.[117] Notwithstanding his claims to an ancient aristocratic lineage, it appears that Zackay eventually left Cochin because he was struggling financially. Initially, he moved to a clerical job in Hong Kong, then to San Francisco, then briefly to New York and London, before moving back to California. His migration, therefore, was similar to that of some other South Asian economic migrants, including Thind, but the exogeneity eventually allowed him certain opportunities that were closed to men like Thind.

These economic mobilities inserted old racial identities into new political, legal, and social contexts. The work that these identities had to perform multiplied. Their deployments and iterations multiplied. Thind's claims about "upper caste" status or "Aryanness" were akin to Zackay's claims to "whiteness." Unlike the nineteenth-century circulation of ideas about the "Aryan race," these claims did not simply land in new contexts and evolve independently. They also fed back to Punjab and Cochin. A successful immigration application in the United States was amplified in Cochin and used to push a different goal. In the end, the meanings given to "whiteness" in San Francisco and Cochin did not coincide, but they both seemed to mutually reinforce, that is, to intensify, these racialized identities.

Exogeneity, it is worth noting, always opened up the door for people to be pushed out of nations, like India, that are increasingly defined by an emphasis on indigeneity and autochthony. While it would be misleading to draw too strong a causal connection between seroanthropological race science and the out-migration of India's smaller minorities, it would be equally erroneous to entirely absolve the history of race science from the emptying of the once-famed Jew Town of Cochin or the many beautiful synagogues of the Konkan and the Deccan.[118]

INTERCHAPTER

Letter 3

Najrul Islam
Kampala, Uganda
5 January 1934

Dear Hemenda,
I cannot adequately express in words my joy in hearing back from you so soon. That you have taken the time to respond so warmly to my letter fills me with gratitude.

I must also thank you for mentioning the story of Numa. I shall endeavor to press the locals further to see if Numa and Tana might, in fact, be the same child. The colony of Neanderthals somewhere in the forests of British East Africa is also fascinating. Do, pray, ask your friends if they have any further details about where this colony might be? My plans here are likely to detain me for a few more months in this part of Africa, and should your friends have any more information, I would be delighted to enquire into the matter myself.

Your comments about the inveterate promiscuity of man—and indeed his myriad ancestors, is also well taken. Not being of a philosophical bent of mind, I cannot say that I had ever given such matters the kind of thought that you have obviously given them. Still, as a man of the world, I can vouch for the general truth of what you say. No human community, in my experience, whatever their race, tribe, caste, or religion, has ever remained entirely carnally isolated from their neighbors. The most stringent laws of endogamy have been sacrificed at the altar of Love or butchered at the feet of Lady Lust!

I had never thought of connecting this simple fact, which I have seen repeated everywhere I have traveled, with the question of evolution. I am eager to see how you will deal with these weighty matters in *Man's First*

Adventure. Needless to say, it will be a singular honor for me to receive a copy of it directly from you, my hero, whenever the book is published. In the meantime, I shall await it eagerly.

You must also forgive me for my unthinking use of the phrase "playing god." I share your conviction that man is not just biology. I am not a writer like you and have none of the felicity with words that you have. I used the phrase to express my unease with the way these future-humans seemed to be remaking themselves. I did not intend to deny the spiritual dimension of man's being. While I cannot say that I have yet had the kind of experiences your physician friend has had, I remain open to the question.

In fact, upon reading Mr. Sen's diary, I find that there is indeed a reference to the continued operation of spiritual forces, even if only in a dark and malignant way, in the kingdom of the future-humans. Mr. Sen writes of a visit to the State Museum where the futuristic kingdom celebrated its achievements. Like the Indian Museum you mention, they too crammed the building with models and fossils, all telling the story of their arrival in Africa and their subsequent evolution. Pride of place in the museum was reserved for a statue of Chadrasen, the first scientist who created this new race from the original shipwrecked group. Yet, Mr. Sen was warned by the locals that no one dared to enter the room with Chandrasen's statue after dark. Upon enquiring about this, Mr. Sen learnt that the statue was supposed to come alive after dark and had been heard moving around with its heavy stone legs. Anyone who met the statue in the museum at the time did not survive to tell the tale!

Mr. Sen does not tell us anything further about this intriguing aspect of this society, but it would seem that underneath the rationalized, reassembled humanity they had created, there still lurked the dark forces of cosmic evil. A man who had sought to play God seems to have ended up being a bloodthirsty *pishacha*!

For me, though, Hemenda, the spiritual dimension of life comprises more than just a set of mysterious forces that we do not comprehend. To my mind, it is also inseparable from man's quest for the sublime. His thirst for beauty, his quest for adventure, his appetite for knowing the unknown, all of this and more, I feel, Hemenda, connects us to the infinite. I do not know of God, at least not yet and not in the way you suggest. But when I am standing face to face with death in the form of a wild beast, where the tiniest sliver of time stands between life and death, or when I am sitting alone on a moonlit night on the banks of the vast Lake Kivu, I can sense being part of something immense, something more infinite than my body. At these moments I can sense the vastness of the universe at the very core of my being.

I digress. Forgive me. But perhaps it is not such a digression. For the more I think of it, I realize that perhaps what disturbs me most about the future world that these strange future-humans—I really know not what to call them—have tried to create is the utter lack of any sense of beauty, pleasure or aesthetic appreciation. How can man reach for the infinite if there is no sense of beauty? no appetite for sublime pleasures?

There is a section of Mr. Sen's diary that describes the strange food habits of these people. Let me paraphrase some parts of the diary for you in Mr. Sen's voice:

"The Chief Scientist and I reached the next room to find innumerable fruits upon two small tables. The Chief Scientist placed his barrel-like body on one side of the table and withdrew his three legs into the shell. Then, turning to me, he said, 'Come, Amal, let us eat. We do not waste too much time on eating.' Having said this, he opened his mouth wide, like that of a huge cobra, and within a minute or two consumed a heap of fruits. Finally, he gulped some water and, turning once again to me, said, 'So, we have finished eating. Let's rise.' I thought to myself, of course you do not waste time in eating, why would you, when you can eat as much as ten people in a mere couple of minutes. But I cannot eat like this. I will have to starve if I take my meals with this Chief Scientist. I had barely managed to eat two apples by the time he had finished. But the Chief Scientist seemed unconcerned. He said to me, 'You know, eating too much blunts the mind. One should not eat much. I only eat the bare minimum necessary. Keeps you sharp. Though, alas, there are some even amongst us who think that eating a lot increases your bodily vitality. All such ideas are tosh!'"

Indeed, Mr. Sen also writes about one Bhombol, a glutton who was looked down upon by the Chief Scientist for his stupidity. Yet, it was this Bhombol who eventually helped Mr. Sen escape. Though he was considered amongst the least intelligent in that futuristic kingdom, it was this Bhombol who seems to me to be most human, most like us.

Mr. Sen's greatest supporter in the strange and cruel place, however, was the Chief Scientist's daughter, Kamala. Though Mr. Sen initially saw her as a beautiful young lady, he later realized that this was only a temporary form that Kamala used to assume. She too, like the others of her kind, was in fact a formless, shell-dwelling creature. Yet, unlike her father, Kamala resented the lack of beauty and love. She had heard stories about how their ancestors used to be beautiful and hence donned that shape. But this angered her father, who saw it all as an utterly foolish form of prodigality. Kamala was alone in her quest for beauty.

The senses, Hemenda, are not just our windows to the world. They are what allows us to cultivate pleasure and beauty. To imagine the infinite. To experience the sublime. How can a people remain human if they have given up on sensory pleasures and aesthetic goals? How are we any different from animals at that point? Indeed, even animals can distinguish good and bad food. There is a modicum of aesthetic experience there. But these people are baser still. They are mere automata!

Imagine for a moment, Hemenda, the kind of danger that Tana, the Daughter of the Gorillas, courted repeatedly to steal food from my tent. Was it because she was starving? No, of course not. All the gorillas had enough food in that region. My food itself was meager: biscuits and jelly that I had carried with me. She came for the novelty and the sheer pleasure of eating these novel things. Reveling in their newness. Even one such as she, whose humanity itself is moot, could delight in sensory fulfillment. What kind of humans would we be once we lose the ability to feel moved by the rich world of sights, sounds, smells, and flavors that surrounds us?

While I am no epicurean, neither have I ever been attracted to the austerities of the fakirs we occasionally see in India. Scenes of men who sought God by lying for years on a bed of nails put me off as much as the lurid accounts of the gurus who live lives of indulgence and excess. Denial and overindulgence, it seems to me, are but two sides of the same counterfeit coin. The senses are what allows us to cultivate beauty and aspire to the infinite. To rise above our limited existence. So, there could be few futures more disappointing to me than one in which sensory exuberance is entirely banished, where eating is merely a way of fueling the body and beauty is a needless extravagance. No, Hemenda, that future is not for me.

I am sending you Mr. Sen's diary along with this letter. But I do hope we can continue to exchange missives. For all the beauty and romance of the African jungle, I must confess that I miss a good Bengali *adda*. I shall look forward to hearing from you.

Your brother,
Najrul

CHAPTER 3

A Taste for Race

> O Gods, sources of bitternesses, eternal and indifferent, you condemn us to unceasing returns!
>
> —Frantz Fanon, *Parallel Hands*

Praveen Kumar Seth of the Punjab University introduced his innovative study in 1962 by reminding his readers that "the ability to taste or not taste PTC [phenylthiocarbamide]" was a "sort of honorary blood group."[1] The comment had originally appeared in Robert Russell Race and Ruth Sanger's influential textbook, *Blood Groups in Man* (1954), where they had written that, "blood groups apart, it [PTC taste sensitivity] is the only *normal* character the manner of whose inheritance has been established with reasonable certainty."[2]

From the 1950s seroanthropology began to diversify its earlier focus on simply mapping the frequencies of the main ABO and MN blood group systems. The institutional homes for such studies had diversified, with the expansion of the Anthropological Survey of India; the rising number of anthropology departments in universities; and the increasing presence of research institutes like the Indian Cancer Research Center in Bombay, the Hematological Unit of the Calcutta School of Tropical Medicine, and, somewhat later, the Human Genetics Unit at the Indian Statistical Institute in Calcutta. Along with institutional growth, the range of genetical traits mapped also began to grow rapidly. Seroanthropology itself began to evolve and diversify at this stage into an array of new disciplines, ranging from physical anthropology to human biology and genetics to population studies.

Throughout these evolutions, one aspect of the studies remained fairly stable. The overwhelming conviction remained that India's population was biologically segregated into a number of discrete endogamous units and that their differences could be captured by tracking inheritable serological traits, and these assumptions continued to frame the new studies. One study in 1958, which tracked PTC along with a few other traits, began with the redolent declaration that "classification of human races has been undertaken since the days of Herodotus. The races are still being classified and reclassified." The authors pointed out that basing racial classifications on "biological differences" was a much more recent development. Moreover, such studies of "biological differences" could follow either of two separate "methods": the "morphological methods" of the older studies, such as those deploying anthropometric techniques; and the "genetical methods" of the newer studies.[3] The view that race classification had been given a new, biological basis through the development of genetics was fairly widespread in postcolonial India. Numerous authors, such as the team just cited, explicitly invoked the language of race. Indeed, in 1971, Ranjit Chakraborty, a statistician based at the Indian Statistical Institute, Calcutta, even sought to develop a statistical "theorem for race mixture" based entirely on blood group data.[4] Such explicit comments help to refute claims that the prewar studies of "race" were entirely displaced by the postwar studies of "populations." Rather, race and population—whatever the rhetorical differences in the use of the two terms—constantly bled into one another in experimental practice and sustained a longer genealogy of racialization of human difference.[5]

A wide variety of new traits began to be racialized through the burgeoning studies of the 1950s and 1960s. Most of these were the so-called "antigenic factors," that is, those factors which could be tracked through the body's immune response. The original ABO blood group studies were of this sort. Gradually more factors, such as Rh, Kell, Duffy, and Lutheran, began to be mapped. Subgroups of the recognized groups—for instance A_1 and A_2—also emerged. But alongside these, by the mid-1950s, two nonantigenic traits had also become important for researchers. One of these was considered pathological, namely, the sickle cell trait, about which we will hear more in the next chapter. The other, without pathological implications, was the ability to taste PTC.

In 1931 a brief mention had appeared in the journal *Science* stating that Arthur L. Fox, a chemist at DuPont, had discovered that "the chemical para-ethoxy-phenyl-thio-carbamide is intensely bitter to some persons, but tasteless to others."[6] The following year Fox published a longer account of his discovery. He had been working at the DuPont laboratory next to a colleague, Dr. C. R. Noller. That day, he had been working with phenylthiocarbamide, and as he tried to put it in a bottle, "the dust flew around in the air." Noller, upon tasting the dust accidentally, complained of its intense bitterness. Fox had been closer to the bottle and hence tasted more of the dust. Yet, he tasted nothing. Upon Noller's complaining, he deliberately tasted some of the dust and still did not taste anything. Based on this initial discovery, Fox tested a number of different people and declared that he had "established that this peculiarity was not connected with age, race or sex."[7] Yet, remarkably (given this categorical denial of a racial basis to the trait), in 1966 S. R. Das, a researcher employed at the ASI in Calcutta, published a lengthy survey of the extant researches on PTC explicitly titled "Application of Phenylthiocarbamide Taste Character in the Study of Racial Variation." Not only had PTC's differential tasteability been thoroughly racialized by then, but indeed it had emerged among Indian researchers like Das as one of the most productive traits by which to racially classify people.

The earliest published Indian study to deploy PTC taste as a racializing marker was carried out in 1949 by L. D. Sanghvi and V. R. Khanolkar. The study used PTC-tasting ability as one of seven genetic traits, including mostly blood groups, to genetically compare six caste groups in and around Bombay. The study not only compared the PTC rates of the six groups studied, but in some cases, such as that of the Koknasth Brahmins and the Chandraseniya Kayasth Prabhus, it also compared the rates with other groups like "Whites" and "Negroes."[8] These comparisons demonstrated the tendency to deploy PTC-tasting rates as a way of defining and comparing both Indian and non-Indian groups within a global racial matrix.

Another, similar early study comparing multiple groups and multiple genetic traits was carried out as part of a PhD dissertation at Bombay University by one of Khanolkar's students, Satyavati Sirsat. As we saw in the last chapter, Sirsat investigated several so-called migrant communities of Bombay. Earlier blood group analysis, done by Sanghvi and Khanolkar during the 1949 study, had failed to find any biologically legible difference among some of these groups. For instance, Marathas, who were usually, though highly problematically, considered by biologists as a single, homogeneous group indigenous to the region, had been found to be indistinguishable from the Bene Israel Jews. This is where Sirsat's study came in handy. She found that PTC-tasting, along with the new Rh factor, did in fact show statistically "highly significant" differences between the two communities.[9]

Studies such as Sirsat's demonstrated the value of the PTC trait, even as newer studies increasingly tracked a whole range of traits rather than any single trait. Single-trait studies devoted to PTC, though fewer, continued to flourish as they promised to make previously illegible forms of difference scientifically meaningful. D. K. Bhattacharya of the

University of Delhi, for instance, tested Anglo-Indian subjects living in Lucknow, Kanpur, Delhi, Bombay, and Pune exclusively for PTC taste sensitivity. Based on that alone, he concluded that "the frequency of non-tasters in general shows that the population is intermediate between European and Indian frequencies. They show a rise in the non-taster gene frequency with the rise in dosage [*sic*] of European ancestry in their parentage."[10]

In comparison to these largely local initiatives to study urban groups (often in fact from the immediate vicinities of the laboratories), a slightly different trajectory of research began to map PTC taste sensitivity among more rural and distant communities, often classified as "tribal." Possibly the earliest such work was carried out in 1952 as part of a collaborative Indo-Australian study of 132 members of the Chenchu tribe in Guntur and Kurnool districts in modern Andhra Pradesh. The study was initiated as part of a large number of studies carried out throughout Asia and Oceania by the Australian serologist Roy T. Simmons and his team based at the Commonwealth Serum Laboratories in Melbourne. Simmons's work in turn was part of a larger, more diffuse project that Warwick Anderson has called the "cultivation of whiteness." Scientists such as Simmons sought to prove that Aboriginal Australians were "dark Caucasians."[11] Their interest in the Chenchus derived from a long history of claims about Australian Aborigines being related to Indians, which had begun with T. H. Huxley's comment that the lascars he saw at the East India Docks in London resembled Aborigines. Simmons and his Melbourne colleagues, however, seldom traveled outside of their Australian laboratories. Most of their foreign studies, such as the Chenchu study, were performed by local collaborators who collected the samples and shipped them to Melbourne for analysis. Of the men on the ground in India, G. W. L. D'Sena, a physician at the Christian Medical College in Vellore, accompanied by two other technicians, had been the one to actually collect the samples.[12]

In the case of PTC, the authors mentioned that they could classify 79 of the 132 subjects tested as "tasters," 3 as "weak tasters," and 50 as "non-tasters." But it was not clear how these figures fitted into their general analysis. The authors asserted that "on the basis of blood genetics alone the relationship between the Indians and the Whites appear to be much closer than that between the Indians and any other race, although one might suspect also a slender relationship with the Australian Aborigines." Their conclusion, however, relied almost entirely on the frequencies of the Rh factor.[13] Though this study is perhaps somewhat exceptional in the degree to which it lacked internal coherence and pulled in multiple different directions,[14] the way PTC statistics were deployed suggests the emerging popularity of the marker as well as the lack of clarity about how it fitted into the new postwar race science.

By the early 1960s both rural, fieldwork-based studies and urban studies were thriving. New research centers, away from the traditional urban academic centers at Bombay, Calcutta, and Lucknow, now began to emerge. The universities of Delhi, Chandigarh, and Bhubaneshwar, to name only a few, became centers for genetic research into human difference, and PTC taste tracing became popular among some of the researchers. P. Khullar and Prasanta Kumar Chattopadhyay, for instance, were both based at the University of Delhi and conducted several studies individually.[15] J. C. Sharma and P. Parmar were both based at the Punjab University at Chandigarh and also pursued PTC taste studies.[16] Similarly, K. C. Tripathy, Usha Deka Mahapatra and Priyabala Das conducted PTC taste studies at the Utkal University, Bhubaneswar.[17] Sharma's is a good example of the studies from the 1960s. The study tracked the PTC-tasting trait in people Sharma called "Tibetans," "Spitians," and "Lahaulis," defined by their alleged geographic origins—even though, following the influx of Tibetan refugees in the wake of the Chinese takeover, all of the subjects were actually living in the same area. Sharma asserted that the Spitians and Lahaulis were in fact "a hybrid of Mongoloid and Mediterranean racial elements derived from the Tibetan and Hindu sources respectively."[18] Even when he failed to find any statistically significant differences among the three populations, he still held on to their racial difference by asserting that "the proto-Australoid racial strain in the region," rather than the "Mongoloid element," had lowered the nontaster frequency among the Spitians and Lahaulis.[19] The Bhubaneswar study examined three communities in Assam, namely, Brahmins, Kalitas, and Muslims. It did not further specify how it constituted categories such as "Muslims" or "Brahmins." But based on PTC taste sensitivity, it was able to establish statistically significant biological differences between Kalitas and Muslims.

Numerous similar studies were undertaken throughout the 1950s and 1960s. A rough and incomplete estimate suggests that more than thirty full-scale studies involving PTC tracking were conducted in India in those decades. Some of these, such as P. K. Seth's comparison of Kumaoni Rajputs and Brahmins, remained unpublished and are now known only through citations in other studies.[20] Others were published in relatively obscure departmental journals.[21] Several, however, were published in international journals and continue to be cited in the genetic literature.[22] PTC-tasting came to be firmly established in these studies as yet another tool through which to render human difference legible, measurable, and biological.

Summing up much of the published data in 1966, Das attempted to relate the Indian material to global findings and worked out a veritable global racial map based solely on PTC taste sensitivity. He asserted that "Caucasoid populations" had the highest frequencies of nontasters, while tasters predominated among the "American Indians, Mongoloids, Lapps, Egyptians and Negroes." Polynesians, it seemed, were "quite close to Mongoloids and American Indians," and "Eskimos" resembled the Europeans in their frequencies. Australian Aborigines had the lowest frequencies of tasters, which was extremely close to the so-called "Nishadic" or "Proto-Australoid" tribes of India. "High caste Hindu endogamous groups" were mostly distinct from these tribes and had taster frequencies of between 25 percent and 35 percent.[23]

Calibrating Phenotypes

The distinction between "taster" and "nontaster" was the kind of externally legible characteristic that geneticists called "phenotypes." But it was nowhere as clear-cut as one might assume from the enthusiasm among researchers for using the distinction to classify humans into races. As soon as we look more closely at the practicalities of these studies, we are struck by how difficult the task of sorting tasters and nontasters really was and how the distinction itself did not make sense without the scaffolding of carefully calibrated research protocols to hold the two apart.

The word "phenotype," according to the *Oxford English Dictionary*, was first used in 1910 by botanists and denotes "the sum total of the observable characteristics of an individual, regarded as the consequence of the interaction of the individual's genotype with the environment" or "a variety of an organism distinguished by observable characteristics rather than underlying genetic features."[24] The emphasis in this definition and in most of the early uses was on "appearance" and "observation." The term was evidently conceptualized with reference to the visual sense. PTC taste sensitivity, considered as a phenotype, strained the limits of this concept. Unlike the observable traits of botanical specimens, upon which the concept had been developed, PTC taste sensitivity could not be directly observed by an external observer. The observer had to depend upon what the subject told them.

This dependency and the anxieties it produced were hinted at in the earliest studies. Albert Francis Blakeslee of the Genetics Department at the Carnegie Institution of Washington, DC, was among the first to explore these anxieties systematically.[25] Blakeslee took an early interest in the PTC taste sensitivity trait and began to publish on the issue from

1931. He even co-authored a piece that year with Arthur Fox, who had initially discovered the trait. It was also Blakeslee who first propounded a simple Mendelian mechanism for the inheritance of the trait by way of a single recessive gene. Already by the end of that year, Blakeslee voiced concerns about the "psychological influence of expectation" upon test subjects.[26] He explained that "those who had heard about the test and therefore expected a bitter taste might have scored a lower threshold than those who knew nothing of the chemical."[27] Besides such psychological issues, Blakeslee also spoke of linguistic issues. In his case, since he did not seem to have tested non–English speakers, the linguistic problems arose mainly with respect to children. He found that "young children [we]re not as familiar as their parents with bitter tastes." Hence, he decided to exclude from his study all children below the age of ten. But the perception of taste went beyond a strictly "linguistic" familiarity. Blakeslee noted that children brought up in the country usually compared the taste of PTC to that of dandelions, while urban children tended to compare it to some medicine or the other.[28] Another researcher, D. S. Falconer, while presenting an enormous amount of data collected at the Galton Laboratory in London in 1946, observed that "many people pride themselves on the acuity of their senses and would be expected to over-estimate their sensitivity, whereas others through excessive honesty might under-estimate it."[29] Evidently, as soon as the notion of a phenotype went beyond evidence that could be strictly visually observed by the researcher, such as a blood group observed on a slide, a range of anxieties began to gnaw away at notions of experimental objectivity.

Some of these anxieties were visible in the early Indian studies, too. The 1952 Chenchu study by Simmons and colleagues, for instance, despite its reticence about the actual results of the PTC tests, noted that "the field workers assisted by interpreters regarded the results recorded as fairly reliable."[30] That such comments were not considered necessary for the several other antigenic traits tested in the course of the same study betrays a certain underlying anxiety about the reliability of a phenotypic trait that could not be directly apprehended by the eye. S. R. Das's 1956 study of Rarhi Brahmins in the Barisha, Sarsuna, and Behala areas of southern Calcutta similarly mentioned that the study had excluded not just all children under the age of six but also all those who "appear[ed] to be lacking in sufficient understanding for proper application of the test."[31] Another study, carried out by a team of Japanese researchers in eastern and northeastern India, expressed concerns about the allegedly common habit among Khasi subjects in Shillong to regularly "chew very bitter leaves and fruits." The researchers worried that this habit might have led some of their subjects to being "adapted for the bitterness" and therefore "insensitive for the bitterness of PTC."[32]

Overall, though, it is remarkable how little we hear in the Indian studies about these anxieties. Even when the researcher and the subjects were clearly culturally and linguistically alienated, no mention seems to be made of the problems of psychology, language, or taste education. H. N. Agarwal's 1964 study of recent Burmese immigrants in the Andamans, for instance, was pursued across linguistic and dietary differences, and Agarwal almost certainly had to rely on interpreters and translators.[33] Sharma's 1967 study of Tibetans and others along the Indo-Chinese border also, most likely, required translators,[34] as no doubt did Tyagi's study of Oraons and Mundas or K. C. Tripathy's study of various castes in Odisha.[35] Even when the researcher and the subjects hailed from the same region, it is highly likely that local dialects would have varied and hence created some linguistic difficulties. And this is not even to mention the kind of variation that Blakeslee had noted between rural and urban children. Yet, the problems of translating taste were not discussed in these studies.

The existence of complex and frequently localized everyday medical traditions also meant that many Indians were exposed to medicinal bitters for prolonged periods of time. In a recent ethnographic study of metabolic illnesses in Mumbai, for instance, Harris Solomon has described the existence of household remedies that involve the consumption of intensely bitter stuff like *karela* (bitter gourd), *karu* (a bitter bark), and *methi* (fenugreek). One such remedy even involved retaining the harsh, bitter taste of *methi* in the mouth for a whole day.[36] In much of northern and eastern India, similarly, herbs like *chirata* (chiretta), whose taste has been described as an "intense and persistent bitterness," have been used as tonics and stomachics.[37] Likewise, the extensive historical use of quinine since the colonial era for malarial fevers also exposed many Indians to intense bitter tastes for prolonged periods.[38] Undoubtedly, such practices would impact one's sensitivity to bitterness.

Even though the explicit discussion of such impediments to the objective calibration of taste-based phenotypes remained somewhat muted in the Indian context, Indian researchers continued to constantly tweak their experimental designs in pursuit of more stable phenotyping. Blakeslee had already, in the 1930s, proposed the use of a PTC solution rather than crystals. He had argued that making the chemical into a solution would enable better contact with the subject's taste buds and therefore produce more reliable results. The attempt to use solutions, however, produced another unexpected result. Blakeslee found that even among subjects who were able to taste the chemical, there were significant variations regarding the level of dilution at which a person could taste it. Some subjects tasted even weak solutions, while others could only taste strong solutions. This led Blakeslee to propose measuring "taste thresholds" rather than simply dividing subjects into "tasters" and "nontasters."[39]

Blakeslee's findings inspired the preeminent British statistician, Ronald A. Fisher, to take up the challenge of defining "taste thresholds." From 1934 to 1939, when war cut them short, Fisher's team at the Galton Laboratory conducted a series of experiments and collected a large amount of data on taste sensitivity. These experiments were led by Fisher's assistants, G. L. Taylor and R. R. Race. These experiments standardized some of the techniques of mapping the PTC trait and defining "taste thresholds."[40] The methods used in the London experiments were elaborate. For example, they involved a number of different solutions of different solvents, only one of which contained PTC. (The others used quinine, sugar, acetic acid, and seawater.) Moreover, the subjects were told that multiple samples would be without taste. A further attempt was made to standardize the solutions themselves. Thus, first a solution was made using 0.16 grams (gm) of PTC in 100 cubic centimeters (cc) of tap water. This was designated P11. The whole solution was then diluted with an equal amount of water to make a P10 solution. The P10 solution in turn was diluted with an equal amount of water to make a P9 solution, and so forth down to P1.[41]

The intervening war meant that Fisher's experiments were not published until 1946. Once they were available, however, other researchers sought to further develop the methods. H. Harris and H. Kalmus added an extra sorting mechanism to the tests. The subjects were first given a series of solutions, beginning with the most dilute and going upward until the subject first perceived a bitter taste. At this point the subject was presented with eight glasses of water. Four of these contained PTC, and four did not. The subject was told this was the case but was not told which was which. Their task was to correctly identify them. Once this was successfully completed, the same thing was repeated with the next lower concentration of PTC, and so on until the subject stopped being able to correctly identify the glasses.[42] Harris and Kalmus's methods proved the most popular among Indian researchers, though most of them—following a Norwegian-Danish researcher, Jan Mohr—chose to replace tap water with distilled water.[43]

Remarkably, and contrary to the popular fable that moving from "essence" to "frequency" dismantled "race" as a scientific concept, in the case of PTC it was precisely the move *away* from a clear and dichotomous division between "tasters" and "nontasters," and toward a more graduated spectrum of "taste thresholds," that allowed a more full-fledged racialization of the trait to emerge. Thus, for instance, a 1937 study by William and Lyle Boyd, which followed a dichotomous model of tasters and nontasters in order to map racial differences between a motley assortment of Welsh, Ukrainian, Georgian, Spanish, and Egyptian subjects, never became very popular in India.[44] By contrast, a study by N. A. Barnicot, which deployed Harris and Kalmus's methods and found significant "taste deficiency" among "African negroes" and Chinese subjects was widely cited by Indian researchers.[45] Several Indian researchers in the 1960s, such as Tyagi, Agarwal, P. Khullar, R. P. Srivastava, and P. Parmar, cited Barnicot's study.[46] This was especially striking since Indian researchers did cite some of the Boyds' other publications.

Taste, however, remained a truant marker of race. As Das pointed out in the mid-1960s, unwanted variations and anxieties still haunted experimental practices. Researchers, for instance, continued to differ about whether the PTC solution was best given to subjects in liquid form or as paper strips (this method also had been in vogue since the mid-1930s). There were disagreements about precisely how much solution to give the subject. Harris and Kalmus had been generous with the amounts given, but others argued that 5–10 milliliters (ml) would be enough, provided it could be ensured that the solution spread evenly over the tongue and was swallowed in its entirety. Still others found that taste sensitivity varied with the temperature of the solution when given to the subject and sought to determine the optimal temperature. Further concerns centered on the material of the containers in which the solution was stored and the usability of well or river water in the absence of tap or distilled water.[47] Having located a so-called "honorary blood group," therefore, researchers struggled to find a way of reliably standardizing the phenotype.

Yet, interestingly (and notwithstanding such elusiveness), having adopted a graduated scale of taste sensitivity, most of the Indian researchers eventually once again sought to derive a dichotomous distribution out of these measurements. Several Indian researchers of the 1950s and 1960s used statistical tools to recompose the graduated threshold data back into a dichotomous distribution between tasters and nontasters (fig. 3.1).[48] P. N. Bhattacharjee, in the course of a project to map taste sensitivity variation among Rarhi Brahmins and Muslims in Bengal in 1956, clarified that "an arbitrary line of demarcation is deemed necessary for the determination of taster and non-taster phenotype frequencies."[49] Bhattacharjee achieved this end by taking two alternative arbitrary lines and then running statistical homogeneity tests using chi squares before choosing one of them. The general idea was that, while it was unreli-

R. P. SRIVASTAVA 269

TABLE 3

ASTE THRESHOLDS FOR P.T.C. FOR HINDUS AND MUSLIMS

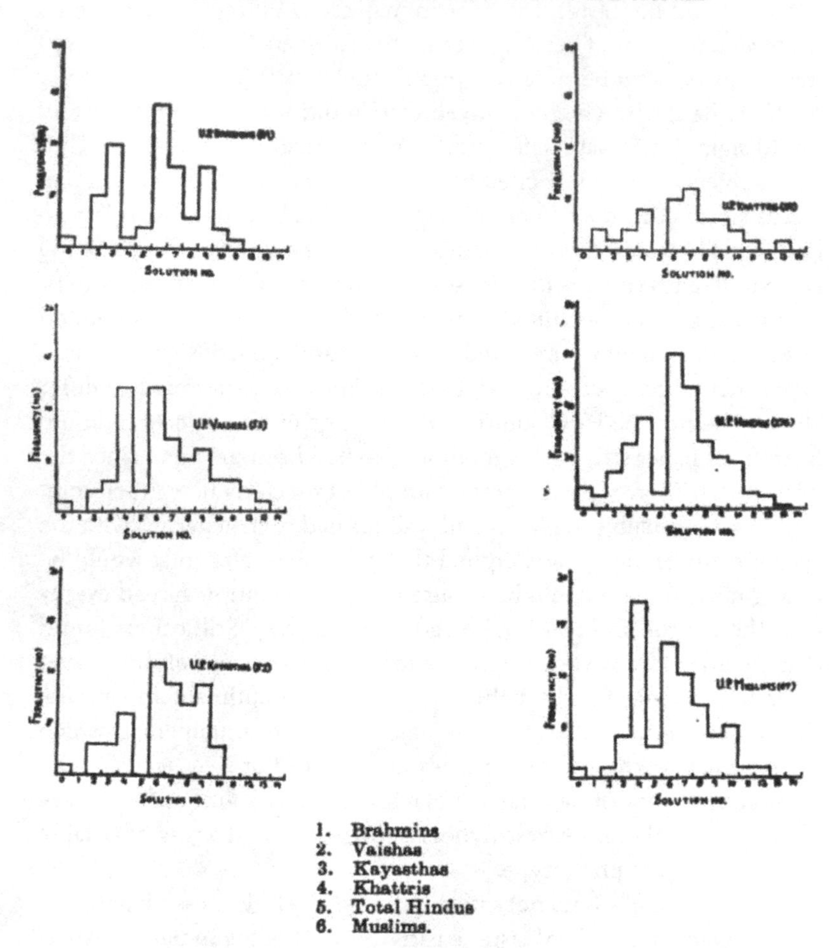

Fig. 3.1. Taste thresholds of Hindus and Muslims. Source: Srivastava, "Measurement of Taste Sensitivity to Phenylthiourel (P.T.C.) in Uttar Pradesh," 269.

able to base such distinctions directly upon experimental findings, one could use statistically constituted dividing lines for separating the tasters from the nontasters. Thus, Das, who emerged as the most influential and consistent researcher of the PTC trait in the 1960s, stated that "the threshold frequencies when plotted usually give a bimodal distribution curve with two maxima corresponding to two phenotypes and a minimum in between."[50]

It was only once subjects had been placed in dichotomous groups that a racialized frequency was worked out for each group, which could then be used to compare and relate them to other groups using chi square values.[51] Tripathy's study of Odiya castes, for instance, compared the frequencies of tasters and nontasters, and came to the conclusion that Odiya Brahmins had PTC-tasting frequencies similar to those of the Rarhi Brahmins of West Bengal and the Vadnagar Brahmins of Bombay, but distinct from those of the Karanas and the Khandayats. The Karanas, in their turn, were closer to the Chandraseniya Kayasths of western India and the Kayasths of the north. Tripathy also noted that all three Odiya castes, notwithstanding their differences, fell within the range of variations noted among the "white populations of Asia and Europe."[52] Mahapatra and Das similarly separated out tasters and nontasters so they could then compare various Assamese castes. They found that Assamese Brahmins had a larger frequency of nontasters than the Brahmins of Bengal, Bombay, Uttar Pradesh, and Odisha, matched the Audich Brahmins of Gujarat. Likewise, the Kalitas of Assam had numbers identical to those of the Rarhi Brahmins of Bengal and the Koknasth Brahmins of Bombay. Assamese Muslims had a higher frequency of nontasters than the Muslims of Uttar Pradesh. These researchers noted that the frequencies of all three Assamese groups fell within the range for "Caucasoid groups of people" and were distinct from the "Mongoloid people."[53] This latter distinction was, of course, implicitly significant given Assam's location next to countries like China and Myanmar that were traditionally imagined to be peopled by "Mongoloid races." While individual results varied, the general thrust of such studies was clearly to produce a single number— the frequency of nontasters—that could be compared across groups in and beyond India. This number was a chi square value (X^2). The measurements of graduated thresholds, ultimately, were not only recategorized into simple dichotomies, but also rendered into a single figure that allowed different groups to be compared easily. Therefore, what had changed, albeit gradually, was not the pursuit of racial identifications, but the extent to which such racial comparisons were mediated by statistical protocols. The new, stronger claims to objectivity were largely founded on the use of statistical tools, such as chi squares and modal distributions.

In part, the reliance on statistical objectivity no doubt derived from the direct involvement in race science of leading and influential Indian statisticians, such as P. C. Mahalanobis. Indeed, we might glimpse broad similarities between the effort to resuscitate dichotomous phenotypes through statistical processes and Mahalanobis's own efforts to resurrect colonial-era anthropometric data sets through statistical reanalysis.[54]

What seemed to be shared was a notion that phenotypes might be difficult to pin down by observation, but that statistical alchemy could render such unreliable data worthy of trust once more. Strikingly, the emphasis remained on finding ways of reestablishing clearly distinguished phenotypes while recognizing that their production required active manipulation of the observable reality.

Here it is crucial to notice how the notoriously elusive character of the sense of taste was rendered, via statistical translations, into a visual trace. A series of ubiquitous tables and, occasionally, bar graphs transformed what was troubling to the researchers, precisely because it was not directly observable, back again into a set of clearly visible inscriptions. As Das's comment points out, eventually the stable phenotypes derived from these visible traces and not from the fleeting ephemerality of what the subject perceived on their tongue.

Ghost of the Gendered Subject

Racialization of the PTC-tasting trait depended as much on the careful scaffolding of phenotypes as it did on the notion of endogamy. Without endogamy, the siloed, racialized characterization of any of these identities as Mendelian populations would collapse. Contrary to this fictionalized and ahistorical account of perfect endogamy laying the basis of stable, biologically distinguishable breeding populations, Dr. B. R. Ambedkar had pointed out as early as 1916 that the "the peoples of India form a homogeneous whole," and that endogamy, which he considered to be the "essence of caste," was in fact a historically imprecise imposition upon an earlier underlying system of exogamy.[55] Dalit feminist scholars have justly taken inspiration from Ambedkar's comments to argue that caste critique and gender critique have to be intertwined.[56] Extending these insights into the history of science, I will argue that Ambedkar's observations are also crucial critical resources for unpicking genetically organized Indian race science.

Two aspects of Ambedkar's observations are important for our discussion. First, despite insisting on the centrality of endogamy to the operation of caste, he also insisted that it was a historically late imposition. Second, and more important, he maintained that the system of endogamy had eventually failed to create stable biological differences, so that the "peoples of India" did in fact form a "homogeneous whole" and the "various races of India occupying definite territories ha[d] more or less fused into one another."[57] Indeed, Anupama Rao has pointed out that this was one of the key ways in which Ambedkar's analysis broke from the disciplinary discourse of colonial sociology. For Ambedkar,

Rao writes, "conflict within the caste system, from sexual violence to structural exploitation, was a conflict between intimates rather than between biologically distinct races, as colonial sociology would have it."[58]

This second point, regarding the eventual failure of endogamy to maintain stable silos of difference, might have resulted from Ambedkar's recognition of the rampant sexual exploitation of Dalit women. This was the other side of the artificially imposed endogamy. As historian Shailaja Paik points out, upper-caste and elite discourse constructed lower-caste women as "public property" and "legitimized upper-caste men's access to the sexual labor of Dalit women."[59] As Durba Mitra succinctly states, the presence of "nonconjugal reproduction represented a fundamental failure of a structuralist theory of caste as endogamy."[60]

The idealized, enabling fiction of endogamy was built on the tacit erasure of the historical specificity of gendered subjects and their actual sexualities. In the case of PTC-taste research, however, this erasure of gendered subjects became blatantly literal. In 1937, William and Lyle Boyd published a study demonstrating significant gender differences within racialized groups. "The discovery of this sex difference, of course, puts a different complexion on the question of the anthropological application of 'taste blindness,'" they mulled.[61] Even more important, in certain cases they found that the racially differentiated frequencies "would disappear if the tests were made on females only."[62] Gender now not only provided an independent axis of comparison, it also threatened to dismantle racial boundaries that PTC research was in the process of reifying.

The Indian studies were quick to adopt gender differences as a topic of data collection. Yet, in all but a handful of cases, these studies then insisted that the gender differences were in fact "statistically insignificant." Bhattacharjee's 1956 study comparing Rarhi Brahmins and Muslims in Bengal found that Muslim men had a slightly higher threshold than Muslim women, while there was a greater frequency of tasters among Rarhi Brahmin women than among Rarhi Brahmin men.[63] Das and Bhattacharjee in 1964 similarly found that among the Rajbanshi caste in Midnapore, West Bengal, the frequency of male nontasters was higher than for female nontasters.[64] Mahapatra and Das found even clearer evidence of gendered variations. They reported that in Assam, among Brahmins and Kalitas, the frequency of nontasters was higher in males, but the opposite was true among Muslims. In all three groups, however, there was clear evidence to show that the frequencies of taster and nontaster varied by gender.[65] Yet, all of these differences were judged to be "statistically insignificant."[66]

"Statistical significance" is nowadays a contested and problematic tool. A recent statement published in the journal *Nature* and signed by 800 scientists called for "statistical significance" to be "retired." They asserted that such calculations were a "waste of research effort and misinform policy decisions." "Statistical significance" also led to the now pervasive problems of experimental results often being irreproducible. They explained: "the trouble is human and cognitive more than it is statistical: bucketing results into 'statistically significant' and 'statistically non-significant' makes people think that the items assigned in that way are categorically different." The problem, as these scientists understand it, arises from the very process of "dichotomization," and is bound to arise in any statistical approach that seeks to dichotomize, "whether frequentist, Bayesian or otherwise."[67] As a result, statistically significant results are "biased upwards and potentially to a large degree," whereas "non-significant results are biased downwards."[68]

This is not the place to delve deeper into the history of "statistical significance." I cite the recent scientific pushback against it merely to flag the artificiality of this measure and the problems with it that are being recognized even within historical traditions of statistically informed experimental practice. For our purposes, what is important is simply that the measurement of "statistical insignificance" allowed for gender differences to be repeatedly subordinated to larger racial differences, even when gender differences were clearly observed.

More remarkable still were the very small number of studies where this statistical technology failed to lay the ghost of the gendered subject to rest. One such study was undertaken in 1958 by G. M. Vyas and his team of Bombay-based researchers. They found that gender differences in PTC-tasting abilities between two groups, Leva Patidars and Talavia Dublas, did in fact rise to the level of statistical significance. Confronted with these results, the researchers declared that they were likely a "discrepancy" caused by "unreliable answers from the subjects tested under most trying conditions of field studies in mofussil [that is, rural] areas. It was difficult to persuade the tribal subjects (T[alavia]. D[ublas].), particularly, the females who were very shy to come forward for this investigation. Hence, the possibility of wrong answers was greater."[69]

The irony, however, was that, despite making such tortuous efforts to reject any overall statistical significance for gender differences in PTC-tasting, most researchers still noted discrepancies between genders. Indeed, Vyas and his colleagues had noted that the Kapol Vania of Gujarat had a higher number of male tasters than female tasters.[70] Khanolkar and Sanghvi's pioneering 1949 study had also found that among the Koknasth Brahmins and the Chandraseniya Kayasth Prabhus, male PTC tasters slightly predominated over female tasters.[71] We also saw how Das, Bhattacharjee, Mahapatra, and Deka all noted gender differences in PTC-tasting. Some, such as Khullar and Agarwal, also presented these differences in visually conspicuous tabular formats.[72]

The question that naturally arises, then, is this: if the researchers were so keen to reject the significance of gendered differences in taste sensitivity, why would they continue to note and remark upon these differences?

Drawing upon the writings of sociologist Avery Gordon, I would suggest that the gendered subject was a "ghost" in these studies: a trace of the "worldly contact" that resisted neat scientific abstractions.[73] Gordon writes that in science's quest for generalizations, it has to suppress the traces of its necessary contact with a messier and fleshier world. The ghost, Gordon continues, is not simply a missing person, it is a particular "social figure" in which something that is "lost or barely visible" is accessible to us.[74] One of those missing social figures that obstinately haunts the scientific imagination, in Gordon's analysis, is precisely the "gendered object of professional desire."[75]

In the case of seroanthropologists, the gendered object of desire is the patriarchic myth of ahistorical, complete, and totalizing endogamy. The social figure that cannot be acknowledged within this framework is the gendered subject whose historically specific subjectivity cannot be entirely exhausted by the fiction of endogamy. She is potentially a doubly destabilizing force. At one level, the history of her sexual labors might subvert the neatness of the Mendelian gene pools; at another her sheer tasting abilities might unspool the apparent firmness of racial boundaries. She remained therefore, in Gordon's terms, a ghost that repeatedly appeared and disappeared in the PTC studies. It was as if the seroanthropologists could not trust the ghost of the gendered subject to remain quietly invisible and felt compelled to ritually exorcise it in one study after another by simultaneously noting her existence and then denying her substantiality.

Bittersweet Affects

An even more ghost-like presence in the PTC studies than the gendered subject were the subjects who turned the whole project on its head by tasting PTC as a sweet substance. While the existence of such subjects had sometimes been suspected in the earlier research, their existence was first noted by S. R. Das in the course of a study on Rarhi Brahmins of Bengal in 1958. He estimated that the number of these sweet-tasters "may not exceed 1–2% of the mass data normally in a population. At least that has been the experience so far in this laboratory."[76] Das argued that, although the actual percentage of such sweet-tasters might turn out to be negligible, before attempting to determine actual percentages a better approach would be to undertake "family studies" and work out the inheritance pattern.[77] The single recessive gene–based model for bitter-tasting that was then popular could not explain the existence of such sweet-tasters.

The following year, Swedish researcher Gunnar Skude published a paper devoted specifically to the sweet-tasters.[78] While Das and others had already noted the existence of such subjects, it was only with Skude's focused study that they emerged as an established phenotype among taste researchers. Skude followed this up with a series of publications on the topic throughout the early 1960s, rendering the sweet-taster phenotype a minor, but still recognizable, presence.[79] Skude also approvingly cited Das's call for family studies and appended a number of pedigree charts for the sweet-tasters he had located.

By 1966, when Das published an encyclopedic survey, the sweet-tasters had become an interesting curiosity that did not merit too much attention. He noted several possible reasons for the variant taste, including hormonal and nervous causes, as well as the simple possibility that the same molecule might produce different tastes depending on which part of the tongue it stimulated. All this, however, was mainly an aside since their numbers were low and "most of the workers in this field have been able to derive quite reliable results unaffected by the taste quality variation. And it should not be regarded as any handicap at all."[80] Having been among the first to note their existence, Das now seemed satisfied to relegate these sweet-tasters to the background in the interests of efficiency.

Notwithstanding Das's loss of interest, the sweet-tasters do open up other and silenced histories of sense, sensibility, and difference. Upon perusing historical dictionaries in Bengali—the language that Das himself and the subjects of his 1958 study used—we notice that the overlap between "bitter" and "sweet" had possible older, linguistic antecedents. Abalakanta Sen's 1892 *Anglo-Bengali Dictionary*, for example, glossed "bitter" as *katu* (pungent), *tikshna* (sharp), *kashtakar* (painful).[81] "Bitter" was also equated with "acrid," for which the Bengali alternatives were *katu*, *tibra* (quick), and *tikta* (bitter).[82] Gopee Kissen Mitter's 1868 dictionary glossed the word *tikta*, nowadays the standard Bengali word for "bitter," with the words "bitter, aromatic, fragrant."[83] Mitter also offered a second, much more uncommon Bengali word, *kadua*, which he glossed as "pungent, rancid, bitter."[84] Sir Graves Haughton's three-way Bengali, Sanskrit, and English dictionary, published in 1833, gave an even larger range of associated words. *Tikta*, according to Haughton, meant both "bitter" and "fragrant." There were two further words that were said to be "corrupt" forms of *tikta*. One of these, *tito*, had the same meaning as *tikta*. The other one, *tita*, meant "bitter, moist, wet or damp."[85] There was also *kadua*, which meant "pungent, rancid, bitter or a disagreeable taste."[86]

The dictionaries, though already abstractions of actual usage, still signal an enormous affective excess. Not only does bitterness seem to blur into other tastes, such as acridity, pungency, rancidity, and sharpness, but even more perplexingly there seems to be a simultaneous synesthetic expansion into the domains of tactile and olfactory perceptions. Fragrance is clearly a form of olfaction, while dampness is grounded in tactility.

Instead of seeing these earlier linguistic taxonomies as mere consequences of some inexplicable category confusion, we should recognize that the "neurological condition of synesthesia" is only an "extreme case of a more general phenomenon," namely, "sensual interconnection." Affect theorists have consistently pointed to "the difficulty of establishing and studying discrete sensual, experiential, and cognitive modes."[87] Ben Highmore has therefore insisted that "substances and feelings . . . matter and affect" are always simultaneously and centrally implicated in our "contact with the world," and that it is these "sticky entanglements" that we need to trace.[88] In the case of food, eating, and tasting, in particular, Harris Solomon has recently argued that we need to locate them within a "reticular perspective" where "food," "the body," and "the environment" are themselves the contingent products of heterographic acts of "absorption" rather than self-enclosed "source[s] of cause[s] and consequence[s]."[89]

The idea that there is a fixed number of discrete tastes and that these could be biologically and universally defined was in itself an assumption that undergirded the work of seroanthropologists. Historically, tastes were not only far more difficult to isolate from each other, but also frequently were entangled with other, nongustatory, senses, such as those of smell and touch. Furthermore, these affective worlds were organized around concrete objects, foods, and habits of eating, each saturated in a complex, historically fashioned habitus rather than in abstraction. Indeed, Highmore offers "social aesthetic" as an "umbrella term" for pursuing a "fleshy sociology" that can "register links between perception, senses, affect and emotions." This, he suggests, is much preferable on intellectual and political grounds to the "bureaucratic business of sorting categories and filing phenomenon."[90]

Intimations of what such a social aesthetic might look like are available in a fascinating genre of Sanskrit and Bengali texts that circulated in early modern Bengal. Known as *Drabyagunas,* these derived from an older Sanskrit encyclopedic genre called *Nighantus.* The *Drabyagunas,* though frequently discussed by scholars as texts on dietetics, in fact included much that was not about foodstuff, such as the shade of a parasol or baths in different types of water, and, in one case at least, even reflections in mirrors. The texts offered a flexible guide to how these diverse "things" affected the body by incarnating one of six possible *rasa,* or "flavors": *madhura* (sweet), *amla* (sour), *labana* (salty), *katu* (acrid), *tikta* (bitter), and *kashaya* (astringent). The actual flavors that were attributed to any specific foodstuff or thing varied over time and in relation to the kinds of material contexts that the authors of these texts were immersed in.[91] Without getting into that history, what these texts offer this discussion is a kind of affective horizon that might help us register the way bitterness was implicated in rich sociomaterial worlds that defy its neat black-boxing as a discrete taste.

One of the most influential among the *Drabyaguna* texts was an eighteenth-century text attributed to Raja Rajballabh and redacted by Narayan Das Kaviraj. Several printed editions of this text circulated in the nineteenth century, proving that the ideas in them retained some purchase in the region well into the era of colonial modernity. One such printed edition discussed *tikta-rasa,* or "bitter-flavor," in several places. In one of the clearest and most explicit discussions, the text stated that "bitter-flavor destroys bile [*pitta*], phlegm [*kapha*], pus [*kleda*] and itches caused by poison [*vishakandu*]."[92] This definition did not identify any specific foods that might actually be "bitter-flavored." Instead, it presented *tikta-rasa* as a therapeutic principle that could destroy *pitta* and *kapha,* which were two of the three para-humors around which much of Ayurvedic pathogenesis was organized.[93] Later in the text a more complex schema was offered, which, though retaining the framework that saw tastes/flavors principally as therapeutic principles, listed three seemingly distinct types of "bitterness": *swadu-tikta-rasa* (sweet-bitter-flavor), *kashaya-jhaal-tikta-rasa* (astringent-spicy/hot-bitter-flavor), and *katu-tikta-kashaya-rasa* (acrid-bitter-astringent-flavor). Each of these affected the para-humors differently. But no concrete example of any of these flavors was provided.[94]

Notwithstanding the absence of examples directly cited alongside these definitions, the text is replete with specific examples of things said to embody *tikta-rasa.* This separation suggests a possibly deliberate attempt to define an abstract principle that would resist reduction to one or two specific sensory perceptions. This possibility is further affirmed by the fact that two of the iconic foods that are usually associated with intense bitterness, namely, *korola/karela* (bitter gourd) and *chirota* (green chiretta), are both mentioned without any reference to *tikta-rasa*.[95] Instead, *tikta-rasa* is identified in such items as any fish cooked with watercress, *vidanga* (Embelia ribes), the tips of rattan canes, and so on.[96] In several of these specific cases, however, we notice the same sensory interpellations that we saw in the dictionary entries. *Kalkasundya saak* (leaves of Cassia sophera) were said to be *swadu-tikta-rasa* (sweet-bitter-flavored).[97] *Kakalidraksha*, a kind of grape with small stones, was said to be bitter and fragrant.[98] A small fish, very popular in Bengal, known as *puntimaachh*, was described as being bitter-acrid (*tikta-katu*).[99] Whereas the abstract principle was left assiduously underspecified, the concrete examples more often than not were presented as a mix of flavors and sensory perceptions. The framework, therefore, was one that remained supple enough to register the complex combinations of the material and the sensory that resisted any easy sorting into distinct, crisply defined tastes.

The most elaborate instance in the text of the way bitterness was embedded within an overall social aesthetic occurs in one of the longest descriptions in the book. In describing the virtues of betel leaves, the text mentioned that "upon the addition of spices etc. [*mashladi jukta tambul*] it is acrid-bitter-sweet-astringent flavored. Its potency is hot [*ushnavirya*]. It destroys wind [*vayu*], phlegm [*kaph*], and worms [*krimi*]. It effectuates the libido [*rati-shakti*]. It is an ornament when speaking to women. It is lovable [*pritikar*] and vanquishes sorrow [*dukkhya-nashok*]."[100] Plainly, for the author, the taste of betel, its therapeutic actions, the associated erotic sociality, the emotions that it evoked were all intimately interconnected. In other words, resisting any neat specification of bitterness in isolation, the *Dravyagunas* presented bitterness within a framework where the "senses, the sensory and the human sensorium" came to be "densely interwoven."[101]

The mismatch between this densely interwoven affective world and the seroanthropologist's search for a discrete sensory taste was not simply a mismatch of words—a mismatch where one person's bitterness was another's sweetness—it was a mismatch of worlds.

Masking and Ageusia

It was in the gap between these two worlds, that of the *Dravyaguna* and that of seroanthropology, that there flickered the bright green leaves of the betel plant. In the former world, it was on the wings of these leaves that bitterness rose to its affective heights. Betel vanquished sorrows and ushered in love, it adorned erotic dalliance, rid patients of worms and phlegm, and added vigor to their sex lives (fig. 3.2). To seroanthro-

Fig. 3.2. An infamous nineteenth-century sexual scandal depicted here through the exchange of betel between the adulterous lovers. For more on the scandal, see Roy, "Tracking the Ephemeral." Source: Wellcome Images.

pologists, however, it proved to be an obstinate hurdle that impeded research by masking a person's true genetic, and hence racial, traits.

In 1959 J. C. Sharma of Punjab University, Chandigarh, investigated four specific dietary habits among Punjabi subjects, namely, betel-chewing, tobacco-chewing, smoking, and drinking alcohol, for their potential to mask or hide the actual taste-phenotype of research subjects.[102] Subsequently, in 1962, Praveen Kumar Seth, also from Punjab University, once more investigated three dietary habits among Kumaoni

Rajputs and Brahmins. The three traits Seth looked into included betel-chewing, smoking, and nonvegetarianism.[103] These were also the three traits investigated by K. C. Tripathy in 1969 among Odiya subjects.[104] Some of the other researchers, too, made passing references to the problem of masking and especially the role of betel-chewing.

While most of the researchers also considered other traits, such as smoking, as Seth pointed out, "workers outside of India have not attempted to find out the effect of betel chewing—a characteristic of Indians and some Far Eastern peoples—and non-vegetarianism."[105] Naturally, "vegetarianism" is also an interesting object of study here, especially since what foods are included under this label and what are excluded vary widely, even within South Asia. There is widespread disagreement across India, for instance, about the status of eggs. Moreover, the associations of vegetarianism with particular forms of Hinduism are well known and its occasional deployment in the exclusionary claims of Hindu nationalism also well established.[106] It was thus simultaneously an experimentally underspecified and ideologically overloaded phenomenon to include as one of the characteristically "Indian" features of research into masking.

Betel, in contrast, both because of its specificity and its widespread use across religious divides in precolonial, colonial, and postcolonial India, was much more significant. The rich affective tapestries it evoked in texts such as the *Dravyaguna* also stood in remarkably stark contrast to its reframing by seroanthropologists merely as an intransigent impediment to genetically determined sensory normalcy.

Seth, however, was not correct in saying that only Indian researchers had attended to the relationship between betel-chewing and PTC-tasting. At least two studies had already systematically noted betel-chewing among test subjects and asserted that such information was crucial. Both studies, which appeared in 1955 and 1957, were conducted by researchers based in Singapore using Malaysian subjects.[107] The omission of these studies might have followed either from implicit assumptions about the historical relationship between Indian and Malaysian populations, or equally probably from a simple lack of awareness of the Singaporean research. Whatever the reasons for the omission, the existence of the research evokes another historically produced set of sensory affiliations between South and Southeast Asia: an affiliation built on shared cultural habits rather than genes, one that could only be registered by seroanthropology in the negative, as a problem.

Betel, in this negative vein, came to engender one of the crucial anxieties for PTC researchers. Those anxieties revolved around the "masking" of the true genetic trait of a subject due to nongenetic reasons.

Betel-chewing, smoking, or vegetarianism were not alone in threatening to mask the true genetic character of a subject as either a "taster" or "nontaster." In 1963, S. R. Das, the researcher who came to be most closely associated with sensory genetics, along with colleagues published a study on PTC distribution among two groups in the Koraput district of Odisha. In discussing the research protocols for the study, the researchers drew upon a 1958 publication by Kalmus to explain how they had introduced an additional filtering mechanism in the fieldwork to exclude those who might suffer from a "general impairment of taste due to pathological reasons."[108] The following year, Das published yet another study on taste distribution of three other groups living in Koraput. Invoking his own publication from the previous year, he once again wrote about the need to filter for pathologically impaired subjects. This time, however, he also added a name to the condition he had previously referred to, namely, *ageusia*.[109] Das's major encyclopedic publication in 1966, which comprehensively summed up the extant PTC research, once again invoked ageusia and the need to control or filter for those suffering from ageusia.[110]

The category of ageusia, an impairment of the ability to taste, had existed in medical circles since the nineteenth century. As early as 1938, it had already entered PTC research. G. Albin Matson of Montana State University used it in a racial study of various Native American groups in Alberta and Montana.[111] Remarkably, in Matson's usage "ageusia" was the genetic condition that rendered a subject incapable of tasting PTC, in other words, the "nontaster phenotype." By contrast, for Das and his colleagues, ageusia was the condition that masked the genetic ability. It was an acquired, pathological condition, rather than an inherited, constitutive trait.

Ageusia could be caused by a variety of factors, including betel-chewing. The crucial fact was that subjects suffering from it needed to be excluded from research data by following strict filtering mechanisms. They posed an implicit threat to the determination of proper racial frequencies by showing up as false nontasters. Ageusia, in other words, effectuated a separation of taste-as-sensory-perception and taste-as-genetic-trait.

No longer was the taste-as-sense an indicator of taste-as-trait. The two became partially independent from each other. Sarah Tracy, in a pellucid study of the universalization of *umami* as the "fifth taste" by geneticists, has insisted that "the sensory is molecular is affective is political."[112] Indian researchers in the 1960s seem to have progressed along a slightly different path of transformations. They increasingly separated the sensory and the affective. The sensory was geneticized and rendered into a trait upon which rested racial identities. The affective, by contrast, was progressively marginalized as pathological and/or as an impediment to research that had to be assiduously filtered out by robust selection protocols deployed in choosing test subjects.

Chemosense to Genosense

Tracy speaks of the "chemosenses" that animate contemporary taste research, that is, the sensory conceptualized in exclusively chemical terms.[113] Helpfully, she also connects this chemosensory paradigm to increasingly industrialized megafood production and distribution networks. Indeed, she terms contemporary taste research "Big Food Science."

The Indian researchers of the 1950s and 1960s had no discernible connections with Big Food and operated within state-sponsored national research networks. At a time of postcolonial nation-building and biometric nationalism, they had neither the need nor the ambition to chemically reproduce or manipulate tastes. Rather, they were devoted to tracking tastes as purely genetic traits that could be used as markers of racial identities. We might say of them that in contrast to contemporary researchers, they operated with "genosenses."

Genosenses were built orthogonally to affective worlds. Recent scholarship exploring how humans "became genetic" has identified a whole range of "clamorous contemporary formations" that have engaged, and continue to engage, in the "geneticizing of human life, history, relationships, and identity."[114] The "geneticization" of human sensory contact with the world, however, has received relatively little historical attention. The translation of the human senses from constitutive aspects of thick affective worlds to increasingly reified genetic traits that could, at least on occasion, be entirely spliced off from their affective milieus was crucial for revitalizing biologized ideas of racial difference.

The emergence of such genosenses was overdetermined by the disciplinary evolutions of the mid-twentieth century through which race science found new academic homes, identities, and traditions. It was perhaps not coincidental that Arthur Fox, whose accidental discovery launched the whole line of PTC research, was in fact a chemist. As Lissa Roberts has pointed out, chemistry was one of the most "sensuous" scientific disciplines. Smells, tastes, and colors continued to matter in chemical research much more than in most other traditional scientific disciplines.[115] Early modern natural philosophy, of course, had been much more sensuous in its engagement with the world.[116] The increasing privileging of visual data over other sensory information began toward the end of the eighteenth century. At the time, the "death of the sensuous chemist" seemed impending. But that death foretold did not come to pass, and chemists remained much more attuned than other scientists to nonvisual sensory data.[117] It is doubtful whether Fox, had he been a biologist or a physicist rather than a chemist, would have paid much attention to the laboratory accident that launched PTC studies.

In colonial South Asia the general sensory detachment of modern scientific disciplines had engendered a critical response. Among those who led the charge against this sensory alienation was Sir J. C. Bose. Scholars of South Asian histories of science have repeatedly returned to Bose's inquiries in this regard. Though he started out as a physicist, Bose got deeply interested in bridging the seeming gap between the "animate" and "inanimate" aspects of the world and sought to craft a new science that could transcend disciplinary boundaries. "Sensation" became a key component of his researches through which he accessed objects as diverse as plants and metals. By 1920 Bose had succeeded in creating an institutional home, named the Bose Institute, for his new, transdisciplinary vision of science that would connect the sensate world with the insensate.[118]

By the 1940s Bose's transdisciplinary vision had been disciplined into biophysics. In 1942 a young man with a master's degree in pure physics from Calcutta University joined the Bose Institute as a research fellow in biophysics. His name was Sudhir Ranjan Das. S. R. Das, as we have seen above, would go on to become the most prolific, influential, and prominent researcher into the racial variations of PTC-tasting ability. As a student, Das does not seem to have been much interested in human biology. He did, however, develop an interest in chemistry and published a couple of papers on sulfur and selenium. One of these was read at the International Union of Chemistry in Münster, while the others were cited in chemistry textbooks of the time.[119]

Barring a brief, eighteen-month stint as a lecturer in physics at Kabul University, Das remained at the Bose Institute until 1948. That year he was hired by the Anthropological Survey of India as a radiologist-anthropologist. The ASI was at the time increasingly engaged in conducting x-ray–based anthropological studies, often within its general racialized framework.[120]

Das commenced his tenure by conducting radiological studies on skeletal growth. He grew interested in genetic factors affecting skeletal growth and increasingly moved away from radiology and toward genetic studies. This was the time when he immersed himself in PTC researches. His studies within the ASI fit in well with the ASI's general model for studying human difference as racial differences, and also allowed him to interest other ASI anthropologists, such as P. N. Bhattacharjee and D. P. Mukherjee, in PTC research.[121]

When Das finally retired from the ASI in 1968, he was immediately recruited by the Indian Statistical Institute to head its Anthropometry and Human Genetics Unit.[122] Here, Das got the opportunity to train a large number of students in his mold. Recently, a memorial volume in Das's honor, though remembering him mainly for his growth and development studies, rather than his studies in sensory genetics, compared him with Franz Boas.[123]

Through Das's move from the borderlands of physics and chemistry, first into a transdisciplinary, critical version of physics that was increasingly interested in putting physics back into conversation with biology, and then through genetic anthropology, into statistical studies, we notice the increasing reification of the "sensation." Whereas Bose's project had been partly to resensitize a world of insensate matter, by the end of Das's career the sensory itself had been rendered into a statistically defined genetic object—in other words, into genosense.

Recently, Shivani Kapoor has perceptively argued that "caste . . . is a multisensorial ordering of spaces, bodies and objects."[124] Colonial science, she further states, not only experienced this "multisensorial nature of caste" but also partially failed to control it.[125] Genosense emerged from this failure as an instrument by which national science could claim technocratic authority over the multisensorial nature of caste by inserting it into an emerging global history of race.

The sensory history of race, focused mainly on the American South and on slavery, has largely avoided contact with the scientific study of sensory phenomena in the context of race science. Mark M. Smith has therefore been correct to allege that most "modern discussions of 'race' and racial identity" are in fact "hostage to the eye."[126] However, his view that attempts in eighteenth- and nineteenth-century America to scientifically study racialized ideas about the senses were "pseudoscientific" is clearly ahistorical and uninformed by the historical studies of science.[127] To quote Michael Gordin, "there is no such thing as pseudoscience, just disagreements about what the right science is."[128]

Remarkably, even such disagreements did not arise with regard to PTC research as a form of race science. No one questioned the effort to classify races according to their ability to taste PTC by labeling it a pseudoscientific project. It was and mostly remains part of a mainstream scientific inquiry into genetic differences. In its continuing status as science, it vividly attests to the simple fact that race was more than just a bitter aftertaste of empire in postcolonial India.

INTERCHAPTER

Letter 4

Hemendrakumar Ray
21, Pathuriaghata Bylane
Calcutta, India
26 January 1934

My dear brother,
You say that you are not a writer or a philosopher, and yet you have such a way with words! As I have said before, you, more than any of my bookish friends, are able to feel the deeper throbbing pulse of our times. In your letters I can glimpse both the dying embers of an age that valued action, pleasure, danger, and mystery and the rising wind of another age that seeks to plan, to rationalize, to avoid risks, and to banish mystery. I do not know which of these two ages will ultimately shape our descendants. If Mr. Sen's diary really gives us a foresight of the future that awaits us, I am afraid you and I will perhaps both be misfits in that future.

Please accept my heartiest thanks for sending me Mr. Sen's diary. I have only skimmed through it, but am planning on reading it closely in the coming weeks. It is such a singular document and full of such tantalizing descriptions of these future-humans who live in a barrel—and, yes, like you, I too am not sure what we might call this strange, futuristic race.

In my last letter, it being my first communication, I had omitted to ask about the bout of malaria you said you had suffered. Given that our relationship has now evolved from that of an author and a reader to one closer to that of an elder and a younger brother, I feel I must enquire about your health. From your letters it does not appear as though you gave yourself enough time to convalesce before you proceeded on your shikar *again. I know young people like you, people driven by the unquenchable thirst for adventure, romance, and mystery, find it difficult to be chained to your beds for long. But, my dear*

Najrul, there are few diseases that are more obstinate or treacherous than malaria. I have said as much in print, too!

I know that ever since old Ross-sahib of the P.G. Hospital identified the mosquito as the villain that transmitted malaria, scientists have thought they have figured out exactly what the disease is and how it is to be prevented or cured. Every few months I read in the papers about some new discovery about the disease. And yet, if they really know so much about the disease, why do so many people still suffer from it? You have only to look around you to see just how many people are plagued by this terrible disease.

Might there be some other agency involved in it? Might there be something that exceeds the powers of old Ross-sahib and the legions of his followers?

A few years back I heard of a strange incident involving a man I knew only very superficially through some common acquaintances. He lived, not far from me, at Baghbajar, and hence we would often run into each other. About six years ago, this man had on a whim taken the tram to the Dhakuria Lake to enjoy a stroll by the water. He had lingered a little too long by the water's edge and let the sun depart on the western horizon, when he suddenly felt a chill come over him. Before he knew it, his teeth were chattering like one of those irritating clickers children buy at village fairs and he was shivering so badly that he could hardly stand straight. To stop himself from falling over, he lay down on the grass and wrapped himself in the shawl that he had with him. He felt unbearably cold, and, in a while, a thick fog descended upon his mind. His ears rang and his eyes watered. Eventually he sank gradually into a fevered stupor.

By the time this man regained consciousness, the lakeshore was deserted. He still felt cold and unsteady, but he also feared that unless he somehow tried to get home, he might just die there, either from the damp late-night cold or simply at the hands of the many goondas *who operate around that area after dark. Somehow, and with immense effort, he raised himself and headed towards the tram stop. Swaying from side to side and having stumbled more than once, he somehow got to the tram stop at the head of Russa Road and leaned on the lamppost waiting for a tram to show up. Even in his foggy and disoriented state, he realized how late it was and how the city had gone to sleep, and he prepared himself as best he could for the night tram to come by.*

He was somewhat surprised when instead of a tram he saw a bus approaching. This was a godsend, he thought. It would get him home quicker, but wasn't it too late for buses to be plying? Yet, when the bus finally stopped near him, he hopped on. Once on it, he was surprised at how fast the driver, a tall Sikh with a flowing beard, was driving. Somewhat irritatingly, the driver also kept honking the horn for no reason. The speed and the noise began to bother him. He also noticed that the stray dogs sleeping on the footpaths seemed to be scared of the bus and ran away as it got nearer. He tried to get the driver to slow down, but neither the driver nor the conductor responded to him. He thought maybe paying the fare would attract the conductor's attention and called him to take the fare.

As the man extended his arm to accept the fare, my acquaintance noticed how the conductor's face had a sallow, malarious look. His eyes were sunken and unblinking. As their hands touched, my acquaintance was shocked at just how cold the other man's hand was. With a shudder he pulled his hand back and looked away. As he did so, to his utter astonishment, he noticed that where only a moment ago there had been the three of them—the driver, the conductor, and himself, there were now at least ten to twelve other passengers. Yet, the bus had not stopped anywhere since he got on. Amongst the passengers were Bengalis, up-country men, and sahibs. But all of them had the same dead, unblinking look in their eyes and the sallow visage that usually characterizes those who have long suffered from malaria.

As you can imagine, my acquaintance was terrified and tried to get the bus to stop. But neither the conductor nor the driver would pay any attention. As he started shouting, the strapping Sikh driver finally turned to look at him, and he noticed that underneath the neat turban and the flowing beard was the face of a skeleton. In place of eyes, mouth and nose were mere gaping holes, and the rest was pure bone that shone in the streetlight they rushed past. Frightened and hysterical, he jumped out of the bus without thinking much.

Next morning people found him lying unconscious by the Russa Road crossing where he had boarded the bus. Many, including his father, said the bus ride was the figment of a fevered mind. But then, why did they find a bus ticket clutched in his hand?

I hesitate to dismiss the bus as a malarial delusion, the product of a fevered brain. Perhaps, I wonder, it might be possible that the overheated brain opens up channels of perception that are otherwise blocked to us? Perhaps the teeming millions of men and women of all races who have fallen prey to malaria here in Calcutta do in fact run a night shuttle transporting fellow victims to their final abode? But then, why did my acquaintance survive? How was he able to get off the malarial bus?

You are a man of action, Najrul. Perhaps you find these rambling reflections indulgent and pointless. Indeed, I cannot defend them by appealing to any higher purpose. But I cannot get rid of a lurking feeling that there is more to disease than our scientists might allow. In our villages, as you know, people have long worshipped Jwarasur, the fever demon. Was all that simply fantasy? Were our ancestors so far deluded and that too for so long?

More than that, however, I worry about your health. Malaria is too stubborn to let go of you easily after once it has sunk its teeth into you. It is also a stealthy disease. It has a tendency to sneak up on you, just as you imagine that you have left it behind. These modern doctors underestimate its doggedness and its stealth. In their hubris they think they are dealing only with a virus, an entity that has no intention, no cunning, and is entirely reducible to its chemical components. I tell you, brother, this disease has not survived so long and harvested the souls of so many by merely being an unthinking chemical structure. Its malignance is not devoid of malice. Its virulence is not innocent of vice.

Be careful, brother. Do not take it lightly. I pray for your health.

Hemenda

CHAPTER 4

Medicalizing Race

> Just as tracks must be created, so too must one fight against leprosy and malaria, against the natives; nature must be changed in spite of itself, violence must be wrought on it; the native must be brutalized, have good done to him despite himself.
>
> —Frantz Fanon, *The Meeting between Society and Psychiatry*

"There you are!" exclaimed the man peering down the microscope at the Pasteur Institute in Coonoor in southern India.[1] The man and the woman standing next to him could feel his infectious excitement. The year was 1952, and this was the first time the "sickling gene" had been discovered in a population that was clearly not African.

On 26 January 1952, India celebrated its second Republic Day. It had gained its independence from British rule only five years ago, and two years ago it had stopped acknowledging the British monarch as its head of state. The first elections to the country's Parliament were still under way. In the midst of all this, two British scientists traveled to India. The older of the two, Hermann Lehmann, was a German Jewish refugee, then based at St. Bartholomew's Hospital in London; he had previously been posted in India a decade ago when India was still a British colony. His assistant, Marie Cutbush, was a serologist who worked for the M.R.C. Transfusion Unit in Britain. The two of them were at the Pasteur Institute at Coonoor to test the blood of some local tribes for seroanthropological reasons. Their brief trip had been financed by the Nuffield Foundation.

On their first day, as was the norm in seroanthropological work, they wanted to test their blood sera on some of the laboratory staff before beginning their work in earnest. P. K. Sukumaran, a young laboratory technician at Coonoor who was eager to learn the new serological techniques, had "paraded some of the staff at the laboratory" and made them line up for the foreign researchers. Blood was drawn from pinpricks on the fingertips of each of the employees who lined up. It was Cutbush and Sukumaran who looked on as Lehmann examined the blood samples they had drawn from the laboratory workers. It was in one of these samples of blood drawn from a laboratory technician that Lehmann first found the tell-tale, sickle-shaped cells that made him exclaim with excitement. Sukumaran would later recall that the man whose sample it was belonged to the lowly Badaga caste.[2]

The unnamed man himself would have made little of the discovery. The presence of a single allele of the sickle cell gene does not usually produce any perceptible ill-health. Sickle cell anemia, which is a painful affliction, arises only in individuals who possess two sickle cell alleles. The former category, heterozygotic individuals (which is what the Badaga man most likely was), only became legible through blood tests. Without the tests to detect the "trait," they would have been indistinguishable from any of their neighbors.

Sickle cell has been recognized by several scholars as being much more than a "simple gene disorder." Rather, it is something of a "cultural icon."[3] Predictably, its iconic nature has led to a robust and rich historiography on the subject.[4] This history, equally understandably, focuses on the United States and to a lesser extent on African countries. Only recently has Elise Burton's pioneering work begun to open up the history of sickle cell and race in Asia.[5] Barring incidental discussions of sickle cell in India in these works, however, there is very little by way of critical historical scholarship on sickle cell on the Indian subcontinent. Most accounts of both the gene and the disease in India are by practicing scientists who generally present the story within a simplified, teleological narrative of progress, which almost entirely omits the disease's complicated entanglements with the politics of race.[6]

The "standard story of sickle cell anemia," as Keith Wailoo reminds us, begins in 1910 with James B. Herrick's discovery of "peculiar, elongated and sickle-shaped red blood corpuscles" in a "Negro patient with severe anemia."[7] The work of Verne Mason and John Huck subsequently further clarified the disease (named sickle cell anemia), its alleged latent stage (sickle cell trait), and, perhaps most important, its racial identity (African). By the 1920s sickle cell not only was firmly identified with those of African ancestry, but was used by racist propagandists in the United States to legally oppose intermarriage between African Americans and whites. It came to be seen as a disease of "Negro blood." So powerful was the association that even when the trait was detected in a visually and socially white family, it was alleged that they had some

African ancestry that was either forgotten or deliberately hidden.[8] Indeed, Duana Fullwiley calls sickle cell an "all-too-reliable weather vane for the climate of race thinking in the United States."[9]

What sustained this racialized framework was a particular technology of testing. "Emmel's test," developed by Victor Emmel at the University of Washington in 1916, was a simple yet visually persuasive test. A ring of petroleum jelly was drawn across a sterile glass slide. A drop of the sample blood was placed at the center of the jelly and covered with another sterile glass slide. After a few hours like this at room temperature, the slides were examined under a microscope and demonstrated "a striking picture" of sickled red blood cells.[10] This led to a framing in which even having the single sickle cell allele came to be seen as a kind of submerged or unexpressed disease. Thus, though the person concerned might suffer absolutely no ill-health, they could still be figured as the carriers of a deadly disease, similar to asymptomatic carriers of viral infections.[11]

The belief that the sickle cell allele might be a dominant one, which according to Mendelian laws would render even heterozygotic individuals ill, further consolidated the idea that any bearer of the sickling trait was a potential social danger. Any transmission of the allele then came to be seen as potentially pathogenic, even if it was passed on to a heterozygotic child. This in turn became the rationale for US racists to try to legally prevent the marriage of African Americans to whites.

All this only began to change by the late 1940s when Linus Pauling and J. V. Neel developed a new identity for the disease, based on a new testing technology. The rise of electrophoresis displaced Emmel's test and relocated the disease in the hemoglobin. Soraya de Chadarevian describes how the new identity of the disease developed by Pauling and Neel clearly distinguished "sickle cell anemia" from "sicklemia." The former was a condition seen in homozygous individuals, while the latter was observed in heterozygotes.[12] By 1954, Anthony Allison's work further demonstrated that heterozygotes with the sickle cell gene enjoyed a natural immunity from malaria.[13] Very soon after this, Vernon Ingram's work on sickle cell hemoglobin further clarified the mechanism of inheritance by locating the difference between sickle cell and non–sickle cell hemoglobin in a single amino acid.[14] This in turn demonstrated that the allele itself was not a dominant one and hence began to reconstruct the narrative around marriages between those with and those without the gene. Cross-marriages between bearers and nonbearers were now seen as being likely to reduce the incidence of sickle cell anemia and also to provide protection from malaria in early life. In place of paranoiac calls for segregation, therefore, from the 1950s on the disease in the United States increasingly became linked to imaginaries of "human engineering."[15]

While the molecularization of the disease from the 1950s certainly helped to break down the racial framework within which the disease was earlier conceptualized, it also progressively off-staged the clinical anemias. The prominence of laboratory analysis and the abstract molecular ideas about the condition meant that sickle cell anemia as a clinical condition began to recede into the background. As Wailoo puts it, "in replacing one technological identity with another, medical science had revised the former, harsh symbolism of the disease and yet played down the centrality of pain, infection, and infant death."[16]

Lehmann worked very much within this "standard story" of sickle cell. So much did he see himself within this general historical outline that he had even, in 1950, on one of his earliest visits to the United States, attempted to meet James Herrick in order to "pay homage."[17] But his own experience in India and Uganda, early in his career, and his anthropological interests made him keen to develop a test that would work well in the "tropics." When he developed the paper electrophoretic method for separating hemoglobin variants together with Elspeth Smith, he mentioned how the method could work well in the tropics, since it could work on a simple battery and was much less cumbersome than the previously used Tiselius apparatus.[18]

Instead of locating Lehmann's eureka moment in Coonoor as a mere footnote to the "standard story," I argue, we need to relocate him at the intersection of multiple distinct histories. The history of the Pasteur Institutes themselves was, for instance, crucial to creating the network and the very space—the Coonoor laboratory—within which Lehmann worked. Given his own invocations of tropicality and race, it is also significant how these Pasteurian laboratories, throughout the colonial era, had sustained, reproduced, and even bolstered racialized views of the "tropics" and its denizens.[19]

More prominent still, especially from the perspective of this book, is the history of seroanthropology in India. Until recently, the role of seroanthropology in Lehmann's research has been underappreciated. Histories of Lehmann's discovery have usually presented the anthropological background somewhat incidentally, focusing instead on the laboratory work.[20] Yet, Lehmann's own early communications about his Indian discovery clearly attest to his deep and abiding interest in "racial history."[21] De Chadarevian has pointed out that, especially from the 1950s, Lehmann himself located his own work more firmly within the seroanthropological tradition of the Hirszfelds. Indeed, she points out that Lehmann is considered to be "the founder of molecular anthropology."[22] Burton, too, has described how Lehmann's work on sickle cell distribution in the Middle East was closely linked with the racialized frameworks of physical anthropology.[23]

Burton also details the anthropological context that propelled Lehmann to undertake his 1952 trip to India. More than just the destabilization of the older, racist model of sickle cell pathogenesis, the immediate context of the trip involved the highly racialized narratives generated by the discovery of sickle cell in the blood of Greeks and Yemenite Jewish migrants to Israel. Lehmann's specific interest in looking for sickle cell in India arose from a comment by the Cambridge anthropologist, A. C. Haddon, that he had heard back in the 1930s. Haddon had suggested that South Asians might have migrated to Africa in pre-Neolithic times. Lehmann now thought that if he could find specifically African traits, including sickle cell, in South Asian groups considered "primitive," his findings would illuminate this anthropological hypothesis.[24]

Veddoid Blood

Lehmann quickly modified his original assertion about the origin of the sickle cell gene in India. By 1953 Lehmann came to argue that the sickle cell gene had first emerged in southern Arabia among the so-called Veddoid race and that it was these Veddoid migrants from Arabia who had spread the gene to both Africa and India. He refuted theories of multiple origins as a form of environmental adaptation to endemic malaria. Instead, he argued that while endemic malaria certainly influenced the success and frequency of the gene, it did not explain its presence in the first place. Noting further that in India, Africa, and Arabia, the gene seemed to predominate in groups that were socially disempowered, Lehmann argued that the endogamy enforced by social disempowerment and the selective pressure applied by malarial environments together had sustained and amplified the gene's frequency, after its initial entry into the group through racial inheritance.[25]

Lehmann's initial discovery in 1952 was followed by a number of other reports of both the heterozygous sickle cell trait and sickle cell anemia in India throughout the 1950s and 1960s. Several of these authors discussed the so-called "Veddoid blood." The first report to appear after Lehmann's was published in the *Indian Medical Gazette* in September 1952. The two authors were both doctors who had encountered a small number of cases of what they believed to be sickle cell anemia—though it seems highly likely that at least some of the cases were misdiagnosed and might actually have been heterozygotic individuals with no disease—among the workers in the tea gardens of Assam.[26] They seemed unaware of Lehmann and Cutbush's publications announcing the presence of sickle cell in India and were not overly interested in historical or anthropological debates about the gene. They did, however, comment on the impossibility of there being any "negro blood" in any of the families they had seen.[27] Since the nineteenth century, tea garden labor had been recruited, through a series of highly racialized and immensely exploitative recruitment practices, from the so-called tribal regions of the Chota Nagpur Plateau.[28] As a result, the families of the patients found in Assam had all migrated there within two to three generations from Bihar and Odisha. Unlike US doctors of a slightly earlier era, the doctors in Assam did not suspect any secret admixture of African genes and instead flatly commented that the presence of "negro blood" was not "considered possible."[29]

Subsequent researchers were more forthright in offering hypotheses that could explain the gene's presence despite the absence of "negro blood." One of these studies was undertaken by Ernest Büchi, the Swiss anthropologist employed by the ASI. He made two field trips, in the winters of 1953–54 and again in 1954–55, with the precise purpose of examining whether "sickling [is] a Weddid trait."[30] Büchi's use of the term "Weddid" in place of Lehmann's "Veddoid" signaled the distance between them.

Both terms signified what has loosely and more recently been referred to as the "Proto-Australoid element" in the Indian population. This category was rooted in T. H. Huxley's observation in 1870 that South Asians and Aboriginal Australians might be related.[31] Huxley's comments were impressionistic and his reference to South Asians was highly imprecise. The only South Asians he had personally seen, and on whom he based his observation, were the Indian lascars working as coolies at the East India Docks in London. But based on that evidence, he surmised that "the only people out of Australia who present the chief characteristics of the Australians . . . are the so-called hill-tribes who inhabit the interior of the Dekhan [*sic*], in Hindostan."[32] Huxley called the "type" constituted by these South Asians and Australians, taken together, the "Australoid type." Leaving aside the conundrum over locating the Deccan in Hindustan, Huxley notably did not mention any specific South Asian group by name. He merely floated the idea that there might be a common racial group that included both Aboriginal Australians and some South Asian peoples.

The French physiologist Louis Lapicque coined the name "Pre-Dravidians" to designate a similar category in 1895. This term was quickly adopted by A. C. Haddon, whose comment had first alerted Lehmann, in his famous work titled *Races of Man and Their Distribution.*

The Pre-Dravidians were a much more specific group than Huxley's "Australoid." Haddon included in this group the Veddahs of Sri Lanka; the Kadirs, Kurumbas, Irulas, and "other Jungle Tribes of the Deccan"; the Sakai of the Malay Peninsula and Sumatra; the Toalas of the Celebes; and the Australian Aborigines.[33] In 1921, when Haranchandra Chakladar translated the work of an Italian anthropologist, Vincenzo Giuffrida-Ruggeri, into English as a textbook for the newly formed Department of Anthropology at Calcutta University, "Pre-Dravidian" was used interchangeably with the hyphenated name "Australoid-Veddaic."[34] The "Australoid-Veddaic" were explicitly distinguished from other proposed groupings of some of the same tribes, such as "an Oriental extension of the Mediterranean race," the "Brown Race," and the "Indo-Erythrean race." This group was sandwiched between the "Negrito" and the "Dravidian" and included such tribal communities as the Panyans, Kadir, Kurumba, Sholaga, Irula, Mala Vedan, Kanikar, Paliyan, Chenchu, and Urali. Significantly, the Badaga, among whom Lehmann found his first evidence of sickle cell, were explicitly excluded from this group and included among the Dravidians.[35]

Throughout the first few decades of the twentieth century more names were proposed and their precise boundaries continued to vary. Ramaprasad Chanda, writing in 1916, for instance, proposed the term "Nishada," while the German raciologist Egon Freiherr von Eickstedt proposed the term "Weddid." The American Roland Burrage Dixon proposed "Proto-Australoid," and a fellow American, Earnest Hooton, later modified Dixon's term to "Pseudo-Australoid." Among the early Indian anthropologists, B. S. Guha, who had been taught by both Dixon and Hooton as a student at Harvard, retained the term "Proto-Australoid" but thought that some of the tribes of the Deccan were an older, and hence truer, prototype of this race than the Veddahs of Sri Lanka. S. S. Sarkar, in contrast, argued for the retention of the basic type "Australoid" and dropping of prefixes like "Proto-" and "Pseudo-."[36]

Lehmann's label of choice, namely, "Veddoid," seems to have emerged initially in anthropological studies of Indonesia. It was the Dutch anthropologist J. P. Kleiweg de Zwaan who first used it, in 1925. He himself built upon the earlier work of two Swiss anthropologists, the cousins Fritz and Paul Sarasin, who had described the denizens of the Celebes.[37] Unlike in South Asia, where the "Weddid" and the "Proto-Australoid" were distinguished from the "Negrito" and the "Dravidian," in the Malay Archipelago the "Veddoid" were located in-between the "Negrito" and the "Malay." Both the need to make locally relevant distinctions meaningful within the vocabulary of translocally shared racial typologies and the specific groups that had to be accommodated within these vocabularies rendered terms such as "Veddoid" ambiguous and open to contestation.[38]

The general model for racial distribution was often borrowed from the discipline of geology, and racial groups were imagined as existing in the form of "strata." The origins of these strata were imagined as successive waves of migration and conquest, which pushed the older layers to the bottom or the geographic fringe.[39] But beyond this broad structural consensus, the specifics of which existing social group fit into which racial category, alongside which other local group, and on the basis of which criteria remained a matter of unending disagreements. For example, some researchers, such as von Eickstedt, made the Veddahs of Sri Lanka the prototypes of the "Weddid" group, while others like Guha not only rejected the label "Weddid" but also thought that some of the tribes, such as the Chenchus, were the prototype for a "Proto-Australoid" group and the Veddahs were a derivative, later group. Still others, such as the British anthropologist R. R. Gates, argued that the term "Veddoid" should be narrowly limited to the Veddahs alone and not applied to other groups.

It was within this overarching context of disagreements that Büchi confronted Lehmann's argument. Superficially equating von Eickstedt's "Weddid" with Lehmann's "Veddoid," Büchi argued that Lehmann had erred in homogenizing the "Weddid" race. Büchi pointed out that the "Weddids" were clearly divided into further subgroups or "sections." "Weddids" living beyond the Nilgiris and Wynad, Büchi reported, did not show sickling. He labeled these other "Weddid" sections "Gondids" and "Malids," respectively. According to Büchi, the sickle cell gene was most likely to have first appeared in the Badaga or Toda tribes, since these two tribes also "show the 'Negro chromosome' cDe(R_0) in remarkable frequency."[40] From there the gene allegedly passed to the "Paniyan-Irula-Kurumba group," most likely through a documented practice of ritualized cross-community sex between Toda and Kurumba couples.

As to the original appearance of the gene, Büchi rejected Lehmann's early theory of an Indian origin, proposing instead two alternative possibilities. First, adopting a theory proposed by Prince Peter of Greece that the Todas had originally come to India from Mesopotamia, he proposed that the Todas might have acquired the gene from "Negro slaves" in Mesopotamia. Second, Büchi thought the importation of "Negro slaves" directly to India either at the dawn of the Common Era or around 1437 might have introduced the gene directly into the country. Of the two, he seemed to prefer the first alternative.[41]

The following year, 1956, P. K. Sukumaran, the man who had helped

Lehmann and Cutbush during their initial investigation, published two brief studies. The first of these was co-authored with Lehmann himself, while the second was co-authored with two other eminent Indian colleagues. Sukumaran's career had been most dramatically transformed by the Lehmann discovery in 1952. A high school dropout, he had been a low-level laboratory technician at the Pasteur Institute in 1952 with no academic qualifications and few prospects. But his work with Lehmann catapulted him into an academic career. By 1956 he was employed as a researcher at the Human Variation Unit of the Indian Cancer Research Center in Bombay. The 1956 publications were themselves a testament to his meteoric rise. It was remarkable, first, that he would co-author an international publication with the already fairly well-known Lehmann, and second, that his name would appear ahead of his other, more qualified colleagues in the second publication. His co-authors, L. D. Sanghvi and G. N. Vyas, were both highly educated scientists who had obtained PhDs from highly reputable institutions. Sukumaran, by contrast, never earned any higher academic degree. Yet, he would rise to the top of the profession and eventually become one of the most respected global authorities on hemoglobin variations. Throughout his life, Sukumaran endearingly referred to Lehmann, with whom he continued to collaborate on numerous projects, as his "guru."[42]

Both of Sukumaran's 1956 papers were brief to the point of being terse. The first, written with Lehmann, further substantiated the 1952 findings by adding more data to illustrate that four "Veddoid tribes of Southern India," namely, Badagas, Irulas, Kurumbas, and Todas, all possessed the sickle cell trait. Yet, neither the Veddahs of Sri Lanka nor the Kothas of the Nilgiris possessed it.[43] The paper did not seek to explain this, asserting merely the presence of the gene among the so-called "Veddoid" tribes.

Sukumaran's second paper that year was similar in avoiding any anthropological explanation. This paper, written with Sanghvi and Vyas, simply reported the fairly high incidence of the sickle cell gene in several socially disadvantaged groups in western India, such as the Bhils, the Dhodias, the Dublas, and the Naikas, who had not previously been tested. They also reported that several, more privileged groups from the same area, such as the Anavil Brahmins, the Leva Patidars, and the Marathas, along with one lower-status group, the Kolis, lacked the gene.[44]

The following year, another brace of publications appeared on sickle cell in various western Indian groups. Authored by pathologists at Nagpur Medical College, these studies were much more forthright in articulating what was only vaguely hinted at in Sukumaran's studies.

Both studies reported the detection of significant sickle cell gene frequencies in three lower-caste groups in and around Nagpur, namely, the Mahars, the Kunbis, and the Telis.[45] The authors asserted that though there was clear evidence of migration from Africa into western India, such migration had only resulted in "local mixing," and it was unclear "what kind of Africans they were" since some East African groups, such as the Somalis, had been found to be entirely free of the sickle cell gene. Hence, they declared on rather obscure grounds that "the presence of sickle cell in various regions of India . . . cannot be due to mixture with Negroid blood."[46] The only source from which the gene could have come, they were certain, was "Veddoid blood."[47]

To support this contention, the researchers emphasized that the Mahars—a numerous "scheduled caste" according to the new Indian Constitution and the caste from which the chief architect of the Constitution and leader of the Dalits, Dr. B. R. Ambedkar, himself hailed—were a "mixed community occupying a position midway between the Marathas and the primitives."[48] Partially contradicting themselves, however, the researchers then added that the two other castes, the Kunbis and Telis, were "different castes of the Marathas." Having thus set up the dominant local group, the Marathas, as a distinct group, they immediately undermined the assertion by labeling the Kunbis and Telis as "castes" within the "Maratha caste."

Prachi Deshpande has tracked the transformations of the category "Maratha" from a "historical, military ethos to the bounded marker of a caste group" during the colonial period.[49] The transformation left unreconciled tensions between an elite, warrior identity and a more dispersed, caste-based identity open mostly to dominant peasant groups. It was precisely that tension that became visible in the attempts to genetically stabilize "Maratha," "Mahar," and "Kunbi" as distinctive biogenetic groups.

The researchers mined the conceptual tension between corpuscular caste identities and historically contingent, open-ended ones in their interpretative work. This strategic exploitation is clearly visible in the way the Mahars were simultaneously identified as a bounded caste group and a hydraulic system through which "Veddoid blood" could flow from one group to another. Researchers asserted that "the habits and customs of Mahars do not preclude the possibility of such infiltration of Veddoid blood [from the Badagas and Todas]."[50]

This dual project of both espousing clearly bounded identities and identifying a common connective group was advanced further in the studies of R. S. Negi. Negi, then employed at the ASI, undertook three studies that built directly upon the Nagpur studies. Negi described the presence of the sickle cell gene among a wide variety of tribal groups in the Bastar region of Central India.[51] He then went on to propose two possible mechanisms for the circulation of the gene. One, which he suggested as a weak possibility in the last of his three publications, postulated that the traditionally nomadic Banjaras might have been the historical "agents of dispersion."[52] The other explanation, which he seemed to favor and repeated in each of his publications with greater emphasis, once again zeroed in on the Mahars. According to Negi, the Mahars themselves absorbed the gene from Badagas and Todas, before further disseminating it to other groups. He argued that the lowly Mahar caste of Bastar was likely related to the Mahars and that it might be from them that other groups, such as the various Gond tribes, had acquired the trait.[53]

It is illuminating to compare this mode of racialization of sickle cell with what was happening in other colonial or recently decolonized contexts. Writing about French West Africa, Fullwiley describes how, "despite the apparent similarities in race logics that were helped by sickle hemoglobin on both sides of the Atlantic, in the then French West African capital of Dakar, Senegal, the scientific emphasis was less on this trait as a biological index of 'Negro' blood and its perceived social threat to whites vis-à-vis 'admixed' progeny. Rather, key colonial physician-anthropologists . . . stressed how ostensibly different and multiple groups of black Africans varyingly exhibited this so-called African pathology across the geopolitical terrain of French West Africa." [54] This is remarkably similar to what we notice happening in newly decolonized India. The measurement of sickle cell gene frequencies become a new technique to stabilize a range of complicated and often ambiguous social identities upon a biogenetic register. As Fullwiley writes, "the pattern of that variation, it was assumed, would congeal ethnicity and race into a metonymic configuration where mental telescoping would allow researchers to scan the geography of race for ethnicity within it."[55]

But this was not all. In India, perhaps in keeping with its Nehruvian credo of "unity in diversity," there was also an attempt to find a mechanism that could string together the discrete and disparate tribal and caste identities. This binding thread, interestingly, could not be an exotic factor. Thus, Lehmann's Arab "Veddoids," for instance, were largely ignored in India. Likewise, the numerous possibilities of transmission through Muslim or African migrants were strenuously denied on the flimsiest and most unconvincing grounds. Both the claim that Gujarati communities had limited "local" intermixture with African communities and the even weaker argument that not knowing where the African migrants came from automatically foreclosed the possibility were clearly disingenuous. Indeed, we know that there are several long-settled Afro-Indian communities in Gujarat and western India, and some of these groups even founded flourishing regional kingdoms ruled by Afro-Indian royals.[56]

These elisions allowed for the emergence of an "indigenous" historical connective tissue that could string together the distinct tribal and caste communities. "Veddoid blood" became that connective tissue and Mahars its main agents of dispersal outward from the Nilgiris into western and central India. It was in this combination of "Veddoid blood" and Mahar mobility that the imaginaries of national cohesion were engendered, even as the colonial racial logic of biologically distinct and discrete communities was being reinvented through the sickle cell gene.

Hemoglobinopathies

From 15 September to 21 September 1957 both Lehmann and Sukumaran attended a major symposium on "Abnormal Haemoglobins," held in Istanbul. The gathering had been organized by the Council for International Organizations of Medical Sciences, and under the joint auspices of UNESCO and WHO. Istanbul was deliberately chosen as the venue since the "Near East appear[ed] to be a sort of haemoglobin crossroads."[57] The symposium naturally included participants from many countries across the world and was intended to bring together the rapidly proliferating local studies of a growing number of "abnormal haemoglobins," such as the one responsible for sickling, HbS.

Interestingly, the job of summing up the state of knowledge and research on the subject in India fell neither on Lehmann nor on Sukumaran. Instead, it was Dr. Jyoti Bhusan Chatterjea who reported on the Indian research. Unlike many of the other researchers we have been dealing with, Chatterjea was mainly trained as a physician. He had obtained his MB from the Calcutta Medical College in 1942 and joined the Calcutta School of Tropical Medicine (CSTM) as a research assistant in the Hematology Department. After receiving his MD in 1949, he secured a two-year fellowship through the Rockefeller Foundation to work at the New England Medical Center in Boston under Dr. William B. Dameshek.[58] Dameshek himself was of Ashkenazi Jewish background and had been responsible for the search for sickle cell among the Yemeni Jewish immigrants to Israel at the end of 1949. It was in fact these reports about Yemeni Jewish sickle cell genes from Israel that had prompted Lehmann to rethink his initial hypothesis about the Indian origins of sickle cell.[59] After his return from Boston, in 1956 Chatterjea became a professor of hematology at the CSTM. A decade later, in 1966, he rose to be the institution's director. He won numerous awards, both national and international, for his work.[60]

In his lecture at Istanbul, Chatterjea mostly repeated the broad contours of the story about sickle cell, beginning with Lehmann and Cutbush's discovery in 1952 and ending with the work of the Nagpur team on Mahars. Chatterjea followed this up with the assertion that, "though extensive studies have not yet been made in different states and races, it appears that haemoglobin S is restricted to aboriginal tribal population in various parts of the country and its incidence in non-aboriginal population must be very low."[61]

Yet what was interesting about Chatterjea's talk was that the discussion on sickle cell was buried somewhere in the middle of it. Although it was Pauling's work on sickle cell that had led to the much touted "molecularization of disease" and inaugurated the work on hemoglobins, for Chatterjea these events were part of a different history that had commenced in India in 1937. Indeed, Chatterjea began with the publication in 1938 of a striking clinical report by Dr. M. Mukherji of the Campbell Medical School and Hospital in Calcutta.

The case was that of a two-and-a-half-year-old Bengali boy called Sambhunath Kar. When young Sambhunath's parents brought the sick child to the hospital in the winter of 1937, he exhibited many of the well-known clinical signs of congenital anemias (including sickle cell anemia), such as an enlarged and squarish head, an enlarged spleen, and so on. A battery of the still novel x-rays, which Mukherji called "skiagrams," led him to doubt that the disease was in fact the same as the one that had been described in 1927 by the American pediatrician, Thomas Benton Cooley. As Mukherji recounted, before Cooley's publication all congenital anemias—including, prominently, sickle cell anemia—were grouped together. But Cooley had distinguished a specific type of anemia that came to initially be named after him. "Cooley's anemia" by 1938 had been squarely identified with people of Mediterranean parentage, namely, Greeks, Italians, Syrians, and Armenians. So strong was the identification that the name "thalassaemia" ("thalassemia" in current spelling) had been proposed and popularized in preference to "Cooley's anemia." (Thalassa was the deified image of the Mediterranean Sea in classical Greek mythology.) Just as with sickle cell's African associations, so too with thalassemia. It was supposedly not found east of the Suez. Mukherji published the case of young Sambhunath to rebut this exclusively Mediterranean association.[62]

Chatterjea invoked Mukherji's discovery not only to signal an alternative history, but also to draw attention to the multiplicity of abnormal hemoglobins that were found throughout India and the difficulties of distinguishing between them. Chatterjea's own work in the past had focused extensively on hemoglobin E, which often appeared together with thalassemia and rarely also with sickle cell anemia. As a result, instead of dealing simply with homozygotic and heterozygotic carriers of the sickle cell hemoglobin (HbS), clinicians were frequently confronted by so-called "double heterozygotes," that is, people who bore two individual but different abnormal hemoglobin genes. Clearly, detailed "characterization studies," involving a whole range of pathological tests for a large number of family members, were needed to establish with certainty precisely which aberrant hemoglobins were causing anemias in specific patients.[63]

Soraya de Chadarevian has described how the continued work of Pauling and his colleagues at Caltech led to the redefinition of "anemic conditions" as "hemoglobinopathies" and to the introduction of the "electrophoretic characterization of hemoglobin . . . as a routine diagnostic procedure in the clinic."[64] Chatterjea worked within this framework. Indeed, his talk in Istanbul was titled "Hemoglobinopathies in India," though significantly, rather than starting with Caltech, he chose to commence the account with Mukherji's work in Calcutta in 1938.

The slightly different timeline and different founding figures were not the only ways in which Chatterjea's work was distinctive. Rather, as in the case of sickle cell, Chatterjea's summary—and, indeed, the entire body of Indian research—was distinctive in the ways in which the Indian hemoglobinopathies were overdetermined by older, specifically Indian racial logics. The casting of sickle cell as a "tribal disease," from which most other castes, except the lowest-ranking ones like the Mahars, were free, was part of this narrative.[65] Remarkably, this kind of racialization was not limited to sickle cell. The large and varied palette of hemoglobinopathies with which Chatterjea and others worked were all racialized to varying degrees. Thus, hemoglobin E, for instance, came to be associated mainly with Bengalis.[66] Hemoglobin D, in contrast, was identified mainly with Punjabi Sikhs and to a lesser extent with Gujaratis, but Bengalis, Biharis, and Nepalis were explicitly found to be free of it.[67] Indeed, Chatterjea even included a political map of India locating the distribution of hemoglobins within provincial boundaries (fig. 4.1). What is noteworthy here is that, whereas the American molecularization of sickle cell disease had possessed a broad race-imploding aspect that disaggregated the presence of the gene from racial identities, in India molecularization seemed to reinforce racializing trends.

Chatterjea was certainly not atypical in his racialized presentation. Several other clinical researchers, not to mention anthropologists, produced racialized numbers to demonstrate the differential incidence of

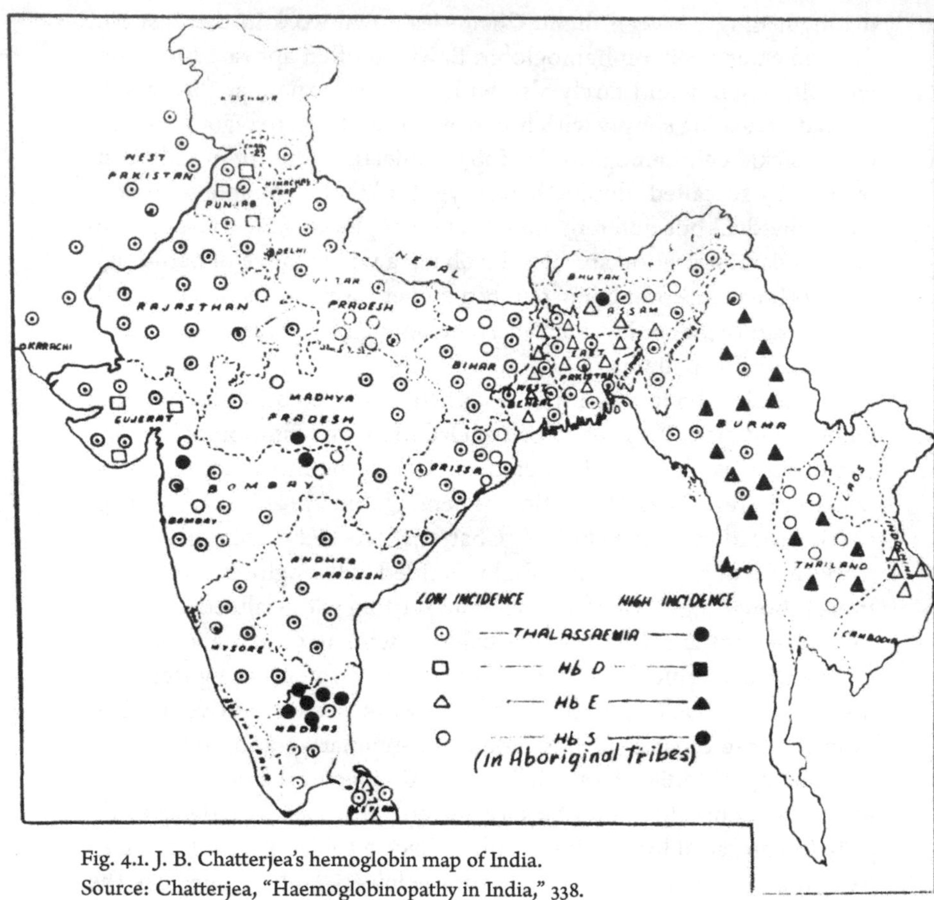

Fig. 4.1. J. B. Chatterjea's hemoglobin map of India.
Source: Chatterjea, "Haemoglobinopathy in India," 338.

specific hemoglobinopathies. Moreover, the redefinition of anemias as hemoglobinopathies served to build interest in hemoglobin variations more generally, even when there was nothing pathological about them. In 1958, for instance, Sanghvi, Sukumaran, and Lehmann co-authored a report on the presence of hemoglobin J among the Lohana caste in Gujarat. They noted that HbJ seemed to have no particular pathological manifestation. Indeed, even though one of the two cases they studied also carried the thalassemia gene, there was nothing to show that HbJ interacted with the thalassemia gene in any way to produce a clinical disease, as was the case, for instance, with the HbE-thalassemia disease reported by Chatterjea and others.[68] The following year the same team of researchers, along with J. A. M. Ager, tracked another new hemoglobin variant, HbL, in the Lohana caste in Gujarat. Once again, they confessed that, "clinically, there was nothing abnormal noticed in the carriers of the haemoglobin L trait which could be attributed to the haemoglobin variant."[69]

Thisline of research blurred the lines between anthropology and medicine. Quite a few studies were undertaken jointly by anthropologists and physicians. In 1967, for instance, the then director of the ASI, Dr. D. K. Sen, personally initiated a detailed survey of the "tribal" populations of Bastar and Koraput for screening for the sickle cell gene. The study was done by a physician, Dr. D. N. Roy, and an anthropologist, S. K. Roy Chaudhuri.[70] The most remarkable testament of this increasingly blurred line between medicine and anthropology, however, was the election in 1967 of Chatterjea, a clinician with a research background and absolutely no training in anthropology, as the president of the Indian Anthropological Society.[71] He was reelected the following year and therefore remained in office till 1969.[72]

Both the new focus on tracking rare, but not necessarily pathological, hemoglobins and the progressive blurring of the line between medicine and anthropology had significant effects on the ways in which identities were racialized in South Asia. One major change was the popularity of "family studies," often as part of "characterization" studies. These studies looked at all the family members of the person who was identified as having a rare hemoglobin or hemoglobinopathy. This kind of study, quite naturally, focused on individuals rather than on the large, explicitly unrelated groups that were the subject of the blood group frequency studies. Instead of seeking to track the differential frequency of genes within a defined racial group, these "family studies" proceeded to actually locate individuals and family groups within larger, more abstract racialized groupings.

The growth of medical genetics in the postwar world led to "family studies" in some other walks of medicine as well, such as oncology and psychiatry.[73] While recognizing this larger field of "family studies" in medicine in the 1950s and 1960s, it is also important to attend to their specificities. The cancer studies, for instance, were largely concerned with establishing the hereditary basis of certain cancers. The hereditary basis of hemoglobinopathies, by contrast, was well established. Family studies pursued other questions, such as characterizing the interaction of different heritable traits.

Two interesting social-theoretical problems emerged from these hemoglobinopathy family studies: diasporas and cross-cultural marriages. I will discuss them here in that order. First, diasporic families were metonymically read as belonging to the localities, provinces, or countries they originated from, rather than the ones where they lived. Structurally, this was similar to the mapping of "Veddoid blood," in which the social body of the nation was understood as being made up of distinct biological layers. The key difference lay in the way that the family studies foreshortened the timescale for such layering. Unlike the "Veddoids," whose distinctiveness was registered on timescales of centuries if not millennia, the timescale of diasporic distinctiveness was usually a generation or two. Yet, in effect the notion of a diaspora was explicitly biologized through such studies.

Such diasporic distinctiveness was particularly clearly visible in the many studies of "Indians" living in countries such as South Africa, Uganda, Malaysia, Singapore, and the United Kingdom. In 1957, for instance, Lehmann's long-term collaborator, Alan Raper, identified a new hemoglobin variant, which he called HbK, at Kampala Hospital in Uganda. The carrier of this new type of hemoglobin was identified as a "Mrs. L. M." who was a "Gujarati Indian."[74] Raper's study was quickly picked up and cited by Indian researchers like Sukumaran, Sanghvi, and Chatterjea.[75] Soon after, in 1959, Drs. F. Vella and P. F. de V. Hart published a case report of a young boy suffering from sickle cell anemia whom they had seen at Malacca General Hospital in June of that year.[76] The boy and all his siblings, as well as his parents, had been born in Malaysia. But the researchers found that the boy's grandparents had immigrated from Odisha, a region with a large "tribal" population and one that by 1959 was already being suggested as a potential sickle cell focus area.[77] This study, too, was quickly cited by Indian researchers.[78] Researchers in Cape Town, likewise, upon encountering HbS carriers, stated that as "pure Asiatic (Indians)" they had carried the gene from India.[79] Diasporic families in each of these studies were identified as biologically "Indian" and thus distinct from the countries they were living in.

Most of these communities in Southeast Asia and Southeast Africa had emerged through the aegis of British colonialism and frequently indentured labor. Until around the time of World War I, these communities had indeed been imaginatively conceptualized by early nationalists as being intimately tied to the Indian nation. Thereafter, in the interwar years, however, the nationalist imaginary was explicitly redrawn along narrower lines to exclude these communities.[80] So it is worth noting how, in the early postcolonial era, the biological claim to the Indianness of these diasporic communities was reasserted. The change from the interwar era is perhaps also glimpsed in the fact that the very first identification of sickling in an "Indian" had actually been made in 1943 in Cape Town. The authors, L. Berk and G. M. Bull, both physicians, had reported on a twenty-two-year-old woman who had been admitted to Groot Schuur Hospital in Cape Town and had shown both clinical and pathological evidence of suffering from sickle cell anemia. The researchers had insisted that the woman's family were "Indian," although both sides of the family had already lived in South Africa—mostly in Durban—for at least three generations.[81] This study was done almost a decade before Lehmann and Cutbush's eureka moment in Coonoor in 1952. Yet, in stark contrast to the postcolonial studies, practically no notice was taken of it. Even after sickle cell in India became a major focus of national and international research, the Groot Schuur study was seldom referred to. By contrast, the studies of Vella, Raper, and others were quickly, and remarkably unproblematically, incorporated into the citations of Indian researchers.

What is important for me in these cases is the way in which seemingly national designations of identity became firmly biologized. No matter how long one lived in Malaysia, Uganda, or South Africa, medically and biologically, one came to be seen as an "Indian" in the late 1950s and 1960s in a way that seemed at least less popular in the early 1940s. As a result, national identities, independently of the political questions of citizenship status, became a biological matter.

Such biologization of diasporic identities was not unique to the national level. We also find some very similar intra-Indian conceptualizations of local identity around the same time. One of the more revealing of these is a 1962 thalassemia study of two families from Palghar in western India. The subjects belonged to a community at Palghar called the Sorathis, whose elders told the researchers that they had migrated to their present location in Palghar, via some other, intermediate settlements, from Surat as a consequence of the invasions of "Mohamed of Gazni." The researchers identified this lore as referring to the invasions of Sultan Mahmud of Ghazni in 1024 CE and sought to link the families back to Saurashtra, rather than Maharashtra.[82]

The second, and not unrelated, social-theoretical problem emerging from the family studies involved cross-cultural marriages. Naturally, one of the most common social consequences of human mobility is the cross-cultural marriage. Such marriages, much more than changes in language, food, or dress, threatened to dislodge the neatly racialized grids that researchers were cultivating by the late 1950s. The scale of the "family studies" easily picked up on such marriages, but the studies then craftily tried to work around them by parsing apart different members of the same family unit. In consequence, racial identity became even further consolidated and biologically legible, in ways that were more individualizable and independent of even the most intimate forms of social grouping.

Two examples will illuminate this process. The first example comes from Chatterjea's research and speaks to the problems of intercultural marriages within the boundaries of the nation-state. Chatterjea had repeatedly emphasized the nonexistence of HbS (sickle cell hemoglobin) among Bengalis. Instead, HbS was identified as a trait found among the "aboriginal tribes."[83] These terms, however, were deceptively transparent. The political boundaries of the province of "Bengal"—whatever they may have been at prior historical junctures—were repeatedly redrawn in the twentieth century. In 1905, for instance, its eastern districts were spliced off and added to Assam to create the province of East Bengal and Assam. In 1911, this was annulled and the eastern districts rejoined to the western half, but the province of Bihar and Orissa was now carved out of the western half. In 1947 the province of Bengal was once again divided between the newly independent nation-states of India and Pakistan. In the 1950s, the process of the Linguistic Reorganization of States further changed the boundaries between West Bengal and Bihar. Moreover, throughout this period substantial numbers of people classified as "aboriginals" and "tribals" lived within the shifting boundaries of Bengal and spoke languages and dialects long connected to standardized forms of Bengali. So, ethnonyms such as "Bengali" and "tribal" are complex, historically contingent labels.[84] Similarly complicated are distinctions between a "Bengali" and an "Assamese," "Bihari," or "Odia" identity, especially along what are nowadays border districts. Yet, Chatterjea felt it plausible to repeatedly invoke these categories as neatly corpuscular groupings.

Chatterjea's neat characterizations were challenged when he first came across a nine-year-old girl who was carrying both the HbS and the thalassemia genes. He had already clearly identified the former as an "aboriginal tribal" trait and the latter as a trait often found among "Bengalis." Yet, here was a child who possessed both traits simultaneously. Chatterjea solved this seemingly unresolvable contradiction to his neat schema through a family study. He explained that the father, from whom the child had inherited the HbS gene, was in fact a Bihari, while the mother, from whom she had received the HbE gene, was a Bengali. He does not tell us on what basis he made such an identification. It is almost certain that the grandparents of the child, if not the parents, would have been born at a time when Bihar was part of the larger Bengal presidency. The statement is further complicated by the fact that he followed the statement with another reasserting the "tribal" roots of HbS.[85] It seemed that he was suggesting that somehow being "Bihari" and being "tribal" were the same thing, while being "Bengali" was clearly separate from both these identities.

The second example comes from Lehmann. Along with J. A. M.

Ager, a lecturer in clinical pathology at St. Thomas Hospital in London, Lehmann published a report on HbK in a forty-three-year-old Londoner and his son. The man was being treated for tuberculosis when the variant hemoglobin was discovered in his blood. A family study revealed that one of his sons had the same hemoglobin. Though the man was a Londoner, Ager and Lehmann found that he was born and raised in South Africa, from where he had migrated to Britain. Even before that, however, his grandfather had apparently arrived in South Africa from "India (Madras)." We are not told who his grandmother or his parents were and where they might have come from. All that we are told is that "there is no evidence in his appearance" or in the family history he gave of "any admixture of non-East Indian blood."[86] In his son's case, however, there was clearly some "admixture." The man had married a woman of "Scots and West Indian origin." The son, therefore, clearly had "Indian," "Scots," and "West Indian" roots, if not "South African" as well.

Wailoo has evocatively offered his own personal example to point out the ways in which genetic truths remain partial and contested. He points out that although he is "phenotypically and culturally African American," he is actually a naturalized US citizen born in South America to a family whose own memories point backward to both Africa and South Asia. He points out that, despite the prominent personal and mnemonic connections with South America, genetically the land of his birth would probably be untraceable in his genes. Any answer to the question "who am I?" Wailoo points out, is about choosing a lineage and determining a starting point and a stopping point. Any person can choose to represent their mother's or father's lineage. They can equally choose their mother's father's or father's mother's lineage. At each point in their past, they have to decide which routes to emphasize. They have to decide where to stop and where to begin. "Genetic testing offers," writes Wailoo, a "specific yet highly limited entry point into a complex past."[87] Wailoo's personal account resonates with the complexities of the identity of the unnamed Londoner from the 1950s. Ager and Lehmann chose a lineage and determined a set of starting and stopping points for the man that almost certainly did not correspond to the ways in which he might himself have thought of his own identity. In the process, despite never having set foot in India and, given his wife's identity, perhaps not even speaking any Indian language at home, this man and his London-born son became biologically legible as "Indians."

Especially in the case of the son, this "Indianness" emerged despite the clear fact of his Scottish and West Indian ancestry. It did so because Ager and Lehmann were able, just as Chatterjea had been, to separate out the familial unit and mark individuals according to their racial origins. Nadia Abu El-Haj has illuminated how the new, geneticized discourse on race is no longer premised on the old "one drop" framework, but rather rests on the simultaneous articulations of almost universal admixture and the ability to still parse apart that admixture and calibrate exact numerical quanta of racial traits.[88] Hence, in contemporary genetic narratives of identity there is both an insistence that we are all mixed and a promise that we can find out precisely what percentage of what racial ancestry we possess.

What marks this kind of genetic reinscription of diasporic identities in cases such as that of the London man is that these genetic truths were themselves artifacts of the kinds of technologies and the particular categories being used. These identities were not grounded in any phenomenological or lived experience whatsoever. Hemoglobin K had no clinical or everyday manifestations. It was purely something that was engendered through the specific techniques of paper electrophoresis. Defined as a "fast moving" hemoglobin, using paper electrophoresis in an alkaline pH, it appeared as a band in front of what was recognized as the "normal" HbA. Yet, even then, it "barely separated" from the HbA. Both the hemoglobin and the racialized truth about personal identities based on it were therefore constituted entirely in the "dialectic of mutual construction of technology and disease."[89]

Race as Risk

Tracking a person's race through racialized frames of parental contributions to their genetic identity, especially in a medical context, transformed race into risk. By the 1960s several researchers in India began to talk about the genetic risks of marriages. While such talk that twinned hereditary disease and marriage had certainly existed earlier, in the 1960s two new elements distinguished it. The first was explicit racialization. Unlike the prewar, often amateur, eugenicists, the scientists of the 1960s linked risk more concretely to biologized community identities, namely, race. The second was the linking of risk far more firmly and clearly to specific genetic diseases, such as sickle cell anemia.

Between 10 September and 1 October 1967, L. D. Sanghvi, one of the foremost geneticists in India, delivered a series of four talks over All India Radio titled "The Mystery That Is Heredity." The lectures were subsequently published in the *Journal of Family Welfare*. This journal was a publication of the Family Planning Association of India, an organization started by Lady Dhanvanthi Rama Rau in 1947 to promote planned parenthood. The lectures are remarkable, therefore, in illuminating the convergence of scientific ideas about race as risk with planned parenthood activism and the state's media apparatus.

In the lectures, Sanghvi spoke almost entirely of the potentially detrimental aspects of genetic heredity. With very brief explanations of the mechanisms of genetic heredity, he focused on the medical and pathological aspects of heredity. He discussed at length defective genes and the "blood diseases." Predictably, thalassemia and sickle cell disease featured prominently as examples of these diseases. These two diseases had indeed become exemplary in making a larger point about race as risk for Sanghvi and his peers.

Sanghvi asserted that "thalassemia is common in some populations of Italy and Greece. In India, it is common in the Lohanas of Gujarat as well as in Saraswat Brahmins of Goa. Sickle cell anemia is quite common in Africa and in the tribal populations of India."[90] Notwithstanding such a clear and categorical statement, by the time of Sanghvi's radio address a lot of evidence had already complicated these simple racial frames. Sickle cell, for instance, as we have seen above, was clearly not limited to "tribal populations." Also, it was not at all clear what being "quite common" meant. Even leaving aside complications about how communities were conceptualized or how frequencies were measured, it was clear that the vast majority of people belonging to tribal communities did not suffer from sickle cell anemia. The blanket ascriptions of "quite commonness" in "tribal populations" were precisely the sort of slippage that allowed the carefully crafted rhetorical difference between "race" and "population" to collapse in practice. It also shows how difficult it is to make a clear distinction between specialized scientific discourse on race and genetics, on the one hand, and more vernacular notions of these topics, on the other. That Sanghvi, who was a globally recognized scientific authority on the subject, could deploy such "imprecise" definitions in a nationally broadcast public address patently undermines any clear distinctions between the carefully calibrated world of expert knowledge and lay discourse.

Given the thrust of Sanghvi's first talk, he naturally turned extensively to the discussion of "family planning" initiatives in his later talks. He pointed out that while the main focus in Indian family planning had been on quantity, there needed to be more emphasis on "population quality." He further subdivided the question of "population quality" into two parts. One aspect of it was "dependent on environmental factors such as nutritional levels, education and health measures, job opportunities and social and cultural settings." This aspect was already, according to Sanghvi, being dealt with under the Five-Year Plans. The other aspect, however, was neglected. This second aspect of population quality was "dependent on the genetic endowment or innate capacities and weaknesses with which we start our lives."[91]

To address this aspect of population quality called for "genetic counseling." Such genetic counseling services, Sanghvi felt, ought to be made available by the state as part of family planning. Interestingly however, he expected this genetic counseling to happen not before marriage, but rather after the birth of a "defective child." He argued that parents who had a "defective child" were usually "anxious to know the risk of having a similar child again."[92]

Having thus located the anxiety about medical risk squarely within a familial and parental domain, rather than at the level of nations and populations, Sanghvi quickly moved to racialize it once again. Following up immediately with a discussion of anemias, he turned once more to the high rates of "risk" to particular communities. He mentioned that thalassemia was "relatively more frequent" within the Lohana, Khoja, and Saraswat Brahmin communities and that 1 percent to 2 percent of all live births in tribal communities suffered from sickle cell anemia. Children born with such conditions, he continued, "are for several years a source of intense anxiety to their parents."[93]

It was only after these preliminary discussions of risk and race that he broached the topic of premarital counseling. He mentioned that a district in Italy had apparently started providing premarital genetic counseling, which included laboratory tests to detect the aberrant gene, so as to avoid marriages between carriers of the sickle cell gene. Apparently, Cameroon too was moving toward such a system. Sanghvi also spoke at some length about the need to avoid "inbreeding" by way of consanguineous marriages that were the cultural norm in many southern Indian communities.[94]

He dubbed the program he was advocating "preventive eugenics," which he defined as "reducing the load of deleterious and harmful genes in the population by creating a more enlightened attitude towards marriage and reproduction."[95] He distinguished this from "progressive eugenics," which was concerned with improving the frequency of desirable traits in a population. The "problem" with the latter was arriving at a scientific consensus on what these desirable traits might be. But in the case of preventive eugenics, clearly such a consensus was easy to arrive at. He further emphasized that such preventive eugenics was all the more important because, given the state of genetic knowledge, it did not seem "possible in the foreseeable future" to "correct" any defective gene, "such as the one for sickle cell." Therefore, the country's family planning program would have to help people avoid situations where such correction would become necessary.[96]

More ominously, Sanghvi invoked the specter of "differential fertility," whereby the genetically weaker elements in a society would proliferate faster than the better elements. He argued that, as urbanization gathered pace in India, those who had "greater physical and mental strength which might enable them to withstand the hardship of urban life" would be more exposed to family planning propaganda and would reduce their own fertility, while their genetically weaker relatives left behind in villages would breed more.[97] In this strange twinning of rural-urban migration and genetic fitness, Sanghvi most clearly articulated how the carefully restrained rhetoric of pathological genes could easily slip into something altogether more patently prejudicial.

Sanghvi's views were hugely important both because of his stature as a leading geneticist in India and internationally, and because these views were being aired on All India Radio. Nor was he alone in expressing such views. Genetically inspired new eugenicist views flourished in India in the 1960s. In September 1959, for instance, the *Times of India* carried a long editorial by one Sapur F. Desai titled "Place of Eugenics in a Democracy." Desai expressed fears about differential fertility that were almost identical to Sanghvi's. He asserted that those who were least fit to reproduce were breeding more and those who were most fit, through greater exposure to family planning programs, were breeding less. Desai argued that, notwithstanding its motivations for creating a better nation, the family planning program was effectively functioning "anti-eugenically" by discouraging the more enlightened from breeding. He advocated a more calibrated approach that used both "positive" and "negative" eugenics. The former would work through encouraging certain people in society to have more children, while the latter would operate by voluntary abstinence from marriage, sterilization, and so on targeted at those who were "diseased, deformed, deranged, anti-social and such." Eugenics and democracy, Desai felt, were mutually complementary, and no proper democracy worth its name ought to neglect eugenics.[98]

Opinions such as Desai's, though not universal, continued to appear in the press—particularly in Bombay, where Sanghvi was based—throughout the 1950s and 1960s. By the second half of the 1960s, when Sanghvi delivered his radio talks, such views were beginning to enter political discourse as well. In late 1969, for instance, when the federal government introduced a new bill to rationalize the legal position on abortion, the minister concerned stated that abortions would be legally permissible on three grounds. The first and the second of these dealt with the mother's health and the origin of the pregnancy in cases of rape and such. The third ground, however, was explicitly described as "eugenic," that is, "where there was a substantial risk that the child, if born, would suffer from deformities and disease."[99] Earlier that same year, a nominated member of Parliament and family planning activist, Miss Shakuntala Paranjpye, daughter of the famous mathematician, Sir Raghunath Purushottam ("Wrangler") Paranjpye, introduced a bill in Parliament that would mandate compulsory sterilization for certain medical conditions, such as tuberculosis, leprosy, and insanity.[100] Paranjpye, not incidentally, had been a member of the Maharashtra Legislative Council before entering Parliament. Though her bill was eventually defeated, it demonstrated both the popularity of medically driven eugenics and its resonances in and around Bombay in particular.

Sarah Hodges has drawn attention not only to the continuities between pre-Independence and post-Independence eugenicist thought in India, but also the fact that postcolonial "population control efforts were often far more 'eugenic' than eugenic advocates in colonial India had ever been."[101] She also describes the involvement of nationalist women's movement activism in promoting eugenicist thought in both the late colonial and postcolonial eras.[102] These observations help us contextualize the positions of leaders like Paranjpye as well as the larger context of scientists like Sanghvi. Yet, Hodges has emphasized the role of neo-Malthusian frameworks, promoted by US development agencies, in this "more eugenic" postcolonial eugenics. While this is of course true, an overemphasis on neo-Malthusianism and its obsession with "population quantity" risks missing the medically and genetically inspired emphases on "population quality." This medicalized qualitative focus was also much more capable of racializing caste and tribal identities than the neo-Malthusian quantity discourse, which mainly targeted Muslims.[103]

Indeed, the racialization of caste and tribal identities, as has been illuminated by Luzia Savary, had long been prominent in the "vernacularized eugenics," or *santati-shastra*, that had developed in late colonial India. Drawing eclectically on older South Asian and Western sources, this internally heterogeneous assemblage had, by the 1930s, developed both a set of rationalities of human difference and a conceptual vocabulary that braided indigenous terms of difference with more exotic fare. These vernacular discourses were also much more clearly in conversation with medical notions of fitness and unfitness than were the neo-Malthusianists.[104] The mechanisms by which such "vernacular eugenics" might have influenced the genetically structured eugenics of the 1960s, as well as the extent of that influence, remains to be explored.

Though the parliamentary bills did not explicitly refer to hemoglobinopathies, choosing instead to focus on such redolent afflictions as tuberculosis and insanity, hemoglobinopathies were in fact frequently discussed in family planning circles. In December 1969, precisely the month when the abortion bill and its eugenic rationale were being debated in Parliament, the government organized a National Conference on Population Planning in New Delhi. At the conference, B. L. Raina delivered a lengthy address on the medical aspects of family planning. While some of his lecture focused on contraceptive devices, much of it focused on the genetic burdens of disease. Once again, he invoked sickle cell as a key example.

Explaining the usual theories about the magnification of sickle cell traits in malarial regions through environmental selection, he added that "it is believed [heterozygotes] are more fertile in general. This may lead to the perpetuation of genetic load."[105] He therefore invoked the problem of "differential fertility" in a new context, whereby harmful mutations are augmented through the mechanism of enhanced heterozygotic fertility, thus increasing the genetic load of deleterious genes in the population as a whole.[106] Raina went on to state, in a tone that was almost lamenting, that modern civilization had reduced the effectiveness of natural selection. Simultaneously, he pointed out that certain selective pressures, such as malaria, that had given harmful genes some functional value in the past were now redundant, through the control and eradication of diseases like malaria.[107] Yet, these genes could no longer be weeded out by nature now, owing to the triumph of modern civilization. In this context genetic counseling became a necessary component of family planning, essential for reducing the genetic load of harmful genes.[108]

Like Sanghvi, Raina was quick to distance himself from progressive or positive eugenics, clarifying that he did not want to suggest that "cattle stock breeding principles should be applied to human beings," but merely to make the point that suffering could be prevented or mitigated by genetically enlightened premarital counseling.[109]

Raina's advocacy of premarital genetic counseling, which seemed on the face of it to be targeted to individuals, soon turned to the familiar racialized framework. He began by pointing out how consanguineous marriages among the Parsees in India had created a substantially diseased population. He mentioned that 13.6 percent of Parsee men suffered from glucose-6-phosphate dehydrogenase deficiency, which, in turn, produced a type of hemolytic anemia as well as jaundice upon taking certain drugs. The Mahar community also suffered from such deficiencies, and in certain "isolated districts" up to 38 percent of the Mahar community were said to be affected. Finally, there was sickle cell. Though Raina did not name the communities, he mentioned that some communities in Bombay and Bastar had as much as 20 percent and 40 percent sickle cell incidence.[110]

What is significant in this discourse is the way race is reproduced as risk. Community identities are racialized, that is, depicted in purely biologically inheritable ways, but this racialization is seemingly socially nonhierarchized. It placed Saraswat Brahmins and Mahars, for instance, at the same footing. It distinguished them not in terms of their ritual or social standing, but only with reference to the quantum of pathological risk they carried. This risk was inheritable. Hence, someone marrying a Mahar or a Saraswat Brahmin had to contend with the risk of giving birth to a "defective child."

Robert Aronowitz has described the emergence, in the postwar United States, of a new type of medicine he calls "risky medicine." There were three new features to this "risky medicine": first, the "market-driven expansion of risk interventions"; second, the "converged experience of risk and disease"; and third, the progressive understanding of medically effective intervention as something that reduced risk.[111] What we see in the same period in India is both similar and different.

Market-driven risk-reduction interventions, for instance, were almost entirely absent in India. But the convergence of risk and disease was fairly explicit. Additionally, however, this converged understanding of disease and the risk to get the disease was entangled with a racialized view of communities. We might say, therefore, that in India race, risk, and disease all began to converge. The conceptualization of effective intervention as that which reduces risk, rather than actually tackling disease, was also visible in India.[112] But it was packaged alongside family planning initiatives—not only attempts at genetic counseling and popularization of genetic ideas of heredity through public radio, but indeed research projects themselves being presented as a public health initiative. That so many hospitals, physicians, and pathology laboratories became involved in quasi-anthropological research into the calculation of risk speaks to this last aspect of Indian risky medicine.

Indian versions of risky medicine, we might therefore say, were most conspicuously legible in discourses and practices that sought to enumerate risky races.

INTERCHAPTER

Letter 5

Najrul Islam
Kampala, Uganda
28 February 1934

Dear Hemenda,
I am touched and grateful for your concern for my health. Please do not worry. I have indeed recovered my health since that unfortunate bout of fever.

Also, please rest assured that I share none of the hubris of our modern doctors towards malaria. Indeed, in the winter of 1925 and 1926 I had suffered greatly from this disease. Repeated courses of the usual medicines had failed to return my health to me. Eventually, upon the advice of a close friend, I performed Kali puja for the first time in Nadia. Somehow, I got it into my head that the terrifying mother, Kali, would vanquish the malaria. Until today, I have not been able to make up my mind as to whether my subsequent rapid recovery was indeed a consequence of the puja or of the medicines I had continued to take. I like to believe that Mother Kali played her part in my recovery. And so, Hemenda, I have continued to pay my obeisance to her in both illness and health. Naturally, I did not forget to ask for her help during my recent convalescence, either.

However that may be, and whether it was through the grace of the Mother Kali or the box of Paludrine I had carried with me from Calcutta, I have now fully recovered from the malaria I suffered on the banks of Lake Kivu. Happily, the bracing air, the fresh game, and the nourishing water of the mountain streams have allowed me to rapidly beat back the debility, weakness, and other vexatious sequelae that usually follow this fell disease like so many foul, carrion-eating beasts following the train of medieval conquering armies.

Recalling the illness reminds me, however, of another incident that you would, I think, be intrigued by. As is often the case with malaria, I had not been the only one in my camp to fall ill. A few of the native coolies too had come down with the fever. When we eventually managed to get the doctor's native assistant at one of the distant village dispensaries to come out to test our blood for the disease, however, these coolies flatly refused to allow their blood to be drawn. As you might imagine, weakened by my own fever, I was in no position to argue with them but urged their *sirdar* to explain the procedure to them. But he too refused. The insolence and insubordination were galling, I tell you! Yet, in that condition I could do precious little and left them to their own devices.

By the time I recovered, too much water had flowed down the Ruzizi, and it did not seem worth the effort to discipline them. Yet, I remained curious about their refusal and inquired why they should have behaved in such a fashion. They kept saying to me that the native doctor's assistants, such as the one who had come out to our camp, were all *batumbula*. The word, I believe, is derived from the Luba word *tumbula*, which has come to mean something akin to a vampire!

I would have laughed off the quaint fable, had I not recalled your account of Mrs. Kumudini Choudhury. There, too, I recalled how physicians had initially struggled to make sense of the high rates of anemia among the locals and put it down to various hereditary and dietary shortcomings, only for it to finally emerge—and of course almost immediately be hushed up—that the well-respected, wealthy Christian widow, Mrs. Choudhury was in fact a vampire! Remembering your account stopped me from laughing off the stories of the *batumbula*. In fact, I have often heard friends who keep abreast with the latest in the world of science fulminate about the connections between anemia and malaria, especially in some of these wilder parts, and found myself wondering about Mrs. Choudhury's persecutions of the people of the Santal Parganas.

Not just that. You might find me becoming rather fanciful here, but Hemenda, I also caught myself repeatedly drifting back to these vampiric tales again when I was reading Mr. Sen's descriptions of vivisections in his diary. He speaks of the heartlessness with which the Chief Scientist in the kingdom of the future-humans performed these gruesome vivisections. It seems that the Chief Scientist had first conducted these grisly experiments upon a Turkish man who had wandered, accidentally, into their kingdom. The poor man died a terrible death in the Chief Scientist's laboratory as he attempted to further improve their own national race. But even this tragic end did not deter the Chief Scientist from his abhorrent path. Undeterred by either the scandal that followed the death of the Turk or the king's edict forbidding such experimentation, he even tried to vivisect Mr. Sen himself!

The whole description seemed almost like a national or racial vampirism to me. Mr. Sen mentions the Chief Scientist boasting that he wished he could find a steady supply of foreigners on whom he could conduct his experiments so as to enhance the life force of his own nation. Unlike Mrs. Choudhury, the vampire you met in Jhajha, the Chief Scientist did not want to enhance his own vitality. It was his nation's life force that he aspired to augment. Yet, was it any different than Mrs. Choudhury's vampirism? The hapless foreigners, like the anonymous Turk or Mr. Sen, had nothing to gain from this enhanced vitality, and yet it was their bodies, their blood, and their lives that were to be consumed in the quest for this enhanced vitality! How is this anything other than vampirism?

That the Chief Scientist and his two assistants would willingly descend to such butchery was compounded by the fact that they themselves saw little wrong in their actions. They justified their actions in grandiose terms. They invoked the glory of science, the pursuit of knowledge, and, of course, the future of their nation, in seeking to justify their bloodshed. It was utterly shocking for me to read Mr. Sen quoting the Chief Scientist declaring that he will willingly disobey the royal law in order to "uphold the dignity of science" by vivisecting him. I tell you, Hemenda, the sheer moral depravity of these statements boggles the mind! I could not help but be struck by the way what seemed to be laudable, altruistic ideals could authorize such abhorrent hecatombs.

An obsession with abstractions, and with ushering in a future they ardently believed in, seems to have blinded these scientists not only to the violence that they inflicted on particular people but also to how little their experiments actually meant to the people they brutalized. They seemed convinced that science and race improvement were ends for which their victims ought to willingly sacrifice their lives and limbs. At one point, Mr. Sen describes how, when Bhombol gets the king to intervene to stop Mr. Sen's vivisection, the Chief Scientist confidently argues that, far from objecting, Mr. Sen should in fact willingly give up his body since it would become the basis for a future, improved race.

I have always been a man of action and have learnt what little I have from the book of nature and in the library of experience. So perhaps I do not sufficiently value the worthy ends these future-minded men of learning pursued. But I must confess, Hemenda, to being staggered by the perverse morality, the lack of empathy, and the sheer hubris of these scientists of the future. I sincerely hope our scientists of today are not cut from the same cloth!

In any case, by now I hope you have had a chance to read Mr. Sen's diary yourself and I wonder what you make of his descriptions? Do, please, drop me a line when you can.

Your obedient brother,

Najrul

CHAPTER 5

Blood Multiple

> . . . drops of clotted blood in the trough of orbits, set out to storm the firmament.
>
> —Frantz Fanon, *Parallel Hands*

Race science, or any other form of science for that matter, was never merely about concepts and ideas. One of the most productive moves in histories of science has been to link conceptual developments back to practices, instruments, and materials.[1] Those working on the intersecting histories of mid-twentieth-century race, indigeneity, serology, and genetics have also contributed to this trend by drawing attention to the ways in which samples were collected, preserved, and studied.[2] I follow their lead here and look more closely at the specific types of blood that were used to produce seroanthropological facts about race.

"Traffic in blood" is how Elise Burton describes the complex, heterogeneous institutional and infrastructural regimes of "exchange and knowledge control" by which the division of labor between the work of "collection of samples from individual bodies" and the work of examining and interpreting the "disembodied samples in the laboratory" came to be articulated in postwar human genetic research into Middle Eastern blood.[3] This traffic in blood depended upon institutional and infrastructural regimes that had built on colonial legacies of the interwar years, while also increasingly reinforcing many of the postwar nation-making projects of the region.[4] These heterogeneous and mutually articulated infrastructural networks allowed blood drawn from individual bodies to flow from "fields" to "laboratories" despite the myriad frictions between those whose blood was drawn, those who did the drawing, those who studied the blood in laboratories, and those who finally aggregated the data in their publications. It was the infrastructural networks that ensured that the traffic in blood remained smooth, despite these multiple frictions.

The smoothness of the flow eventually depended upon the unity of "blood" as a category. While the blood revealed distinctive racialized identities, there was a sense in which the substance itself continued to be seen as a single, homogeneous, biological entity called "blood." This ontological unity of the samples is an epistemic assumption, rather than a pregiven ontological injunction. Jacob Copeman and Dwaipayan Banerjee have recently highlighted the "political polyvalence of blood as an aesthetic substance" through their exploration of South Asian "menstrual activism."[5] The South Asian feminist activists whom Copeman and Banerjee write about have been at pains to upend the dominant patriarchal coding of menstrual blood as a specific substance clearly marked off from the broader category of "blood" more generally, as a substantive site of disgust.[6] Such menstrual blood is quite distinct, for instance, from the blood of martyrs, which is often seen as a sacred substance demanding devotion, loyalty, and so on.[7] Likewise, blood transfusions across South Asia continue to run up against deeply held figurations of the bloods of different castes being somehow fundamentally dissimilar.[8] Indeed, as Lawrence Cohen pointed out in 2001, roughly in the period between the 1950s and the 1970s, the biomedical technique of blood transfusion had itself become a redolent Hindi filmic device for an attempted ambitious recoding of the postcolonial citizen's body. Several classic films from the period showed individuals considered separated by caste or religion eventually donating blood to an upper-caste, Hindu protagonist and thereby establishing a sanguinary unity within a larger national framework. In Cohen's words, in this "Nehruvian vision, modern medicine recode[d] . . . the biomoral logic of local transactions across caste, gender and generation into a national logic of distinction amenable to scientific planning and demonstrably supportive of life."[9]

Cohen's observations are particularly important for us in light of Jenny Bangham's fascinating recent work highlighting how central transfusion services were to the emergence of human genetics. Bangham asserts that "blood group genetics depended on the infrastructures and social practices of the transfusion services" and that "transfers and exchanges followed the contours of nation, class, friendship, institution, and ethnicity—and these, in turn, made blood groups available for fixing pedigrees and for drawing maps of genetic diversity." In other words, "human genetics was made possible by social relationships forged and articulated through the exchange of blood."[10]

While Cohen's insights, and indeed those of Copeman and Banerjee, inspire us to explore the active co-constructions of substance and meaning, Bangham draws our attention to the ways in which institutional structures and disciplinary protocols shape the biological materiality of the blood that substantiated human genetics. Together, they push beyond the findings of anthropologists that "blood is a dense metaphor in most cultures."[11] Rather, Cohen and others point out that the biomedical/biological category of "blood" is itself shot through with equally dense metaphorical charges and, moreover, the very materiality of this blood (and not simply its metaphoric meanings) is itself historically contingent.

Following Annemarie Mol, we might consider the biological entity called "blood" as "an intricately coordinated crowd": the blood multiple.[12] Mol argues that if we defer the ready acceptance of bodily tissues as always already unified, singular ontological entities and instead foreground the actual practices of manipulation through which people engage with these objects, then "reality multiplies."[13] In each practice we find materialized a distinctive, rather than a singular, object. The ontologies of things like blood or other human tissue, she insists, are "not given in the order of things," but "brought into being, sustained, or allowed to wither away in common, day-to-day, sociomaterial practices."[14]

As Bangham points out, the history of human genetics is essentially "a story of the practical links between blood and a formal science of kinship."[15] If we approach it through Mol's reality-multiplying lens, we can resuscitate the extinguished relations that could not be accommodated within the "formal science of kinship." We will glimpse how scientific notions of human kinship eventually relied upon contingent, local relationships between particular human, nonhuman, and even divine bodies. In this chapter, therefore, I will avoid the abstract kinships between groups that the scientists theorized about and look more carefully at the actual bodies whose blood operationalized the theories.

Subject Blood

It was April 1941. D. N. Majumdar, newly returned from Cambridge, was out in search of Korwa blood. The Korwas lived in the Chota Nagpur region, and Majumdar had planned to collect blood in the then-independent princely states of Dudhi, Sarguja, and Palamau. But Majumdar did not simply march into Korwa villages and ask for blood. Experience had taught him that this might not work. Instead, he resorted to subterfuge.

Majumdar dressed himself as an itinerant homeopath and traveled from one village to another, dispensing simple medicines and in the process acquiring blood from the unsuspecting Korwa men and women. Majumdar's strategy was so successful that he followed up his initial fortnight of traveling by another similar tour a couple of months later, at the peak of summer, and again, in April of the following year. The data that he thus collected formed one of the first and most comprehensive serological datasets on the Korwas. What was more, in writing about it, Majumdar unreflexively admitted that this was not the only occasion on which he had resorted to deception to procure the blood of so-called tribal groups. Posing as a homeopath, he wrote, "ha[d] stood [him] in good stead on so many occasions."[16]

Majumdar's tactic of impersonating a homeopathic healer, which he almost developed into a standard technique for accessing so-called "tribal" blood, was not unique. We know that Eileen Macfarlane, working in the decade prior to Majumdar, had used a very similar technique. While trying to obtain blood from groups living in the Deccan, she noted the same difficulties and came up with remarkably similar techniques of access. "Any field workers who have contacted our tribes and aboriginals know . . . that it is difficult to persuade them to co-operate in an unusual project," she asserted. She explained that "most of them are independent minded, as well as suspicious and superstitious folk. Weird rumours might circulate about the motives of a person who want to collect drops of their blood." Preexisting social relations and locally entrenched notions of honor, etiquette, and so forth were also important. Thus, Macfarlane found that some of the Banjaras and many of the Chenchus refused to give blood because she had first bled the Bhils before contacting them.[17] To get around this problem, she mentioned that since the people she was among were already well acquainted with the traveling medical officers who vaccinated and inoculated them, it was easy to "pass as such."[18] This seems remarkably similar to Majumdar's hoodwink.

This recourse to subterfuge bears a passing resemblance to Douglas Botting's disingenuous deployment in 1956 of what he called the "medieval idea" of bloodletting to convince Socotrans to let him take their blood.[19] Yet, it is different. There is no invocation of a "medieval idea" here, despite the constant invocation of the alleged primitivism of the group. Instead, Macfarlane and Majumdar both built on a familiarity with the itinerant practitioners of modern medicines, including homeopathy, who were clearly themselves artifacts of colonial modernity. In 1921, for instance, when unseasonal floods unleashed an epidemic of fevers, the government deployed forty "vaccinators" and two "traveling doctors" to circulate through the Chota Nagpur region, precisely where Majumdar would later travel in search of blood, dispensing quinine powders.[20] Likewise, in 1919 we hear of similar "itinerant doctors" appointed by local district boards, roaming through many of the Bengal districts in response to the influenza epidemic. Thirteen of these peripatetic doctors are mentioned as traveling through the "tribal" districts, such as Bankura and Birbhum, that would later become favored hunting grounds for seroanthropologists.[21] In areas where the medical investment was abysmally low—precisely those areas whose inhabitants usually came to be categorized as "tribal"—it was the onset of sudden epidemics that brought forth ad hoc medical arrangements such as appointment of a small number of peripatetic doctors. This, in turn, created a social memory that the seroanthropologists mined in their own travels.[22]

Macfarlane and Majumdar built on this preexisting pattern of practice of colonial medicine. In so doing, however, they broke with the earliest practices of seroanthropological blood collection. The earliest seroanthropological researchers, such as Malone and Lahiri, had not traveled at all. Rather, they had depended almost entirely upon people whose right of refusal was severely limited. These included patients coming to the Pasteur Institute in Kasauli, prisoners in various jails, and tea-plantation workers.[23] The latter are particularly worth discussing both because they remained a popular source of blood for later researchers and because, superficially, they seem to be freer than the patients and prisoners.

The tea workers used were all recruited by the Tea Districts Labor Association at Ranchi. The workers were hired under draconian employment contracts, and the Tea Districts Labor Association, created by the industry itself in 1878, was intended to systematize the entire coercive apparatus.[24] A major aspect of this coercion was an explicitly racialized recruitment policy that viewed certain "primitive" tribes as fit for plantation labor.[25] It was precisely these racialized recruitment policies that made these laborers objects of seroanthropological research. S. S. Sarkar, for instance, working more than a decade after Malone and Lahiri in the 1940s, still chose to turn to the laborers at the Fulbari Tea Estate in Darjeeling.[26]

This preference for plantation labor, prisoners, and patients was both distinct from and simpler than the later practice of travel and subterfuge. As Macfarlane put it candidly, "in hospitals and prisons everything is simple."[27]

It is difficult, however, to plot this shift from unfree bodies to apparently free bodies, through the development of traveling practices, along any simple line of movement from coercion to persuasion. There were many forms of traffic in blood that seemed to blur the lines between "persuasion" and "coercion." This hazy demarcation is perhaps best exemplified in the research projects of S. S. Sarkar. In the winter of 1937 and 1938, Sarkar pursued seroanthropological work in the Santal Parganas. He was accompanied by Macfarlane for some of this research, though the result was published exclusively under his authorship. Sarkar found it most convenient to set up a blood collection booth during the weekly market days. Yet, he found that "it was a matter of great difficulty to get the aborigines to consent to the pricking of their fingers." What did the trick, he found, was relying on the help of forest officials in the *damin* areas and the local police in the non-*damin* areas.[28] The distinction between the *damin* (or, to give the full designation, *damin-i-koh*) and non-*damin* lands was "not so much as the perimeter of a country as a line dividing two kinds of conceptual terrains—the wild land of 'primitives' and the permanently settled land of the mainstream Indian peasant."[29] In the *damin* lands, the forest officials enjoyed significant police powers, while in the agrarian lands the regular police were in power. The whole official apparatus created by this division was further stabilized by a regime of widespread indebtedness, often enforced by these very forest and police officials. Where this regime of debt and power came together was, significantly, at the weekly markets. Prathama Banerjee points out that the colonial government had established a number of new markets and enabled the entry of a range of consumer goods. It had even insisted that the Santals, initially excluded from so-called civilized society by their being settled in the *damin* areas, now "enter the market in order to engage in exchange with 'civilised' peoples."[30] Finally, it was mostly the Hindu Bengalis who functioned both as moneylenders and often as the petty officials.[31] Santals dubbed these people *dikus*, or "outsiders," who held them in their thrall.

Sarkar, as a Hindu Bengali, would have been identified with this oppressive group, and his direct reliance on police and forest officials, as well as his presence at the very markets where Santal indebtedness was produced and enforced, would unquestionably have reinforced his links with the coercive apparatus. The blood that he managed to extract by these means was clearly indebted to the presence of a colonial apparatus of power premised on a subtle calibration of debt and brute force, and worked by officials and merchants who shared social ties with Sarkar.

It would be incorrect to suggest that Sarkar, or indeed any of the other seroanthropologists, was directly involved in or even supportive of the structures of exploitation within which many of his subjects lived. But it is unquestionable that the researchers were more than willing to opportunistically use these structures to extract the blood they needed for their researches. Perhaps the most egregious case of such opportunism can be seen in the studies on Jarawa subjects in the Andamans.

The Jarawas had long avoided contact with outsiders and had remained outside of British imperial control as such. Encroachment on forest land, the closeness of the British to the Jarawas's traditional rivals, the Great Andamanese, and repeated efforts to capture Jarawa subjects had gradually made them increasingly hostile to the British presence. Since 1902 there had been a number of violent attacks and counterattacks, often with the British attacking Jarawa huts under the cover of darkness, killing people and looting their belongings for anthropological purposes. It was within this cycle of increasing violence that in September 1938 a police picket on Bluff Island surprised a small group of Jarawas traveling on a raft. The Jarawas attempted to fight the rifle fire with their bows and arrows but eventually had to retreat. The single Jarawa man, seeing defeat imminent, jumped overboard and swam all the way to Spike Island. The police were then able to capture the rest of the travelers: a pregnant woman, three of her children, and an unrelated young girl. All five were brought to Port Blair.[32]

Dr. Bijeta Chaudhuri, the medical officer of Port Blair, immediately tested their blood types. Born into an elite Brahmin family in Shillong, Dr. Chaudhuri had obtained his medical degree from Grant Medical College, Lahore, before going to London in 1922 to pursue his medical education further. In 1926, he entered the Indian Medical Service and returned to India. He married a niece of the Nobel Laureate Rabindranath Tagore and eventually rose to be the director general of the Armed Forces Medical College, Pune, after decolonization. Dr. Chaudhuri had already been conducting seroanthropological researches on Onge subjects when the Jarawa group were captured. He wrote to the Canadian-born British geneticist, Reginald Ruggles Gates, that "we were extremely lucky to capture three months ago a mother (pregnant) and her three boys and a girl apparently not of the same family. This is the only hostile aboriginal tribe that still exists in the Andamans, but they are of *purest blood*."[33]

Even more remarkably, the group—now including another son born to the captured woman while in Port Blair—remained captive. The mother and her eldest son died a few years later when the Japanese captured the Islands from the British during World War II. After the islands were recovered, the surviving children remained and grew up in captivity. These children, therefore, continued to provide blood for further studies.

In 1952, the very first issue of the *Bulletin of the Department of Anthro-*

Fig. 5.1. Jarawa child born and raised in captivity, 1952. Source: Sarkar, "Blood Groups from the Andaman and Nicobar Islands," pl. 9.

pology, published by the newly created institution that would soon be renamed the Anthropological Survey of India (ASI), carried an article by Sarkar on blood groups from the Andamans, including the Jarawa children.[34] Sarkar pointed out that Dr. Chaudhuri had failed to test the newborn son of the captured woman. The boy was thirteen years old at the time of Sarkar's research, and Sarkar duly tested this boy and the other survivors. Fourteen years after their initial capture, and five years after Indian Independence, these citizens were still living in effective confinement at a mission on Car Nicobar (fig. 5.1).[35]

Writing in 1962, Sarkar still agreed with the government that the "taming of the Jarawa" was a "moot problem."[36] The government had been trying to capture individual Jarawas, treat them well, and let them go with "presents" in a bid to earn their goodwill. Sarkar advised the government to focus on a group that seemed related to the Jarawas but was cut off from the main population on a smaller island to the north. He also advised that, alongside other efforts, such as airdropping pictures of friendly gestures and announcing messages loudly through specially fitted loudspeakers, the government should open a new local substation of the ASI close to the splintered northern group of Jarawas.[37]

Dr. Chaudhuri and Sarkar were both men who spanned the divide between the late colonial and the postcolonial eras. There is nothing to suggest that either of them had any hand in actually capturing or prolonging the imprisonment of the Jarawa children. Yet, they were also clearly excited to be able to test these "aboriginals . . . of the purest blood" and depended on the state's carceral establishment for obtaining access to these bloods. We must say of mid-twentieth-century India, as Bangham writes of twentieth-century Britain, that "encounters around blood were strongly conditioned by the institutions in which they occurred, and by the power relationships between donors, doctors, and scientists."[38] That the institutions and the relations of power, especially between the scientists and those categorized as "primitives," remained largely unchanged by decolonization shaped whose blood was available for seroanthropological research and thus, in turn, what kinds of genetic kinship could be derived.

Known Blood

The instrumental and opportunistic use of subject bodies will likely jar with our contemporary ethical sensibilities. It is crucial, however, not to impose our contemporary sensibilities upon researchers working in very different historical milieus. It will also help to remind ourselves that the researchers also treated their own bodies in similarly instrumental and opportunistic ways.

Seroanthropology did not simply need blood as the object it analyzed. Blood groups "could not be seen and manipulated directly, rather they were defined when samples were mixed—when red cells clumped together, or agglutinated, samples were ascribed to different groups."[39] What the researcher did in practice, therefore, was to mix the particular sample with a series of antibody-containing reagents or antisera. The group to which the sample belonged was identified by tracking which particular antisera made it coagulate.[40]

Antisera were the fundamental tools by which blood was grouped, and their manufacture depended crucially upon blood. Moreover, most of the main grouping reactions, such as grouping into A, B, O, and AB, required antisera made with human blood. Until the very end of the 1930s, leading British laboratories frequently used the blood of their own workers to create the necessary antisera.[41] It was World War II that suddenly and dramatically changed the institutional basis of blood grouping. As Bangham explains, "in just a couple of years, transfusion shifted from a set of small-scale donor systems with few formal institutional connections between them, to a nationwide wartime service underpinned by the routines and procedures for the mass storage and management of blood."[42] As a result, the reliance on laboratory employees and tight circles of donors soon became insufficient, and the laboratories were forced to look for larger, external sources of blood for the production of large quantities of antisera.[43]

Though blood banks and transfusion services did commence in India in the 1930s and there were wartime drives to collect blood for the needs of war, India did not experience the sudden and widespread infrastructural growth that transformed blood grouping in general and antisera production in particular in Britain. In 1956, more than a decade after the war ended and nearly a decade after decolonization, several states in India, such as Bhopal, Coorg, Himachal Pradesh, Kutch, Manipur, P.E.P.S.U., Saurashtra, Tripura, and Andaman & Nicobar Islands, did not have even a single blood bank.[44] There was therefore no centralized production of antisera. By 1956 the King Institute, Guindy; the Haffkine Institute, Bombay; and the Central Research Institute, Kasauli, were the principal, though by no means exclusive, producers of grouping antisera.[45] By the end of the decade all of these were reporting higher demand. There were also new pressures to align local antisera production with emerging international standards. The Haffkine Institute reported in 1960 that "there is an increasing demand of Anti-A and Anti-B sera from all over the country. The Anti-A and Anti-B sera prepared by the Haffkine Institute compared favourably with the WHO International standard. 3,784 blood donors were bled for the preparation of dried plasma."[46] Whereas the Haffkine Institute relied on donors to fill the increasing demand, the King Institute took another route by setting up a "mobile Blood Bank Team" to travel to the district jails all over Madras state and collect blood from prisoners for the production of grouping antisera that was then to be supplied to various state blood banks.[47]

The variable manufacturing practices and the lack of standardization made the choice of antisera a tricky issue. Though seroanthropologists did not always explain the reasons behind choosing one antiserum over another, they were explicit about the need to choose and occasionally hinted at their lack of trust in certain antisera. In 1945, for instance, when Majumdar had to use antisera prepared by the Haffkine Institute, he retested some of his sample "against antiserum of known titre with capacity to react with known A, B cells made in Lucknow."[48] Lucknow was Majumdar's hometown, and he had long experience of working with antisera prepared by the pathology department of King George Medical College, Lucknow.[49] In 1960, S. R. Das and his colleagues at the ASI hinted once more at some skepticism about Haffkine antisera: "High titre anti-A and anti-B sera of sufficient avidity were chosen out of the supplies obtained from Haffkine Institute, Bombay and the anti-A_1 (absorbed anti-A) serum was manufactured by the Ortho-Pharmaceutical Corporation, U.S.A."[50] By the end of the 1960s most seroanthropological researchers based in Calcutta—variously at the ASI, the ISI, or the Bose Institute—all chose to use commercially made anti-Rh sera, usually manufactured in the United States, rather than the sera produced at the Haffkine Institute. All of them used antisera manufactured by Dade Reagents of Florida.[51]

The choice of Dade antisera was not automatic. We know that the Haffkine Institute was manufacturing anti-Rh sera since at least the late 1950s.[52] We have also seen that most seroanthropologists did in fact use the anti-A and anti-B sera made by the Institute. Yet, they chose not to use the anti-Rh serum made by the Institute. This clearly shows that using antisera was far from a mere mechanical act of using whatever was most easily available. It was moreover these questions of choice that also made the anthropologist's own blood and those in his immediate circle important.

Seroanthropologists constantly needed to test their sera, not only during the initial selection of the sera but through subsequent periods of usage as well. Such constant vigilance over the antisera became particularly necessary in view of India's widely varied climates and the lack of stable refrigeration infrastructure throughout our period of study. These general challenges were further accentuated by the very nature of seroanthropology. Since the researchers often targeted subaltern groups that lived in remote regions, the communication infrastructures they dealt with were, as a rule, worse than those in the cities where their own laboratories were based. Under such conditions the threat of antisera deteriorating and becoming useless was ever present.

To meet these challenges, the researchers developed two key strategies. First, they tried to complete as many tests as quickly as they could whenever fresh antisera arrived. In 1938, during the survey of blood groups in the Santal Parganas, for instance, the antisera arrived from the Haffkine Institute in Bombay and the researchers—S. S. Sarkar and Eileen Macfarlane—rushed to do most of their tests within a day of the arrival of the antisera and to finish the entire work within two weeks of that date.[53]

Second, they regularly checked the potency of the antisera against the blood of members of the research party whose blood groups were already known. They gradually developed a protocol whereby they would, at fixed intervals, check the antisera against a sample taken from someone whose blood group was already known. For instance, D. N. Majumdar, in performing the large serological survey of castes in United Provinces (Uttar Pradesh) in 1941, mentioned the use of three "controls." The use of the word "controls" in this context needs to be delineated a little further. Other researchers working with blood groups, particularly those exploring possible relationships between blood groups and particular diseases, often used the word to designate groups. For example, "control populations" referred to a group of people unaffected by the disease under investigation who provided the statistical foil against which the specific frequency observed in the tested group was evaluated.[54] Though loosely connected, this was clearly very different from the practice of using one's own blood, or that of someone closely connected, to "control" or "evaluate" the antisera being used.

In the summer of 1936 when Macfarlane carried out her study of Cochin Jews, she mentioned that "my own blood, Group AB, was used each day to check the potency of the test sera."[55] By the early 1940s the system seems to have become both more established and more elaborate. Instead of checking the sera with a single "known" blood sample, researchers now used multiple controls for different grouping antisera. In Majumdar's 1941 survey of UP castes, as we saw, he used three "controls": all members of his own research team. Majumdar's assistant was the control for group A, he himself was the control for group B, and his peon was the control for group O.[56] The following year, when S. S. Sarkar was testing Oraon subjects, he too mentioned how he tested his sera every single day against his own group O blood.[57] Majumdar's own subsequent study of the blood of Korwa subjects in Uttar Pradesh once more utilized his own B group blood and the O group blood of his peon as controls.[58] Later in the decade, during the winter of 1948–49, when Sarkar and Dilip Kumar Sen studied blood groups in the Santal Parganas, once more they mentioned the use of "one of our workers" to provide a daily control blood belonging to group AB.[59] This latter project was particularly important, since it provided the basic methodological protocols for other, similar projects carried out by the newly formed ASI.[60]

This was certainly not the only way in which the sera could be tested, however. R. P. Srivastava, for example, did not use his own or his colleagues' blood as "controls." Instead, he used the blood of previously tested subjects themselves to assure himself of the continued potency of the sera.[61] Likewise, Sarkar mentioned during his Oraon study that since he did not have a team member with AB group blood available for calibration purposes, he had to persuade one of his subjects to become his control. This young man, Sarkar wrote, "was much cared for and bribed."[62] This practice of using one of the subjects themselves for control purposes, however, seems to have been less popular than self-calibration. Difficulties in repeatedly accessing the same subject might have worked against this practice. Indeed, Sarkar's comment hints at precisely such difficulties. Using a team member or oneself was a much more reliable way of providing a stable, regular source of known blood against which to check the sera.

Self-calibration was not unique to South Asia. There are a few stray examples from a slightly earlier period that suggest that the practice had been in use among physical anthropologists elsewhere as well. Its extent and longevity, however, remain an open question and will require much further research. One of the earliest instances of this practice comes from the early 1930s in the United States. In 1933 the American anthropologist Robert A. McKennan narrated a minor incident during his fieldwork among Chandalar Kutchin subjects in Alaska that evoked the commonplaceness of self-calibration. Before leaving for the field, he had taken a brief course in "typing technique at the Dartmouth Medical School" and proceeded with sera supplied by the same institution.[63] In the field, he wrote that "my failure to observe any agglutination led me to suspect that something might be wrong with my sera, and since my own blood was type O, I was unable to use it as a control in the field."[64] It was only after McKennan reached Fort Yukon and was able to test his antisera on other blood samples that he was satisfied that his antisera were in fact unimpeachable. This anecdote, though early, demonstrates that researchers usually would rely on their own blood as controls, at least on occasions when there was some reason to suspect the antisera's continued reliability. It is also worth noting that while McKennan's actual anecdote dates from 1933, he only published it—as part of an academic article—in 1964.

Even more significant, perhaps, a key technical manual, published in 1942, seems to refer to the practice. The manual was coauthored by one of the most influential seroanthropologists, William C. Boyd, and included a foreword by Karl Landsteiner, the scientist who discovered human blood groups. Though the reference was more to laboratory practice than to fieldwork as such, the principle was clearly stated: "Once test sera anti-A and anti-B are available, determine the blood groups of your immediate associates in the laboratory. In this way standard donors for experimental purposes are obtained to whom one can have recourse in case of necessity."[65] To be sure, this statement referred to the use of "standard donors" for the production and initial calibration of the antisera in the laboratory, but it would clearly be useful in the field as well when sera needed to be retested.

It might also be possible to locate self-calibration within a longer, and more foundational, genealogy of anthropological "autoexperimentation." Henrika Kuklick has made the intriguing suggestion that anthropology's disciplinary interest in "participant observation" might itself have derived from a culture of autoexperimentation that was prevalent among physicians who doubled as anthropologists before the professionalization of the discipline.[66] Emily Martin, building on Kuklick's work, has similarly suggested that the iconic Cambridge Anthropological Expedition to the Torres Strait Islands in 1898 had developed a set of ethnographic methods that involved the anthropologists themselves becoming standardized "embodied instruments." Martin argues that this was an extension of Wundtian experimental psychology.[67] Indeed, we may cast our net even more broadly and locate the practices of self-calibration of antisera as a somewhat special case within a more expansive history of modern "self-tracking," that is, the trend toward translating bits of our own bodies and minds into data, and especially numbers.[68]

Self-calibration was a way not only of ensuring continued potency of the sera, but also of ensuring against any accidental mislabeling, which in turn might skew the results. This latter anxiety was manifested when Ernest Büchi—the Swiss anthropologist temporarily employed by the ASI who trained a generation of Indian anthropologists—confronted differences between his own data on Tibetans and the data Macfarlane had collected earlier. Büchi explained the difference by inferring that Macfarlane's sera might have been accidentally mislabeled. "This assumption seems the more plausible," he continued, "because Macfarlane does not mention anything about having checked the sera with known blood."[69]

The use of known blood as "controls" against which to calibrate and thereby ensure the reliability of the results became firmly established as one of the key research protocols in seroanthropological studies. Büchi's comment also clarifies that it was important not only to take such precautions, but also to include them in the published reports. As a result, practically all researchers of the Nehruvian period, many of whom were directly trained by Majumdar, Sarkar, and Büchi, included in their reports explicit comments about the use of controls. Yet, interestingly, by the mid-1950s, such comments about "controls" became anonymized; that is to say, one could no longer learn from the publications exactly whose blood was used as the "control." We merely learn that "known bloods" were used as "controls."

Notwithstanding such anonymization, the instrumentalization of the researchers' own blood did not stop, and indeed expanded further in the 1960s. Instead of merely being used to test the antisera, by the early 1960s some researchers were occasionally using their own blood to make their own grouping antisera. Thus, Mukul Chakraborty, who studied the blood groups of Zeliang Naga subjects in Nagaland in 1962 and 1963, mentioned that the antisera he had used were prepared from his own blood and that of a colleague. Chakraborty himself had group A blood, and his colleague, Dr. K. R. Datta, the medical officer of the nearby Peren Government Hospital, had provided group B blood.[70]

Such practices might have developed at least partly in response to the Indian government's restrictions on the use of limited funds for buying expensive sera from foreign manufacturers. Writing in 1969, one researcher commented that the development of genetic studies of caste and tribal groups had "remained somewhat restricted because of the inability of the country to spare foreign exchange, as large amount of blood group sera (at present mainly imported) is required for these studies." Yet, he enthused that, "despite this, the urge for trying to investigate fundamental problems did not get lost."[71]

There has been a lot of excellent recent work in the history of science that highlights the role of tacit knowledge in making scientific experiments work.[72] Much less account, however, has been taken of the instances in which the scientist's own bodily tissue is mobilized as part of a scientific apparatus. In overlooking such mundane, everyday practices, we fail to realize just how historically contingent the scientific findings themselves were.

At its most basic, typing depends upon the agglutination produced by mixing the sera with sample. But, as Schiff and Boyd's 1942 manual had indicated, "different sera of the same group do not all react equally strongly; there is a good deal of difference in content of agglutinin."[73] Thus, the authors recommended working "with known blood groups and strongly agglutinating sera."[74] Moreover, they also pointed out that the aftereffects of certain illnesses, such as carcinomas and high fevers, as well as particular life-stages, such as pregnancy, produced distortions to the process.[75] Even without going into later genetic explanations of why agglutination varied among people with the same blood group, it is clear that the relationship between the researcher's blood and the antisera fundamentally underpinned the results obtained. Two researchers' blood, even if they belonged to the same group, might have yielded somewhat different outcomes. More tellingly still, Schiff and Boyd's manual pointed out that, while two unknown blood samples could be distinguished from each other based on their agglutinating with different antisera, there was no way to determine which was A and which was B unless the antisera were either previously labeled or tested with known blood. In the absence of either, they suggested that "orientation is usually possible from the fact that group A is commoner in America and Middle and Western Europe than group B."[76] Remarkably, then, if one could not regularly cross-match the sera used in different parts of the world, it would be impossible to ascertain whether what was considered group A in one place was indeed the biological equivalent of what was considered group A in another place.

The attempts to correlate local antisera with international standards that we saw the Haffkine Institute attempting in 1960 were motivated by this need to create a standard reference point that in turn would ensure the objectivity and comparability of the blood groupings. Bangham has lucidly described the way in which a specifically London-based infrastructure came to underpin the international standardization of blood grouping antisera in the postwar world. The Blood Group Reference Laboratory based in Chelsea had mainly made and distributed antisera to the British National Blood Transfusion Services after the war. But it also "made blood-based standards that other laboratories used as metrics for assessing local blood grouping reagents." Soon it was also recognized by the WHO as the International Blood Group Laboratory. With the WHO's imprimatur, it "would distribute standardized antisera and check the purity and specificity of reagents from WHO-accredited labs around the world."[77] When, in 1960, the Haffkine Institute was claiming that its antisera compared favorably with WHO standards, it basically meant that Indian researchers working throughout the subcontinent could use Bombay-made antisera and be sure that the reactions they saw were similar to those they would have obtained from London-made antisera. The whole edifice of "aperspectival objectivity" was therefore reliant on a system of institutional technologies for first elevating certain specific antisera as the reference points and then comparing all other antisera to these standard reference points.[78] There was nothing inherent in the antisera that made Majumdar's Lucknow-made antisera less reliable than those made in Bombay, though they probably gave somewhat different reactions with specific samples.

The insertion of the practice of auto-calibration adds an additional and underappreciated level to this process for producing objectivity. Whereas the story of standardization is clearly one that involves large international organizations like the WHO and networks of political power, institutional hierarchies, and such, operationally the system crucially relied upon the bodily substances of individual researchers and their team members. In the field or in small Indian laboratories what reassured the researcher that they had the right results was that the antisera produced the same result when mixed with their own blood as they had seen in the past.

Self-calibration encourages us to rethink our histories of objectivity. We notice how the consistency of the researcher's own bodily tissue and its relationship with the sera provided a uniquely seroanthropological vessel in which to blend more recognizable figures of "objectivity," such as "mechanical objectivity" or "trained judgment," or indeed a "trust in numbers."[79] This objectivity stood in a novel relationship to the individuality of the researcher. It does not adhere to the phenomenologist's insistence that "human knowledge is a product of the use of our bodies."[80] Rather than the sensory mediation or embodied skill that the phenomenologist refers to, here the researcher's blood—already alienated from the body—is a ready-at-hand resource, a "standing reserve."[81] Self-calibration thus puts into perspective our earlier discussion of opportunistic accessing of the blood of subaltern subjects. It reveals just how seroanthropology reduced both the subject and the researcher from being to biology and, eventually, to a "standing reserve" for the production of scientific knowledge.[82]

Caprine Blood

Whatever we make of the apparent reductionism of being to biology, it cannot be denied that casting human beings essentially as agglomerations of biological material also served to open up new connections between humans and a range of nonhumans. The comparative anatomists, the Darwinians, and the animal experimentalists of the nineteenth century and the twentieth-century molecular biologists who worked increasingly with "model organisms": all in their own distinctive ways linked knowledge about humans—by comparison, similarity, or contrast—to the nonhuman animals.[83] The period, roughly since 1800, when the mutual distance and definition between the human and the animal were being worked out was also, not accidentally, the high noon of European imperialism. These overlapping histories were crucial for the scientific constitution of race. As Sujit Sivasundaram points out, "race can be seen as an idea that comes into being at the intersection of the human and animal in post-Enlightenment contexts. The imperial human came to be with animals." Race, as a scientific object, Sivasundaram holds, was constituted through the "material entanglements" of the human and the animal.[84] The skulls of various animals, old and new, for instance, were crucial to the race scientists of the nineteenth century. The nonhuman was, therefore, simultaneously "engag[ed] and hid[den]" in imperial race science and was "productive for imperial and racial subjectivity and postcolonial nationalist renditions."[85] In order to unpick these racialized subjectivities through historical retellings of race science, therefore, it is important to lay bare those hidden assignations between human and nonhuman biologies.

In seroanthropology, the human and the nonhuman became materially entangled through the actual process of blood testing. The antisera required for determining the Rh factor, for instance, were produced by injecting the blood of rhesus monkeys into guinea pigs and rabbits, and harvesting the testing sera from the latter.[86]

The practice of using animal bodies for the production of testing sera built on earlier practices that had entered public health through bacteriology.[87] Antibodies had long been harvested from the blood of rabbits and guinea pigs injected with specific microbes for both the production of diagnostic reagents and "serotherapy," used to bolster a patient's immunity. By the 1920s "institutions responsible for making and distributing animal sera were an essential part of the contemporary public health apparatus, and their standards were coordinated by the League of Nations."[88]

Most of the grouping sera, however, had remained dependent on human blood.[89] This only began to change as the number of factors in the blood being investigated gradually expanded far beyond the basic ABO groupings. As we have seen in chapter 3, from the 1950s more and more factors, such as Rh, Duffy, Kell, Lutheran, and so on, were added to the seroanthropological survey projects. Like the Rh antisera, some of these new factors depended upon nonhuman bodies and bloods. One of the first of these new factors was the so-called anti-H agglutinin.

The anti-H agglutinin emerged in the wake of the discovery that while some individuals secreted their blood antigens into their saliva, others did not. This allowed for a further mapping of the "secretor" factor along racial lines. Fritz Schiff noticed further that the blood of those individuals who were typed as belonging to group O could be easily distinguished into "secretors" and "nonsecretors" based upon the strength of the reaction of their blood to heterogenetic sera. Research into heterogenetic sera had started by the late 1920s, when researchers began experimenting with the differential agglutination of particular human blood groups with the bloods of certain animals. "Heterogenetic," therefore, referred to the heterogeneity of the two species. Unlike the sera used in typing the four main blood groups, which were produced from human blood, the heterogenetic sera were produced from the blood of nonhuman animals. Hence, the "H" in "anti-H" stood for the word "human."

The two main sources of the anti-H sera that distinguished O group humans into secretor and nonsecretor categories were bovine (cow) and caprine (goat) blood. In the former case, the serum from cow's blood had to be absorbed by human AB blood in order to produce the test sera. In the latter case, goats injected with the *Shiga* bacillus produced the necessary sera. Schiff had outlined these sources of anti-H sera by 1934.[90] Widespread use of anti-H sera in seroanthropological work in India, however, did not begin in earnest until the 1950s, after which it grew rapidly, particularly in research projects undertaken by the ASI.

In 1952, Ernest Büchi was busy comparing the blood and secretion frequencies between upper- and lower-caste Bengalis in Calcutta. In the course of his work, he noted that since *Shiga* dysentery was usually very common among Indian goats, one could easily find the necessary anti-H sera at local slaughterhouses. He asserted that if a researcher collected the blood of about fifty goats, they would certainly find *Shiga* bacillus–impregnated blood.[91] The ready availability of dysenteric goats in nearby slaughterhouses not only simplified the process, by obviating the need for artificially producing such dysentery in laboratory animals, it also ensured the availability of fresher sera that gave stronger reactions.

The *Shiga* bacillus, however, is not a single entity. The bacillus was originally isolated in 1898 by the legendary Japanese microbiologist Kiyoshi Shiga from the stool of humans suffering from an epidemic dysentery that prevailed in Japan and was locally known as *sekiri*. Shiga called the bacillus simply *Bacillus dysenterie*. Subsequently, however, a number of similar strains of bacillus were identified. By 1930 an entirely new genus had been identified by researchers and named *Shigella* in honor of Shiga. Today the genus *Shigella* is understood to be composed of four large, and themselves heterogeneous, groups of bacilli: *S. dysenteriea*, *S. flexneri*, *S. boydii*, and *S. sonnei*.[92]

To complicate matters further, Shiga's own later research had pointed toward the production of a form of toxin released by most *Shigella* bacilli. This toxin resembles toxic factors produced by other bacilli that do not belong to the *Shigella* genus, such as *E. coli*. Today, scientists recognize an entire suite of broadly similar toxic factors known as "*Shiga*-like toxins," or SLTs.[93]

All this casts some doubt about exactly what the goats in Calcutta's abattoirs suffered from in the 1950s when Büchi encountered them. Contemporary studies done in other parts of the world have, in fact, failed to find the *Shigella* bacilli in abattoir goats, though they have found other bacteria.[94] This certainly does not rule out the possibility that the goats Büchi used were suffering from "*Shiga* dysentery," but it does raise unresolvable questions about precisely what bacillus was infecting the goats and how that might have affected the test sera produced with their blood.

Though not all bacillary dysenteries are caused by *Shigella* bacilli, most dysenteries in goats are the result of dirty and cramped conditions of living and the contamination of their food and water supplies, especially by fecal matter from infected animals.[95] These conditions were all present in the slaughterhouses of mid-twentieth-century Calcutta. An Ad-hoc Committee on Slaughter Houses set up by the government of India in 1955, which inspected slaughterhouses in Calcutta and other large cities, reported that most of the slaughterhouses were maintained in cramped, overcrowded conditions. Moreover, since goats were almost entirely sourced from the countryside, the journey from such rural locations to the slaughterhouse was itself brutal. The animals were either made to walk extremely long distances or were brought in severely overcrowded trains and lorries. In the process, some animals died on the way, while others arrived in starved and exhausted condition.[96] All this, therefore, contributed to the occurrence of dysentery that became unexpectedly valuable to researchers like Büchi.

Neither can we overlook the facts that both Büchi's turn to goats and the subsequent use of goats by ASI researchers happened in the city of Calcutta. Traditionally a large portion of the Bengali Hindu population is devoted to the goddess Kali. Indeed, some even claim, though controversially, that the very name of the city derives from the old and revered temple of Kali at Kalighat. Goats are usually the favored sacrificial animals of Kali, and as a result many Bengali Hindus used to consume goat meat as sacred food. Shib Chunder Bose, writing in the 1880s, narrated how the sale of goat meat from Kali temples had become so lucrative that a number of new nominal temples to the goddess had cropped up throughout the city simply to slaughter and sell goat meat every morning.[97] More recently, anthropologist Sanjukta Gupta has pointed to the similarity of the food offered to the goddess at Kalighat and the "normal midday meals" of affluent Bengalis in Calcutta, which often feature goat meat curries.[98] This local popularity of goat meat was not uniform across the subcontinent. In Calcutta, however, the almost "ubiquitous presence" of Kali worship in Bengali life, as well as her status as "Calcutta's patron deity," made goat meat consumption popular.[99] Hence, had Büchi or the ASI been headquartered in another city, the number of slaughtered goats in their immediate vicinity would likely have been much smaller.

Finally, the committee on slaughterhouses noted how the consumption of goat meat had been growing exponentially since the departure of the British. Its report stated that "after the partition of the country and especially in recent years due to restrictions placed in many States on the slaughter of cattle, there has been a considerable decrease in the production of beef. The production of goat meat and mutton on the other hand has been constantly increasing to meet the demand."[100] The provincial policies banning the slaughter of cows had emerged as part of the national movement. Since the late nineteenth century Hindu nationalists had mobilized around the issue of cow slaughter.[101] It was also one of the issues that fueled anti-Muslim sentiment and contributed to the animosity between the communities. The introduction of elected ministries in the provinces from the 1920s had brought the issue into electoral politics, and by the time Congress ministries took charge of provincial administrations in the wake of decolonization, banning cow slaughter was a central electoral plank for many of the nationalist politicians. The gap created in the meat market by this postcolonial reduction in beef sales was filled by the expansion in goat meat production. The rapid expansion further burdened the slaughterhouse infrastructures and possibly augmented the incidence of dysentery in the goats.

While all these historically specific circumstances surely enabled Büchi to notice and utilize the blood of dysenteric goats, his own influence on younger Indian scholars in the ASI did much to institutionalize the practice. Practically all the later researchers who used goat blood sera in their racial surveys in the 1950s directly cited Büchi's piece as the source of their method. These included S. R. Das's studies of the Mahars of Maharashtra and the Kondhas of Koraput,[102] Vijender Bhalla's study of Tibetans, Narendra Kumar's study of the Riang in Tripura,[103] and P. N. Bhatacharjee's studies of Kashmiri Muslims and Pandits and the Dudh Kharias of Jharkhand.[104] Each of these studies prepared anti-H sera from goat blood and cited Büchi as their methodological reference. Such studies continued to be done throughout the 1950s and 1960s.

Given this extended and institutionalized popularity of Büchi's methods of using goat blood from local abattoirs, it is also significant to note that the anti-H sera could be produced using the blood of different animals and that the reactions given by these different animal bloods were not identical. One textbook from 1956 identified three animal sources, besides the goat anti-*Shiga* blood, for the anti-H sera: cattle, eel, and chicken anti-*Shiga*. The textbook also delineated differences in the strength of the agglutinations produced by these differently produced sera.[105]

The choice of caprine blood in this context was therefore far from a simple choice. The easy availability of the dysenteric goats in Calcutta

Fig. 5.2. Depiction of cross-species transfusion from a dog to a sheep in a mid-twentieth-century Bengali popular science book. Source: Chattopadhyay, *Payer Nokh theke Mathar Chul* [From Toe Nails to the Hair on the Head], 3 (illustration by Debabrata Mukhopadhyay).

provides only one frame for understanding the turn to goats. Other reasons might have been more deeply embedded in the histories of serology more generally and transspecies colonial serology more specifically. Mitra Sharafi, for instance, has recently pointed out that the imperial serologist's laboratory was already using animal blood sourced from municipal slaughterhouses for forensic and diagnostic tests in the early 1930s.[106]

The relation between animal and human blood has been the subject of fairly long medical, natural philosophical, and, eventually, scientific concern. As a result, it has also produced a rich historiography. This historiography, which includes the discovery and dissemination of blood groups, has discussed at length the role of nonhuman animals in early experiments with transfusions (fig. 5.2). Perhaps the earliest such experiment took place in 1667, when a number of attempts were made in both England and France to transfuse the blood of sheep and calves into humans.[107]

Notwithstanding this earlier history of the role of animals in experiments with human blood, few have pursued this theme in later times and in colonial contexts. The colonial history of transfusion and cross-species contact through blood has been dominated by attempts to trace the source of the transmutation of the simian SIV into the human HIV (human immunodeficiency virus) in colonial Africa. William Schneider has argued that this transmutation happened in the course of blood transfusions that were introduced in African hospitals after World War I.[108] Thaddeus Sunseri, however, has pointed out that this primarily post–World War I and hospital-centric history occludes a more complicated earlier history of transfusions in Africa where experimentation with animal blood was rife. Sunseri writes that "the colonies, as sites of research laboratories, were places where animal experimentation was widespread before World War I."[109] Even more remarkably, Sunseri states that "European researchers in Africa tested human and animal blood for pathogens and possible immune agents, and injected blood of infected or recovered humans into test animals, and sometimes that of animals into people."[110] Even the therapeutic injection of animal blood into humans, of the sort that happened in England or France in the seventeenth and eighteenth centuries, were still being recommended in Africa in the late 1870s.[111]

One area where this kind of transspecies transfusion experimentation and the history of racial science came together was in blood serum therapy. The underlying concept in blood serum therapy, suggested in the 1890s by Japanese scientist Shibasaburo Kitasato (who incidentally was also the inspiration and mentor of Kiyoshi Shiga), held that the race- or species-specific immunity to certain diseases might be transferred to others via blood transfusion.[112] What is directly of interest to our present concerns is that as early as 1895 an English doctor called J. Murray, who wrote a guidebook on tropical health, recommended the direct injection of caprine blood into malaria patients. Murray argued that "goat's blood and serum" were "useful as preventive and therapeutic agents in malarial poisoning."[113]

It is interesting that this suggestion of goats being immune to malaria, one of the classical "tropical" diseases, was very similar to ideas about the alleged racial immunity of several African and South Asian peoples.[114] It might also suggest that there were more subtle and unstated logics that informed Büchi's choice. After all, as Sunseri points out, along with transfusion and serum theory, "blood analysis was a new way to determine closeness of 'races' and species."[115] It was, therefore, highly suggestive that in Schiff and Boyd's technical manual on blood grouping, for about half of the nonhuman animal–derived blood sera the "chief property" was stated to be the differential rates of agglutination between "whites" and "negroes."[116] Though they were not discussed at length, the incorporation of such statements in key technical manuals hints at implicit connections between animality and racial hierarchies.

Yet, it is important to note that the racial connotations of blood sera remained sufficiently suppressed. Even the textbook only mentioned the differential rates, but did not try to explicitly suggest any characteristic similarities between animals and specific races. This suppressed and apparently objective framing was crucial for the kind of enthusiastic uptake of the technique by Indian researchers that we see. For most of these researchers, the goats were merely a form of "animal technology."[117]

Blood Multiple

Attending to the "blood multiple" is, to follow Annemarie Mol, "an act . . . an intervention."[118] It deliberately refuses to conform to the epistemological normativity that determines the horizons of analysis. Instead of being a totalizing alternative retelling, therefore, its purpose is to denaturalize its object by tracking unexplored and invisible histories. The three bloods I have explored here were clearly not the only bloods that went into the constitution of seroanthropological research. We have seen, for instance, that rhesus monkeys, guinea pigs, rabbits, and indeed human donors were crucial to the manufacture of antisera. Tracking these stories would open up other histories. Likewise, we could have explored a different set of sociomaterial practices for each of the three bloods. It would, for instance, have illuminated another set of important histories if we had looked at how the travel infrastructure used by the researchers, the subjects themselves, and the abattoir goats shaped seroanthropology. Cars, buses, trains, trucks, and treks, and the various overlapping, distinctive mobilities they enabled, unquestionably shaped both the practice and the material of seroanthropology.

Rather than traveling down those tracks, I have chosen here to explore questions of access and mediation. Refusing the closure imposed by an image of science and scientists, I have tried to see how nonscientists, and scientists in their nonscientific, personal capacities, were crucial to the production of seroanthropology's racialized truths. Indeed, in accounting for the presence of dysenteric goats, we have seen how seroanthropology relied on the presence of such "unscientific" infrastructure as Kali temples and feasted upon a cultural taste for goat curry.

The "blood multiple" is a device for the multiplication of realities and pasts. It helps us access invisible histories embedded in what I have generically called seropraxis. But it is also a way to unpick seroanthropology's claims to objectivity. On the basis of what we know about goats in Calcutta's abattoirs and the later history of research into the *Shiga* bacillus, for instance, we can ask: if the goats did not really suffer from *Shiga* dysentery, how might that have impacted the results obtained by using antisera produced on the assumption that their blood contained the bacillus? Likewise, we can wonder: if grouping sera continued to be mostly produced in decentralized and localized ways, and were essentially only standardized with reference to the blood of individual researchers, how far are results obtained by different researchers using different sera still comparable?

This is not just to insist, as Warwick Anderson has insisted, that "absolute objectivity, or objectification, is unachievable in science."[119] It is also to point out, after Mol, that objectivity is *enacted* by actively denying and suppressing the "manyfoldedness" of the objects any science studies. Distinguishing this manyfoldedness from "pluralism," Mol points out that the former refers specifically to the multiplicity that is folded into the singularity of biological samples.[120]

What is probably not immediately apparent in this rendition of the "blood multiple" is that it also calls attention to absences. Thus, while seroanthropology was a global phenomenon and anti-H sera were required for research beyond India as well, goat blood was seldom used in those researches. Just as certain antisera, such as those derived from bovine blood, remained absent in India.

But there was one specific type of blood that was conspicuous by its almost complete absence from Indian seroanthropology. Anyone familiar with the extant scholarship on similar research projects in other parts of the world will immediately be struck by the absence of frozen blood in our histories of seropraxis. Joanna Radin has shown how, in the decades following World War II, population genetics increasingly came to be instantiated in frozen blood, especially when pursued in postcolonial contexts.[121] Yet, our Indian researchers up until the 1970s seemed remarkably unconcerned with frozen blood.

The reason for this absence is not difficult to fathom. The "cold chains" that were necessary for the use of frozen blood were fundamentally reliant upon "access to technologies of preservation—refrigerators, freezers, dry ice, and liquid nitrogen."[122] Such infrastructure, as well as the power necessary to run them, remained expensive and inaccessible in early postcolonial India. As a result, Indian seroanthropologists throughout the period continued to work with fresh blood, or blood that had only been frozen briefly (for a few days). The long-term freezing of large samples and their potential for reuse in multiple research projects remained entirely absent.

In revealing the invisible histories and the visible absences, the blood multiple shows us, therefore, that not only the "national styles" but indeed the very biological objects of the sciences of human difference were intransigently contingent historical objects.[123] Only by latching onto the praxes through which these objects were locally articulated and amplifying their innate manyfoldedness can we scrape off their universality and reveal their embeddedness in conditional, localized, and messy sociomaterial worlds.

INTERCHAPTER

Letter 6

Hemendrakumar Ray
21, Pathuriaghata Bylane
Calcutta, India
2 April 1934

My dear brother,
As always, it was wonderful to hear from you. I was particularly relieved to learn of your complete recovery. In the meantime, I have finished reading Mr. Sen's diary. What a remarkable narrative it is!

The world of these future-humans that he describes is both fascinating and disturbing in equal measure. Like you, I find myself deeply troubled by the values cherished by these descendants of ours. I worry whether this is an aberrant evolutionary dead end, which will run its course to extinction like the Neanderthals, or is it indeed what humans in general will become in the future?

It is clear to me that Mr. Sen, too, was revolted by this advanced civilization. His own thoughts, wherever they peek through his descriptions, clearly attest to a deep clash of values. The chasm that divided Mr. Sen's way of looking at the world and that of these future-humans seemed too big to be able to speak across. The two sides do not seem to have any common ground upon which to appreciate each other's positions.

The Chief Scientist and his ilk look down upon Mr. Sen as the representatives of a backward people. They think of him as a brute, devoted to bodily concerns and incapable of appreciating the tremendous technological advances and the new political order in which he found himself. Whilst Mr. Sen, for his part, sees the future-humans as a physically degenerated, amoral, and misguided people lacking even the most basic values that make us human.

The irony of this clash could not, I am sure, have eluded you. We Bengalis, who pride ourselves on our progressiveness, our modernity: we, too, might appear to some, such as these future-humans, as mere primitive brutes! No different indeed from how some of our own people look upon the Santals.

I can see clearly now, Najrul, why you thought that Mr. Sen's experiences spoke to your own dilemma upon encountering Tana, the Daughter of the Gorillas. Both experiences shake us out of our complacency about our place in the world, our system of values, even our understanding of what it means to be human. But neither experience offers us any clear insights. They leave us bobbing up and down in the muddy waters like the once-worshipped idols after they have been immersed in the murky waters of the Hughli. Every now and then we think we catch a glimpse of what we once held sacred and then it disappears again.

I wonder if you have read my little report on the experiences of my friend, Mr. Gupta of the Indian Civil Service? It happened a few years ago when the British were still running our country. Mr. Gupta, at the time, was posted at a place deep inside the Santal Parganas. Anytime he had to travel out of his post to come to Calcutta he had to undertake an onerous trek to a small, lonely railway station to catch a narrow-gauge connecting train. On one such occasion, when he had obtained a short leave to come to see his family in Calcutta, he planned his trek in such a way that he could spend the night in a small hunting lodge belonging to a local Raja so that he could catch the early morning train.

The Raja's employees, though perfectly willing to ingratiate themselves to Mr. Gupta by accommodating him, warned that there were uncanny rumors about the hunting lodge and it might not be a safe place after dark. Upon enquiring further, Mr. Gupta learnt that the locals believed that a Santal god who was worshipped at a simple, nearby forest shrine came into the lodge every night to sleep indoors. I have never been able to decide whether Mr. Gupta himself believed these rumors or not. For, as you know, Najrul, these I.C.S. types, especially the Indian ones, huff and puff endlessly to convince others of their modern, Europeanized outlook. In any case, he decided to brush off the rumors and stay the night at the lodge. It might have also helped that he learnt that the retiring British police superintendent, Mr. Taylor, would also be staying the night at the lodge. The loud-mouthed, blusteringly pompous old British police officer, Mr. Gupta might have imagined, would revolt any god worth its name!

When the day arrived, Mr. Gupta was further surprised to learn that Taylor had got it into his head that he would in fact carry the idol at the center of the uncanny rumors back with him to London! Though he found the old, brightly colored wooden idol grotesque, Taylor thought it would make for a good conversation piece in his London drawing room. Men like Taylor, of course, seldom cared for the sacredness of such idols. They were keener to hoard them as exotic trophies from their time in the far-flung outreaches of the empire or to use them as props for their fledgling ethnographic pretensions in plush retirement.

That night, as Taylor and Gupta finished their dinner and retired to their respective rooms, a massive storm broke over the district. I do not know, Najrul, if you have ever witnessed a storm in the Santal Parganas. To be sure they are rare occurrences. The district is one of the driest in these parts. But when the storm breaks, its fury knows no bounds. Even in the midst of such fury, both Gupta and Taylor heard a steady, monotonous chant in a Santali language. Perhaps it was the tempestuous surroundings, but even the pompous Taylor seemed unnerved by the droning incantation. This was followed soon by repeated attempts to force open the main door. Someone seemed to be kicking at the door vigorously.

Neither Gupta nor Taylor could muster up enough courage to open the door and check what was going on. They simply sat together with their guns, ready for the intruder to burst in. After what seemed an interminable time, the kicking and the chanting stopped. Gupta and Taylor eventually withdrew to their respective rooms and fell asleep.

Next morning, as dawn broke over the rain-washed forest, Taylor's courage seemed to return with a vengeance. He insisted that the incidents of the night must have been the handiwork of crafty Santal priests, and prepared to leave for the station in his two-seater car with the expropriated idol on the seat next to him. Gupta seemed a little less sure, perhaps because, not having a car, and given the state of the muddy roads after the storm, he realized he would have to stay at least another night before the roads were dry enough for him to ride to the railway station. But Taylor bade Gupta goodbye and left quickly to catch the train to Bombay while dreaming of the ship back to Old Blighty.

That night, as Gupta tried to fall asleep and brush off the strong uncanny feeling that had returned to him, once more he heard the chanting and the pushing at the main door. He was utterly unnerved and paralyzed with dread for a while. But then he remembered that the idol by now was well on its way on a train to Bombay. So, this clearly could not be the idol. The thought gave him courage, and he hollered out a warning in the gruffest civil servant voice he could muster. When the noise continued, he spat out another warning in an even gruffer voice. That it persisted despite these warnings infuriated Gupta, and he put his double-barreled gun to the door and fired twice in quick succession. There! That should teach the scoundrels not to try to break into the lodgings of an officer of the Crown, he thought to himself.

A blood-curdling cry followed the shots, and that shook Gupta again. But it was followed by silence, and gradually Gupta calmed down enough to get a modicum of repose during what was left of the night.

The next morning was so completely different from the two previous mornings that it would be difficult for those who haven't witnessed the storms of this area to believe that such a transformation could be effectuated in such a short time. The whole country had dried out during the night, and the sun shone mercilessly again. Gupta was happy that he could finally proceed onward to Calcutta. At breakfast the orderly brought him a copy of the local morning paper. Opening it, Gupta was stunned to read on the front page that the recently retired police chief, Mr. J. Taylor, had been killed the previous day in a terrible car crash. The report blamed the weather and the poor roads for the crash.

Gupta seemed unsure of the reasoning and ran out on a whim to check on the idol at its shrine. As he reached the shady clearing in the grove where the simple shrine stood, his deepest foreboding seemed to come true when he saw that the garishly colored old idol was exactly where he had seen it when he first arrived at the lodge! But this was not all. A further shock awaited Gupta. As he moved closer to take a look, he noticed two significant details. First, he saw that there was mud on the feet of the idol, this despite the fact that it stood upon a humble, but dry, stone altar. Second, and this shook Gupta even more, there were two big bullet holes on the body of the idol. Around the holes and in a trickling pattern beneath them there was a dark stain that seemed to be dried blood!

I do not know, my dear Najrul, what you will make of this curious incident. Gupta has always been reticent to discuss it at any length. But he keeps asking men of science, is it possible for old wood to somehow bleed if it is shot at?

Personally, I am not overly perturbed by whether wood can bleed or not. As long as I don't have to deal with bleeding wood, I really can't say I care that much. What I do care about, brother Najrul, are the chasms. The unbridgeable gaps in values, morals, beings, and worlds.

Taylor, in his pomposity, saw only a grotesque old wooden statue. Gupta, with his usual Hindu disdain for Santali culture, spoke of the idol as an apadevata*, a "counter-god," and a* bhuter raja*, a "king of ghosts." But what kind of a ghost is this? It is surely not the ghosts of dead relatives and friends who speak to us at the spiritualist séances held at the house of my neighbors, the Tagores. Nor indeed is it the fish-stealing ghosts of our children's tales. And what the devil is a "counter-god"? No. There are enormous chasms here. Not just between Taylor and the Indians, but also between Gupta and the Santals.*

It seems both tragic and remarkable to me that even as Gupta keeps enquiring about whether wood can bleed, so far as I know he has never endeavored to find out the meaning of the mysterious incantation that seemed to accompany the visitations. I know you have traveled far and wide, and accumulated many interesting experiences. Perhaps you can tell me what it means? Here it is: logo buru dhirko sinin ghanta bari ma kawad.

With my heartfelt affection,
Your,
Hemenda

CHAPTER 6

Refusing Race

> Others justified their refusal in accordance with specific Koranic demands . . . Indeed, we have to look for what is concealed behind that absence of imagination, that refusal of fiction.
>
> —Frantz Fanon, *TAT in Muslim Women: Sociology of Perception and Imagination*

"These people are not easy to approach and some people have a superstitious objection to losing a drop of blood," wrote Eileen Macfarlane in 1937. She was speaking of "Tibetans, or Bhutias, as they call themselves," in Kalimpong whose blood group frequencies Macfarlane was the first to study.[1] Macfarlane, as we saw in the previous chapter, was not alone. Many researchers faced difficulties in bleeding their subjects. D. N. Majumdar for instance, wrote of the "very great difficulty" he faced in bleeding the Hill Doms in 1942.[2] G. W. L. D'Sena, who studied the Chenchus in 1952, similarly reported that "many difficulties were encountered both in locating and contacting the Chenchu before they disappeared into the forest."[3] Likewise, in 1954, Narendra Kumar, studying the blood group frequencies of the Galong Abor people of Assam, wrote that "the Galong are very superstitious" and that it was "[at] first difficult to collect blood."[4] The ascription of "superstitiousness" was frequently invoked, especially if the group being tested was not an elite group, to explain the recalcitrance of subjects to participate in racializing projects.

"Superstition" has a long and complex genealogy in modern India. Cultural anthropologist Agehananda Bharati has argued that it is an "anti-traditional device" deployed in "urban Hinduism."[5] The usage of the word in Indian English, Bharati argues, most likely originated among Christian missionaries but was then appropriated by non-Christian

Indians who use or are exposed to English. In time it was taken up by Hindu religious reformers who used English, such as Swami Dayananda and Swami Vivekananda. Among these latter it came to function as a boundary marker that distinguished reformed and respectable religious behavior from unreformed and "traditional" forms of behavior.[6] Along a parallel pathway Indian atheists and rationalists have also adopted the term, but they have done so to demarcate a different kind of boundary.[7] What is common to all these usages in Indian English is that the word "superstition" has come connote social hierarchy. Bharati writes that "the user of 'superstition' seems to imply that the thoughts and deeds that come under the heading, are not thoughts and deeds common to people of equal or higher status; and that they occur within a domain typical of lowlier folks."[8]

Unfortunately, there has not yet been any detailed accounting of how this hierarchized notion of "superstition" has informed the human sciences in India. Public health and social science literature, especially, frequently deploy the word in both policy documents and scholarly publications, just as Macfarlane and Kumar had done, but obviously to ends distinct from those of Hindu reformers or activists championing rationality and atheism. The precise nature of the hierarchy constructed, or, more specifically, the nature of the beliefs, practices, and frameworks that were subordinated by the use of the term in scientific works, have thus remained unexamined.

Even the few scholars who have, in passing, remarked on the invocation of "superstition" in medical contexts have interpreted it within a paradigm of "resistance."[9] While the scholarship on "resistance" has indeed been valuable in challenging the high-modernist narratives that simply dismissed such friction as misplaced recalcitrance and ignored claims of "cognitive justice,"[10] it fails to engage meaningfully with the rationalities that inspired such refusal. The resistance framework is largely reactive. The initiative rests exclusively with those acting in the name of science. Others can only react to what the scientists initiate.

One consequence of this reactive framing is that very dissimilar reasons for refusal are grouped together. Widely divergent social groups with significantly different reasons for opposing a particular scientific project are treated as doing the same thing. This is a particularly salient problem in India, where the science is often mobile and pan-regional, orchestrated by centralized institutions, but where the communities under study are much more geographically and culturally bounded. Once again, the histories of public health provide telling instances of such lumping together of communities across the length and breadth of the subcontinent that gives "agency" to those refusing these initiatives but does so by entirely ignoring their specific rationales for the refusal.[11] Initiatives such as quarantines and vaccinations were, and are, rolled out by the state at an all-India level. The techniques and regulations were mostly common to all the communities. Occasionally, some of the staff were also at least somewhat mobile. But the communities that sometimes opposed these initiatives did so from within their very specific and distinctive geographic, linguistic, political, and philosophical frameworks.

Likewise, seroanthropology was clearly the pursuit of relatively small groups of researchers working at a small number of institutions who traveled to and studied widely dispersed and much more culturally distinctive communities. While research projects might therefore have been identical, the refusals to participate in them—often put down simply to "superstition"—were in fact motivated by a range of different rationales.

Recent scholarship in critical race studies has inspired scholars to reflect anew on "refusal." Carole McGranahan has argued that "to refuse can be generative and strategic, a deliberate move toward one thing, belief, practice, or community and away from another. Refusals illuminate limits and possibilities, especially but not only of the state and other institutions."[12] As Audra Simpson shows, the lack of transparency about the grounds of refusal might itself be a tactical move that performatively marks the limits to established hierarchies of power.[13] Refusal to give fuller reasons or to engage in elaborate debates over the rationality of seroanthropological testing, therefore, might itself have been a performance of limits to sovereignty of national elites, the state, or even simply authorized science. Unlike "resistance," whose grounds—at least theoretically—can be made transparent, the grounds of "refusal" are frequently tactically and deliberately opaque.

In fact, there are some precocious antecedents to this kind of approach within the historiography of opposition to public health and state science. Rumor research in particular was a pioneering approach that explored the alternative frameworks behind the motivations and interests of those who opposed scientific projects. David Arnold's fascinating early exploration of the rumors circulating during the British Indian plague epidemics of the nineteenth century, or Luise White's imaginative interrogations of vampire rumors in British East Africa, are good examples of such early attempts to move beyond a simply reactive reading of "resistance."[14] Indeed, in White's *Speaking with Vampires* there is also an explicit effort to see how the same set of medicalized interventions that generate superficially similar idioms of refusal might in fact be grounded in more bounded, local structures of feeling, reasoning, and being.

Instead of either simply writing off the refusal of various subjects to be bled as "superstition" and unwarranted "suspicion," as the researchers did, or trying to recast it as some form of "resistance," therefore, we can explore why the subjects who refused to be bled did so. What were the beliefs, practices, motivations, and sovereignties that briefly flickered into view at the moment of refusal?

Tibetan *Bla tshe*

One of the groups most often described as "superstitious" and hence opposed to being bled for genetic research were the "Tibetans," a label that included many hill communities steeped in a broadly Tibetan culture. Many of the early researches were conducted in and around Darjeeling, and the people studied were more immediately from Bhutan. We have already heard Macfarlane decry their superstitions. Later researchers concurred. More than a decade after Macfarlane's comments, Ernest Büchi in 1951 wrote that "as Tibetans are very superstitious it is very difficult to persuade them to donate a drop of blood."[15] Why did these men and women refuse to be bled? What were these "superstitions" that kept them from being compliant subjects?

While details about the specific people and what they thought or said are now lost to us, we can tentatively reconstruct the grounds of their refusal by exploring Tibetan ideas about the body and health. Of central importance to Tibetan ideas about the body, life, and health are two life forces, *srog* and *bla*. *Srog* is life itself. Its decline leads to loss of lifespan (*tshe*) and its loss leads to death. *Srog* can be astrologically calculated and can be strengthened through specific rituals. *Srog* is supported by *bla*, a "subtle life-essence."[16]

Bla features prominently in Tibetan medical ideas and practices, as well as in lay health beliefs. It is a complex, ambiguous, and polysemic concept that exists in multiple variant forms, and its full historical or contemporary description is difficult to outline. But two aspects of *bla* are important for our present discussion. First, *bla* is not equally distributed throughout the human body. There are specific places (*bla gnas*) in the body through which *bla* moves. Second, *bla* is tied to a number of complex rhythms. Its strength and location in the body vary according to the time of the day, the day of the week, the season, and so forth.[17]

Understanding the precise bodily locations of *bla* and its temporal rhythms suggests an unexpected rationale for the Tibetan reluctance

Fig. 6.1. *Bla tshe* leaving the body through the tip of the finger. Source: Parfionovitch, Dorje, and Meyer, *Tibetan Medical Paintings. Illustrations to the Blue Beryl Treatise of Sangye Gyatso (1653–1705)*, 140.

to be bled. Barbara Gerke has suggested that the medical paintings in Sangye Gyatso's seventeenth-century medical commentary the *Baidūrya sngon po*, is one of the relatively early texts where we can witness the coming together of ideas about *bla*, *bla gnas*, the lunar rhythms of *bla*, and clinical practices.[18] One of the features of these paintings is the way *bla* and *tshe* are conjoined and illustrated together as a "little human-like figure." The conjoint notion of *bla tshe* is glossed as "the life-essence which supports the life-span principle."[19]

One prominent image (fig. 6.1) describes how the *bla tshe* might be "stolen by a demon." What is remarkable about this image for us is the route that it describes for the *bla tshe* leaving the body. The image depicts the little humanoid figure leaving the body of a wild, possessed-looking man by the tip of his right ring finger.[20] In fact, Gerke tells us that this right ring finger, known as the *srin mdzub*, is called the "pure finger." It is believed to be kept in the nostril of the fetus and thus protected from any contact with the maternal fluids of the womb. It is, most important, the finger through the tip of which spirits of all sorts enter or depart the body.[21]

Now, we know that the fingertip was in fact the most popular spot from which researchers usually drew blood. D. N. Majumdar, S. S. Sarkar, and Narendra Kumar, to name only a few, all explicitly mentioned pricking the fingertips of their subjects.[22] Sarkar, for example, mentioned that "blood was obtained by pricking the finger tips with the help of a Franck's needle."[23] Sarkar's comments are particularly telling since he was a collaborator of Macfarlane's, and they had done a fair bit of research jointly. One can assume, therefore, that Sarkar's methods were also those of Macfarlane, and that she too obtained blood mainly from the fingertips of her subjects. Given what we now know about *bla tshe* leaving through the fingertip, it appears highly likely that her Tibetan subjects were opposed to *where* they were being bled rather than to any abstract notion of being bled as such. Such a conclusion is also supported by the fact that these Tibetans practiced bloodletting as part of their own medical traditions.[24]

Of course, we do not know for certain which fingers Macfarlane or Büchi might have pricked. But then, neither do we know how far the specific importance of the right ring finger within scholarly medical discourse might have been generalized into a broader lay discourse about spirits and life forces leaving the body via the fingertips. In fact, though *bla* features prominently in the clinical protocols of Tibetan medical practice, its presence within the scholarly tradition is relatively muted. Its supreme importance derives mainly from its popularity within the larger society. *Bla,* therefore, is in some ways a bridge that connects the expert and lay medical ideas about the body, health, and longevity.[25] Taking seriously this larger context within which *bla* operates, we realize that the textual record is only a partial trace of the beliefs about *bla*. Limited as it is, however, the textual record of notions about life essences that can depart or be "stolen" and their relationship to fingertips forces us to consider it as one of the frameworks within which Tibetan refusals to being bled may have operated.

Moreover, there is a strong belief that any injury to a *bla gnas,* that is, to a place where *bla* flows through, can immediately cut short the lifespan of an individual. For practitioners of Tibetan medicine, who are usually very mindful of not causing unintentional injury to their patients by accidentally performing therapeutic actions at a *bla gnas,* such injury can arise from two types of therapeutic actions, moxibustion and bloodletting, collectively called *gtar bsreg.* The latter, naturally, might easily be related to the act of bleeding for genetic testing. As a result, it might well be conceptualized along the same lines. This is especially important since the admonition to not perform bloodletting at *bla gnas* is one of the few remarks about the whole concept of *bla gnas* that is actually mentioned in the *Rgyud bzhi*, the foundational text of Tibetan medicine.[26] The basic idea that drawing blood from a *bla gnas* might seriously shorten one's longevity is therefore one of the most conspicuous ideas about *bla*.

Moreover, independently of the *bla tshe* being stolen through the ring finger, there is certainly a more generalized concern about *bla gnas* being injured. Depending on the temporal rhythms, various points on specific fingers were certainly *bla gnas*.[27] *Amchis* (practitioners of Tibetan medicine) are acutely aware of *bla* mostly within this context of not inflicting harm to their patients, and contemporary clinical protocols and textbooks in use in Darjeeling, the area from which Macfarlane and Büchi both drew their subjects, emphasize the necessity of attending to the *bla gnas* in determining the place and timing of therapeutic interventions.[28]

Determining the *bla gnas* without any reference to the temporal rhythms is impossible. Gerke noted during her lengthy fieldwork that at Kalimpong, another site where Macfarlane and Büchi had both worked, the annual almanac published by the Men-Tsee-Khang, the foremost Tibetan medical and astrological institution, based in Dharamshala, always "sold out shortly after the Tibetan New Year."[29] Most Tibetans keep a copy of the almanac at home and can read it to determine auspicious and inauspicious days. The almanac contains a specific table to determine the temporally fluctuating qualities of the two main life forces, that is, *bla* and *srog*.[30] Men-Tsee-Khang–operated medical clinics are also supplied with similar "vitality tables" that help determine the temporal rhythms according to which *bla* fluctuates. These tables are used by physicians to determine the proper time and place for particular treatments, such as bloodletting. Monks use the tables to perform longevity rituals. These temporal rhythms are not the same for everyone; they vary from individual to individual. Vitality tables work out individual temporal rhythms with reference to the person's birth signs, organized according to three options, namely, a "life essence day" (*bla gza'*), a "life force day" (*srog gza'*), and a list of "days of inauspicious planetary configurations." The favorable and unfavorable days are worked out with reference to the person's sign as well as the relationship of that sign to the particular days of the week.[31] What might be a good day to bleed one person, therefore, might not be the right day to bleed another.

What was important was that *bla* "wandered." Gerke points out that the adjective used most often to speak about the mobility of *bla* is "wandering" (*'khyams pa*)—a notion that was drawn mainly from the wider Tibetan culture where it was attributed to the two types of nomadism that were popular in the Tibetanized societies across the Himalayas. The first of these were the pastoralists who seasonally moved their flocks around and were called *'brog pa*. The second were vagrant-like groups that did not have stable seasonal homes. These people were called *'khyams pa*. "The notion of *'khyams pa*," Gerke clarifies, "seems to involve some sort of mobile temporalization, either cyclical along an annual route, or errantly from place to place. It bears some similarity to the principle of the *bla* wandering along the route of the *bla gnas*, following the moon cycle, but also—when lost—straying like a vagrant."[32]

Interestingly, Büchi explicitly mentioned that the people he tested were all "connected with trading businesses." These included 43 traders, 39 donkey drivers, 30 porters and coolies, 14 servants, 9 pilgrims, 4 lamas, 1 beggar, and 10 women who were accompanying their husbands.[33] In fact, he is explicit that "their main purpose for entering India was to trade in the frontier trade center of Kalimpong."[34] Macfarlane, too, pointed out that "Kalimpong is the terminus of the main trade route between Tibet and India and throughout the dry season Tibetan merchants and muleteers arrive there every day with ponies and mule trains bringing wool." A few of these families had settled in Kalimpong and Darjeeling, but there were also a "number of wandering professional beggars from Tibet."[35]

Places like Darjeeling and Kalimpong witnessed a gradual but major demographic shift, as Tibetans were gradually overtaken by Nepali migrants as the majority community.[36] As this shift unfolded, there was a certain amount of cross-fertilization of ideas between these two cultures. Nepali culture has a stronger notion of spirit involvement in ill-health. One aspect of this belief in spirit involvement is that injections anger spirits. Before having injections, therefore, one needed to take necessary ritual precautions.[37] Such ideas, though originally Nepali, have gradually been partially absorbed by their Tibetan neighbors.

The refusal to be bled encountered by the researchers was therefore could well be inspired by ideas about either wandering *bla* or spirits. In either case, it was much more than a mere reaction to the seroanthropologists. It was grounded in complex ideas about the body, health, life, and death. What was being opposed was not so much the bleeding itself, which we know from the popularity of bloodletting to not be a problem. Rather, the issue was who was bled, where on the body the blood was drawn from, and on which day. Perhaps they were also concerned about what ritual precautions had been omitted.

Pushing past the language of "superstition" or "resistance" allows us to see why the refusal made sense to those being bled. Instead of the reduction of their rational choice to either unreason or simple socioeconomic calculations, interrogating their refusal allows us to recognize the distinct views about the body and health that separated the seroanthropologist and their subaltern subjects.

Sādā Santal Totemism

In 1937 Sarkar published an extensive account of blood groups in the Santal Parganas. Though the publication was in his name alone, he acknowledged that most of the work had been done jointly with Macfarlane and that she had accompanied him during the field trip. Of the groups whose blood they collected, the largest number of samples belonged to the Santal tribe. Sarkar also mentioned his difficulties in collecting the samples. "The widespread abhorrence to letting blood is not only confined to the aborigines," he lamented, "but to many backward classes of the Hindu community." Most intransigent, however, were members of the "sādā sub-clan" among the Santals. Not a single individual, Sarkar wrote exasperatedly, would permit themselves to be bled, and despite Sarkar's best efforts, he did not succeed in getting any of their blood.[38]

Though Sarkar tells us no more, we can be fairly certain that he would have tried in every way possible to acquire the blood he sought. We have a testament to the lengths to which he could go to get around such unhelpful "superstitions" in his own account of how he acquired a complete skeleton of a woman belonging to the Mālé tribe within a few months of the trip during which he failed to extract Sādā Santal blood.

In March 1938, Sarkar traveled to Guma Pahar, near Durgapur. On a particular morning, accompanied by a teenage local boy from the tribe as his guide, he climbed a small hill to see a tribal cemetery. Once there, he was surprised to find that the rocky soil, and perhaps the local custom of burying people who died from certain diseases like smallpox in shallow graves, had combined to expose a nearly complete, recently buried skeleton. The skeleton was so recent that it still had soft tissue attached to it. Sarkar thought this was a lucky find and wanted to extract the entire skeleton for further study in Calcutta. The boy who had acted as his guide, as soon as he understood Sarkar's intentions, protested with all his power (fig. 6.2). Initially, Sarkar brushed these protests aside and tried to collect the skeleton into a basket. Before long, other villagers got wind of Sarkar's actions and assembled at the perimeter of the cemetery. One woman was particularly upset and began to cry loudly. It transpired that the woman's mother had recently been buried in the

Fig. 6.2. The grave on Guma Pahar with Sarkar's guide standing in front. Source: Sarkar, *Aboriginal Races*, 38.

cemetery, and it is possible that the skeleton Sarkar took was in fact her mother's. Given the attitude of the villagers assembled and the woman's intransigence, Sarkar temporarily relented. He wrote in the third person that "the author had to suspend any further search for the bones, lest anything untoward may happen."[39] Clearly the last phrase suggests an implicit threat of violence.

Yet, Sarkar refused to give up what he had already collected. He had the watchman of the government *dak bungalow* he was staying at carry the bones quickly back to the bungalow via a path in the forest. By the time he himself got back, however, he saw that the villagers had reassembled at the bungalow. The men were all standing to one side, while many of the women were crying. He relied on the forest department officials—who, as we learned in the last chapter, also had substantial judicial powers—to "console" the villagers. He also explained to them that he was not going to use the bones for any "magical purpose" and that the bones would be "properly reburied since they were not properly buried there."[40]

Sarkar himself had noted elsewhere that the draconian and fundamentally racist Criminal Tribes Act had recently been implemented in the area and the Mālé included within its operations. This act prevented any villager from leaving their village for any neighboring village without the prior permission of the police.[41] The act was based on a racialized assumption of hereditary criminality.[42] In the light of this legal regime, the tribe were already under enormous pressures. That they still went so far as to seemingly threaten violence attests to their desperation to prevent the desecration of their dead. That Sarkar braved such threats of violence and committed opposition to still extract the bones attests in turn to his determination to not let uninformed superstitions stand in the way of scientific knowledge production. He was ready to use both white lies that he was taking the bones for a proper burial and the suggestion of counter-violence by bringing in the forest department officials.

Though the incident involved bones, rather than blood, it happened within a few months of Sarkar's tryst with the Santals and in an area that is fairly close. It therefore gives us a clear sense of how committed Sarkar was to his science and its appetite for the biological tissues of these tribal groups. He would certainly have tried his level best, as indeed he said he did, to obtain the blood he wanted. That members of the Sādā sept among the Santals managed to frustrate his attempts bears glowing testimony to the depth of their opposition and the strength of their refusal.

Sarkar seems to suggest that this opposition was rooted in a generalized "abhorrence to being bled," which was shared by both the "aborigines" and the lower-caste Hindus in the region. He does not tell us the reasons for such abhorrence. Nor, indeed, does he tell us how he eventually managed to bleed a fairly large number of other subjects from the area, including other Santals, but not members of the Sādā sept.

We do, however, learn more from a short description of the Santals written nearly half a century before Sarkar by none other than Sir H. H. Risley, the man most often indicted as the promoter of race science in colonial India. The comments, moreover, were made in one of the most authoritative texts produced by the colonial "ethnographic state"—the 1901 decennial census.[43] In noting the customs peculiar to the different sub-septs of the Santal tribe, Risley noted that "men of the Sādā-Saren sub-sept do not use vermilion in their marriage ritual[,] they may not wear clothes with a red border on such occasions, nor may they be present at any ceremony in which the priest offers his own blood to propitiate the gods."[44]

Risley's comments clearly demonstrate two things: first, that the opposition is not part of any generalized belief common to all Santals, leave alone other denizens of the district; second, that the opposition is grounded in the color of the substance rather than its specifically hematic character.

We have other, scattered reports from across South Asia attesting to the importance of color in the bodily imagination of certain other communities. Cultural anthropologist Mark Nichter, for example, has noted that color plays an important role in the ways in which certain southern Indian communities choose between alternative therapeutic options. Writing about communities in south Kanara, Nichter notes that "red medicine particularly in pill form are attributed to be heating and good for wet cough and cold while liquid red medicines are thought to be blood producing." He speaks of such connections as a "color concordance" in which the color red becomes connected to blood.[45]

Closer to the Sādā Santal we find that the color red played a very significant role in Santal culture more generally. Nabendu Datta-Majumder, who had briefly been the director of the ASI in the late 1950s, wrote that "the importance of blood in Santal worship and the fact that a man can forcibly marry a woman by merely rubbing her forehead with some material having red colour, e.g., red ochre, indicates the great importance of the colour red in aboriginal custom."[46] In fact, Datta-Majumder felt that the Hindu practice of using vermilion to mark the foreheads of married women was likely borrowed from the Santals. He also pointed out how Santals, during the spring festival when colored water was used in celebration by the Hindus, avoided all colored water, but most of all red-colored water.[47]

Color, and especially red color, also played a significant part in Santal therapeutics. Paul Olaf Bodding, the missionary ethnographer who collected encyclopedic information about Santal medicine in the early twentieth century, noted the existence of several therapeutic practices that were organized with reference to color. He noted that in diseases where blood was visible, the remedy usually also involved something red. Hence, in the case of "spitting of blood in phthisis the red *Nymphaea rubra* is used, also a bit of a red woollen blanket (preferably of European make). In menorrhagia similar red-coloured ingredients are applied."[48] Another form of Santal therapeutics involved the use of amulets, called *amsam dhiri*. These were essentially "ancient stone-beads found buried here and there." They were used as a kind of protective charm against certain specific maladies, especially various types of dysentery. For the amulet to be effective, however, the color of the bead had to match the color of the evacuations during the dysentery.[49]

Moreover, some of the neighbors of the Santals had further developed strong color-based ideas through relatively recent religio-political mobilizations. Sangeeta Dasgupta writes how the Tana Bhagat movement, which began in 1914 as a messianic new faith among the Oraons, who were neighbors of the Santals, and then rapidly took on a strong political edge, included the avoidance of the color red among its core tenets. Followers of the movement even began to avoid chilies, red paddy, and red potatoes, as part of this generalized avoidance of all things red.[50] To them, the color red stood for the hated British. The abhorrence of the color red also led the Tana Bhagats to oppose vaccination, since it involved bleeding.[51]

None of these anti-red ideas specifically explains the avoidance of the color red by the Sādā Santals. Yet, they do sketch out a larger canvas where color mattered in general and the color red was particularly redolent. Indeed, what might have led Sarkar to hastily assume that the whole district shared an "abhorrence to being bled" might well have been that he encountered a number of different subjects who opposed being bled for different reasons, including distinctive cosmological or political reasons expressed in a chromatic sensibility.

These more general observations allow us to locate color as a mark of distinction, identity, and bodily effects. Whether among the Santals, who refused to use red-colored water during the spring festival of *Holi*, or among the Tana Bhagats, who eschewed all things red, the color red was evidently a way of marking the boundaries of one's group affiliation. Moreover, the use of chromatically calibrated therapies also shows that color was understood to have putative physical effects.

If we combine these insights, we can see in the Sādā Santal aversion to blood both a concern with maintaining their sept's identity within the larger Santal fold and concerns about the bodily effects of the red substance. Indeed, in the context of the neighboring Tana Bhagats, Dasgupta has argued that the movement was as much about internal power struggles within the Oraon tribe as it was about fighting external oppressors. Yet, we have also seen that this complex struggle over internal and external power took on at least a partially physiological aspect when it inspired the followers to oppose vaccinations. Clearly, the concerns about marking group identity and maintaining bodily health and integrity could blur into one another through the mediation of colors in general and especially the color red.

Just as, in the case of the Tibetans, we found their refusals to be likely rooted in ideas about the bodily places and times of bleeding, in the case of the Sādā Santal, we find them rooted in the complex ontology of the color red. Once again, both "superstition" and "resistance" turn out to be inadequate frames for comprehending the refusal to be bled.

Levirate

M. N. Lahiri, as we saw in chapter 1, was the very first Indian researcher to undertake seroanthropological research, in the 1920s. In the same journal issue in which he published his co-authored article on the blood group frequencies of different castes and tribes, he also published another, single-authored article on blood groups. This latter piece clarified the way in which blood groups were inherited and therefore was foundational to the former.

In his single-authored article Lahiri described a single family that had come to the Pasteur Institute at Kasauli to get treated for rabies. They hailed from a village in Shimla. Lahiri had been using patients at Kasauli to test the various rival hypotheses about blood group inheritance and hence persuaded the whole family to have their blood typed as well. Upon doing this, however, he was surprised to find that one of the children seemed to be violating the usual Mendelian laws of inheritance. Confronted with this discrepancy, Lahiri interrogated the family about the child's paternity. "At first", Lahiri wrote, "he [the father] denied the possibility of his not being the father of the child." Upon being pressed, however, "he made a frank confession stating that his present wife was really his younger brother's wife although both of them had access to her as is their custom." One of the two children had been born a year after the brother's death and so was most likely his, but about the elder child there was uncertainty.[52]

The family's refusal to divulge its sexual life to Lahiri is not the same as the Sādā Santal or Tibetan refusal to give blood. Yet, this too is a refusal. As is clear, perhaps in a rather striking way in this case, the pursuit of genetic inquiries depended to a large extent upon additional, intangible information, particularly about sexual relationships. The refusal to share information thus bore some resemblance to the refusal to share blood. Moreover, both the sexual mores of individuals and groups, as well as their reticence to speak about such things to researchers, was often legible to seroanthropologists merely as irrational "custom."

Lahiri's case of the family from Shimla is not the only such case known: we have at least one other case reported from Bombay more than thirty years later. By that time, genetic research, as we have seen in the previous chapters, had evolved away from simply tracking the frequencies of the four main blood groups. One line of research that had become particularly important was the tracking of rare blood groups. Among these rarer blood groups, researchers had begun to find a group they called "weak B." By 1960 a small handful of reports had emerged of such "weak B" blood groups, but most of them had been from outside of India. Only a single case of this blood group had been known in India. Dr. V. V. Vakil was therefore surprised when he accidentally found the "weak B" blood in a fifty-year-old male patient, known simply as K. J., who had been admitted to the K.E.M. Hospital for the treatment of other complaints. Vakil immediately alerted researchers at the Blood Group Reference Center, Human Variation Unit, at the Indian Cancer Research Institute in the city. Three eminent researchers, G. N. Vyas, H. M. Bhatia, and L. D. Sanghvi, undertook the investigation.[53]

Having confirmed that the patient, who they noted belonged to the Dasha Shrimali Vania caste, did really possess the rare blood group, they sought to test his family. It turned out that K. J. was a bachelor, but lived in a large joint family with his parents, his five brothers, and their families. Upon testing the whole family, it was discovered that two other members of the family possessed the same "weak B" blood group. One of these was K.J.'s elder brother, A. J., and the other was A. J.'s daughter, R. A. There were, however, two further difficulties.[54]

First, A. J. suffered from severe congenital mental retardation. So much so that the researchers could not even draw his blood and had to make do with testing his saliva. Second, both K. J. and his niece R. A. were "secretors," that is, their blood group antigens were also found in their saliva. But A. J. was not a secretor. This created a difficult, though not impossible, genetic pathway for R. A. to have inherited her genes from her alleged and mentally incapacitated father. It was more likely that she had actually been fathered by her uncle, K. J. The authors of the study, perhaps out of respect for the middle-class, urban family, chose not to discuss this explicitly. They did, however, insert an unexplained arrow connecting A. J.'s wife to K. J. in the schematic family pedigree that they appended to their publication (fig. 6.3).[55]

Here we might even speak of two refusals: first, the refusal of K. J. and his family to divulge the information about the family's sexual life; second, the refusal of the researchers to explicitly discuss what they likely perceived to be an embarrassing family secret. Naturally, this latter position is in marked contrast to the way Lahiri had discussed the Shimla family. Whether the difference was a consequence of the different social ranks of the two families or the changing protocols of scientific writing is difficult to tell.

What is easier to point out, however, is the sustained importance of the custom of levirate. Irawati Karve, one of the foremost geneticists of

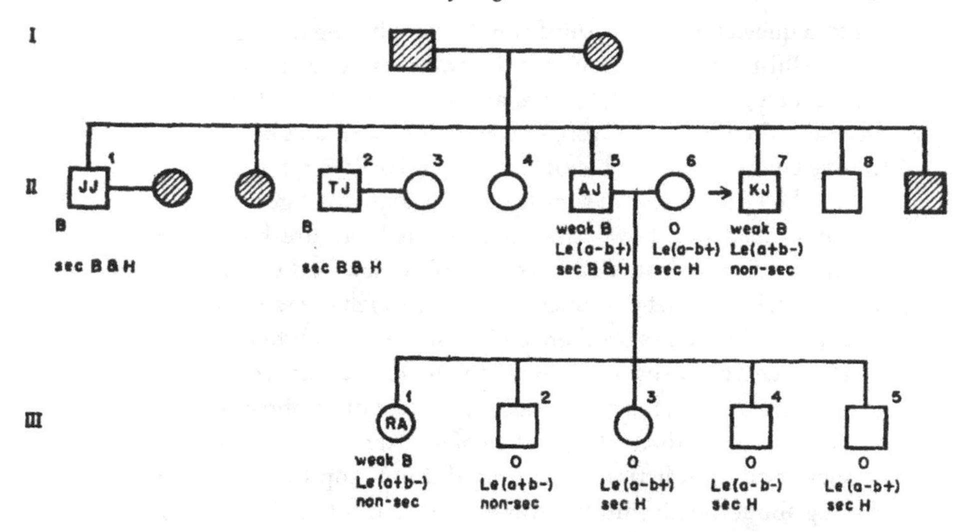

Fig. 6.3. K. J.'s family tree with the unexplained arrow. Source: Vyas et al., "Three Cases," 510.

the period, has discussed this custom at several places in her writings. Indeed, one of her most interesting discussions of this custom arises from her discussions of the genealogies of the preeminent Indian epic, the *Mahabharata*. Karve pointed out that there are several key junctures in the genealogy of the royal house of Kuru, whose members are the main protagonists of the epic, where the royal lineage was maintained through levirate. King Shantanu, for instance, fell in love with a beautiful fisherwoman, Satyavati, in his old age. Satyavati's father agreed to the marriage on the condition that Shantanu's eldest son by a former marriage, Bhishma, would relinquish all claims to the throne. Bhishma did so and vowed to never marry. Shantanu and Satyavati then had two sons. The eldest died in childhood, but the youngest, Vichitravirya, though a sickly young man, became king in due course. Bhishma, his step-brother who had raised him, then ensured his marriage to two princesses. The young King Vichitravirya, however, died young without an heir. His mother, Satyavati, begged Bhishma to marry and continue the royal line. Bhishma, citing his oath to his father, refused to do so. Left with no other option, Satyavati finally revealed that she had given birth to another son, Vyasa, by a Brahmin sage, long before she met and married King Shantanu. This son was then called in to cohabit with the two widowed queens of the dead Vichitravirya and produce heirs. This was in effect an act that could be described as either levirate or fraternal polyandry.

Vyasa not only produced two sons, Dhritarashtra and Pandu, by the two queens but also a third son, Vidura, by the queen's maid. The eldest, Dhritarashtra, was blind and hence overlooked for the kingship; Vidura, being the son of a lowly maid, was also ruled out. Thus, Pandu ascended the throne. He too, however, owing to a curse, was unable to have children and died without an heir. His two queens, Kunti and Madri, then produced five sons through five different gods.[56]

Karve discussed how classical Sanskrit literature had permitted several forms of levirate and fraternal polyandry. In the former, an appointed male surrogate produced an heir through the wife, while in the latter all the husband's brothers and, in some cases, other agnatic relations had certain sexual rights on the wife. Karve pointed out that this position changed later, and by more recent centuries most of the Brahminic law codes forbade the practice of levirate.[57]

Such legal interdictions, however, did not stop the practice itself. Writing about the kinship structures of northern India, Karve pointed out that though the higher castes, namely, the Brahmins, Khatris, Kayasthas, and Vaishyas, denied the existence of the custom, "a lady medical practitioner living in the Delhi region for many years" had assured Karve that "the custom is very wide-spread in villages and is found among all castes."[58] Karve herself claimed to have met with one such case involving an impoverished Sikh family.[59]

While it might appear at first that levirate, if it is strictly limited to brothers, will not seriously undermine the genetic framework of castes constituting discrete Mendelian populations, a further investigation disabuses us of this notion. As is clear from the stories of the *Mahabharata*, levirate—even when limited to brothers—could bring in a host of different contributors to gene pools. Vyasa, for instance, was related to Vichitravirya only through their mother, who herself had been a fisherwoman by birth. Pandu's sons, the main heroes of the epic, were all fathered by gods with absolutely no links to the fraternal line. Indeed, none of the famous heroes of the *Mahabharata*, including the righteous King Yudisthira, had any direct biogenetic relationship to their great-grandfather, King Shantanu, whose throne they claimed.

Moreover, besides the somewhat better documented forms of levirate, there were still other, more localized customary practices. We have histories, for instance, from both Bengal and Bombay in the nineteenth century of young married women having sexual relationships with powerful priests, with this being tolerated or even encouraged by certain communities.[60] Such practices usually became controversial only with the emergence of more "modernized" notions of sexual morality and strict monogamy. While these practices may not have been levirate as such, looked at from a geneticist's point of view, they could have a similar effect.

At issue here is the way descent was conceptualized within a northern Indian, caste Hindu framework. When it came to producing children, sons were greatly desired. But this desire was mainly in order to continue the performance of certain key rituals. Karve speaks of the three debts that any Hindu man owed and upon the repayment of which his ultimate liberation depended. These debts were respectively to the gods, the sages, and the ancestors. Each of these, and especially the last, required the making of regular offerings. These offerings could only be made by a son. Hence, the function of a son was the making of ancestral offerings, rather than the maintenance of a biological or genetic lineage.

The Hindu legal codes, the *Dharmashastras*, recognized twelve different types of sons. Among these categories of sons, the *kshetraja* son was one produced through *niyoga* (levirate). In the oldest legal codes this surrogate fatherhood seems to have been limited to the mother's husband's brothers. By the time of the epics, however, Brahmin surrogates seem to be preferred. In effect, the epics formalized the "open-endedness of possible surrogates."[61] Though *niyoga* does seem to have been as popular as the practice of *dattaka* or adoption, the point is that these and the other categories of sons were all conceptualized within a system where lineage maintenance meant performance of particular ancestral rituals rather than retention of biological continuity. One unintentional consequence of this, especially the open-ended structure of levirate and adoption, was the undermining of the kind of biologically discrete group identities that seroanthropologists evoked.

The refusal to divulge information about a family's sexual practices was therefore not simply motivated by a modern desire to avert scandal. Rather, it was because, within an older customary framework of kinship, "descent" itself worked differently and to other ends. Its functions and conceptions were quite distinct from the biologist's insistence on genetic inheritance. The refusal that left its archival traces in the form of refusals to share sexual information was tacitly rooted in a more radical refusal to accept a narrowly biologized notion of inheritance.

Entelechies

Having dealt with refusals to share blood and information, based in distinctive ideas about the body, life, and inheritance, I will now turn to an instance of refusal to accept the larger cosmological underpinnings of seroanthropology. Studying refusal entails attending to specificities,

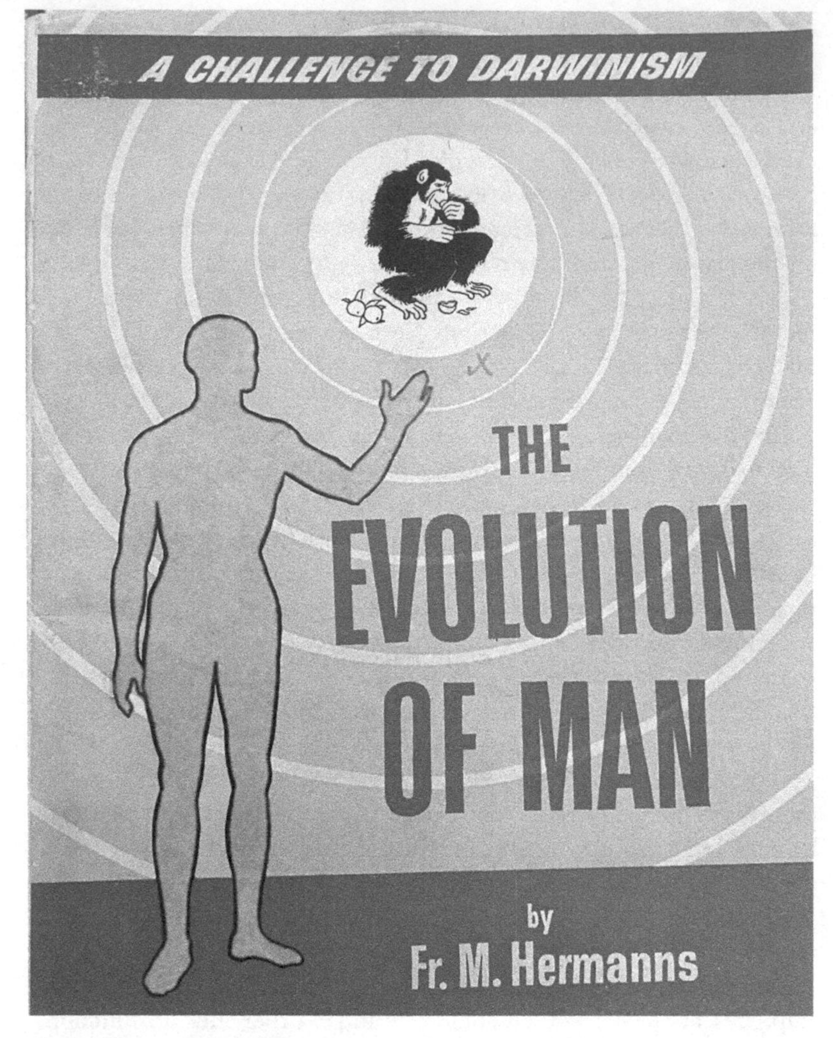

Fig. 6.4. The cover of Mathias Hermanns's *The Evolution of Man*.

rather than grouping diverse responses into the straitjacket of "resistance." This would naturally include looking at relatively singular or even marginal responses. The present instance was clearly much more singular and atypical than the widespread, though largely tacit, practices like levirate.

In 1955 the Society of St. Paul, Allahabad, published a small book on anthropology with substantial discussions of genotypes, phenotypes, the races of man, and so forth (fig. 6.4). The book carried a foreword by Cardinal Valerian Gracias, the Archbishop of Bombay. The humble volume had a rather grandiose title, *The Evolution of Man: A Challenge to Darwinism*. The author of the book was a Roman Catholic missionary named P. Matthias Hermanns (1899–1972), belonging to the Societas Verbi Divini. Father Hermanns had been born in Köln-Niehl, Germany, but moved to the Netherlands in 1914. He was ordained in 1928 and moved to China the following year. He learned Tibetan there and began to study Tibetan communities. In 1947 the advances of the Chinese Communists in the civil war forced Hermanns to return to Europe.[62] He could not go back to China after this. In 1950, when the Anthropos Institute opened a branch in Bombay, he joined it. He also joined St. Xavier's College in Bombay and taught there until 1954. Thereafter he undertook a number of ethnographic studies, studying a wide variety of Indian groups, such as the Bhils, Korwas, Oraons, Mundas, and many others, and in the process doing fieldwork in eastern, central, western, and southern India. He finally returned to Europe for good in 1960.[63]

Two aspects of Hermanns's book make it interesting for our discussion of seroanthropology. First, he discussed the same themes treated in this book, including a detailed account of genetics and race. Second, he studied some of the same groups that our seroanthropologists studied at around the same time. Yet, he drew remarkably different conclusions from the rest and explicitly repudiated much of the theoretical and cosmological framework that informed racialized genetics.

Hermanns explained Mendelian inheritance, or "Mendelianism," as he called it, in reasonable detail. He also described how one can see instances of Mendelian inheritance operating in human beings. Heredity of hair, eye, and skin color, for instance, were cited as following the Mendelian law. But he then asserted that "not all characteristics are inherited in this simple fashion. And many characteristics are not inheritable. There are very complicated processes and it is not easy to find out the dominant genes and the recessive ones."[64] Of particular importance in this latter, more complicated category was the "heredity of psychic characteristics." "Certain talents, for instance," he wrote, "skill in music or in painting. Gifts for poetry or mathematics, are often inheritable and we find great musicians and painters in one and the same family for several generations." Such talents, he thought, presupposed a "special disposition of certain cells in the brain." It was this "special cell-structure" that was inherited alongside the "psychic powers responsible for the talent."[65] Yet, it was "impossible to explain the innumerable physical and psychical co-operating factors" that combined to create "a person." Each person, therefore, Hermanns insisted, was "an individual, not a copy. This individuality is due to man's spiritual soul which is not transferred from the parent to the child but is created individually."[66]

Hermanns argued that much of the evolutionary discourse was confused because different kinds of evolution were mixed up. To bring some order into this alleged chaos, he distinguished four different types of "evolution," which he called "atheistic or absolute evolution," "deistic evolution," "theistic evolution," and "special creation." He rejected the first two and combined the latter two.[67] Central to theistic evolution was the notion of "entelechy." "The body," wrote Hermanns, "is adequate to the animating entelechy; an animal body is adequate to a sensitive, a human body to a spiritual entelechy."[68] In other words, one's psychic, spiritual, or mental powers do not derive from one's somatic makeup, but rather the somatic form always fits itself to the psychic makeup. "Special creation" added to this the idea that the human body possesses a "special, extraordinary position in the kingdom of life which can be explained by a special creation only."[69]

Less expectedly, Hermanns argued that the "special peculiarities" of human life were eventually contained in the "chromosome-gene structure."[70] Slightly later he explained that "man could never be evolved from an animal by a natural process, because man possesses not only a special chromosome-gene-system but also has quite a different spiritual entelechy, a rational soul."[71] The argument, though at times opaque, was that it was man's soul that came first. A unique genetic architecture emerged in response to this unique rational soul.

Having thus grounded human specificity in both the "chromosome-gene-system" and the entelechy of the "rational soul," he peripheralized the importance of phenotypes altogether. He argued, for instance, that just because humans and certain animals in their fetal stages resemble each other, this did not signal any genetic similarity. He wrote that "the phenotype, the outer appearance, only is similar; the genotype, the gene-system, is completely different."[72] It was this separation of the phenotype from the genotype that allowed him to develop his arguments about race.

Hermanns did not deny the existence of different races as such, but he emphasized that "biogenetic law . . . show[s] that man has a special chromosome-gene-system which is fixed and unchangeable; we are forced by physical anthropology which proves that all different phenotypes of man depend on one and the same genotype only."[73] The genetic makeup of humans, rather than allowing us to track racial demarcations as seroanthropologists did, in Hermanns's view, proved our membership of a single, undifferentiated species.

He explained the differences in phenotypes by two sorts of rays, namely, the cosmic and the telluric rays. Among the former he included sun rays, moon rays, and atmospheric rays, while among the latter he included the earth's magnetic and radioactive rays. He explained with allusions to his fieldwork that physical changes ensued not from changes in genetic composition but rather from the influence of these rays. Since human numbers had begun growing and different groups of humans had migrated and settled in different places, they were differentially exposed to these various cosmic and telluric rays, which in turn gradually changed their phenotypes.[74]

Central to Hermanns's framework was the notion of entelechy. As Bohang Chen points out, by the early twentieth century the doctrine of entelechy had become the dominant framework for understanding vitalism. Originally formulated by the German biologist Hans Driesch, entelechies were "nonmaterial, bio-specific agents responsible for governing a few peculiar biological phenomena." Furthermore, Driesch held that "these biological phenomena would never be fully explained by mere physico-chemical mechanisms."[75] Chen has further shown that for Driesch, entelechy was a matter of logic rather than metaphysics.[76] Yet, it is worth noting that the combination of vitalism and logicism did not always appeal to all those seeking a Catholic modernity through mid-twentieth-century genetics.

One famous and controversial figure, whose views on the matter serve as a telling foil for Hermanns's, is Alexis Carrel. Nowadays a very controversial figure, especially in his native France, Carrel was a Nobel Laureate, a believing Catholic, and a notorious Vichy-era eugenicist.[77] In his best known, public-facing work, *Man, the Unknown,* Carrel explicitly dismissed the doctrine of entelechy, calling it a "troublesome reminder of [past] mistakes." "Evidently," Carrel declared, "entelechy is not an operational concept. It is a mental construct."[78] Interestingly, though *Man, the Unknown* was originally published in 1935, a press in postcolonial Bombay reissued it in 1959, four years after Hermanns's *The Evolution of Man* was recommended to readers by the Archbishop of Bombay. Contrasting Hermanns's embrace of entelechy within an explicitly Catholic and noneugenicist framework with Carrel's positions demonstrates both the larger terrain of such quests for a Catholic genetic modernity as well as the diversity of positions within it.

Another figure who worked on the same terrain was Basile Luyet. Like Hermanns, and unlike Carrel, Luyet was a Catholic priest. He briefly overlapped with Carrel at the Rockefeller Institute in New York and shared Carrel's interest in "latent life," wherein life is maintained but metabolism is either completely or almost completely suspended.[79] Luyet pursued his quest for "immortal life" and "latency" through the emerging experimental and technological possibilities of cryobiology. It was his work on refrigeration that, in turn, transformed genetics research, especially in population genetics. Joanna Radin has called Luyet's broader worldview a "cosmology of cold."[80]

Christian missionaries working outside Euro-America have mostly been seen as "go-betweens" in human genetics research.[81] Yet, Hermanns's partial overlaps with both Carrel and Luyet suggest that at least some of the Catholic missionaries were much more than mere go-betweens. They tried to intervene in a global intellectual quest to eke out a new Catholic modernity that could accommodate religious aspirations, emerging genetic knowledge, and ministry amongst culturally diverse populations across the globe.

What lent urgency to such a quest for a Catholic, or even Christian, genetic modernity was the changing circumstances of postcolonial India. On the one hand, the condition of Christians, who were often stereotyped as being insufficiently anticolonial, owing to their shared faith with the colonizers, was deteriorating.[82] On the other hand, "tribal" peoples, among whom missionaries had long been active, were put in a particularly precarious position. First, a resurgent Hinduism now sought to culturally assimilate them by recasting them as "default Hindus."[83] Second, the postcolonial state's appetite for national development led it to often forcefully acquire natural resources on tribal lands.[84] As a result of these diverse new constraints, it became urgent to articulate a Christian modernity in India as well as to explicitly deal with issues of diversity within it.

Hermanns attempted to achieve this by distinguishing clearly between "culture" and "civilization." Confounding these two had, in his view, led to grave errors. Culture, he insisted, was the "cultivation of spiritual faculties and the gift of human nature. It is the development of humanity and spirituality, and the refinement of the human character and personality." By contrast, civilization was "the development of techniques . . . conquest, use and control of nature."[85] While Hermanns's definition is undoubtedly idiosyncratic, his attempt to eschew and denigrate the language of "civilization" is politically potent.

Once again, however, the politics itself is heterographic, occupying different positions in the national and international contexts. Nationally, as David Ludden has pointed out, though the idiom of "civilization" originated in imperialist and Orientalist circles, it was well ensconced in nationalist circles as well.[86] In fact, the immediate context in which Hermanns introduced this distinction between culture and civilization was his criticism of the Hindu philosopher and future president of India, Sarvepalli Radhakrishnan, and his idealized neo-Vedantic exposition of Hindu ideas about life. This neo-Vedantism was part of the intellectual rationale for rendering "tribal" populations as "default Hindus." Hermanns pointed out that this position was remarkably close to what he perceived to be a reductionist and materialist one.

Internationally, however, the distinction between "culture" and "civilization" was reminiscent of European antisemitic rhetoric.[87] Hermanns himself would certainly have been exposed to this darker, European usage of the distinction. Yet, he chose to deploy it in building a critique of Radhakrishnan's neo-Vedantic Hinduism. Perhaps one reason for the choice was that it also allowed him to articulate his distaste for another large, international discourse, namely, that of Soviet-style industrialism, which was also fetishized by the Nehruvian state to the particular detriment of many of the tribal and Christian communities Hermanns worked among. As C. A. Bayly points out, the Nehruvian regime was an ideological "amalgam" that combined earlier ideas about small-scale development with large, Soviet-style industrialization plans and a neo-Vedantic Hinduism.[88]

Hermanns's entelechy is therefore a complex refusal grounded in heterographic politics that is particularly difficult to decode with absolute clarity. Yet, clearly, in his intervention we see a style, level, and geography of refusal that operated very differently from the other refusals we have discussed. Its existence is a particularly strong corrective to any framing that distributes seroanthropology and its discontents along a simple binary organized around a modern, cosmopolitan scientism and romanticism of the bucolic local. Refusals, like the science they refused, operated at multiple levels and upon diverse territories.

Polyphony of Refusals

Histories of race science, even the critical ones, often feel the pressure to assimilate refusals into the straitjacket of resistance. Shorn of the complexity and specificity of their interests, agendas, cosmologies, and geographies, the so-called resistors become cardboard caricatures in histories where the seroanthropologist continues to hold the center stage, even if only as an antihero. The lack of detailed textual archives, as well as the structural imperatives in writing histories of science, all contribute to this reductionist framing. After all, scientists, even when they leave few personal papers, usually leave a slew of publications and other records. By contrast, especially in seroanthropological research, the subjects are frequently anonymous, marginal figures who leave few archival traces that are self-constituted.

Historians of subalternity have long struggled with similar archival absences and narrative pressures for erasure and assimilation. It was precisely in response to such challenges that Ranajit Guha, and historians who worked within the tradition he inaugurated, pioneered strategies of "reading against the grain," supplementing the historical with the ethnographic archive and strategically embraced microhistory in "defense of the fragment."[89] These techniques can be productively employed by historians of science to delineate refusals to participate in extractive and exploitative scientific projects.

It is, however, important to also repudiate the notion that all refusals ensued from marginal groups that lacked the power to produce voluminous, self-constituted archival traces. The textual archive of South Asia is rich and diverse, though its preservation in official repositories is frequently impoverished. Works such as those by Hermanns clearly challenge our notions of who challenged scientific claims of authority. His book, though little known today, was written in English and supported by the Catholic Church in India. Its existence demonstrates that refusals to race science need not always fit into easy binaries that pit the male, Christian European as the exponent of racial science and the hapless tribesman in the tropics as the lone tragic resistor. Decolonization produced a number of strange bedfellows, including the increasingly marginalized Christian missionaries and marginal groups dubbed "tribals," whereas those who increasingly spoke for seroanthropological racialization were middle-class, upper-caste Hindu scientists.

The diversity of this cast of characters also warns us to be wary of the capacity of the "sign of science" to reify a contingent line dividing "science" from "nonscience" as a static, ahistorical reality. The line itself is a politically produced, infinitely variable, and inchoate demarcation, which operated in modern India to simultaneously undo colonial hegemony and constitute the power of nationalist elites.[90] Who speaks for science, and who represents the nonscientific point of view, is far from being an unproblematic matter of fact. Taking the refusals of seroanthropological racialization seriously with all the complexities of such positions, therefore, becomes, at the same time, a refusal to accept the boundary between science and nonscience as anything but a contested and contingent reality.

Finally, it is also crucial to guard against the labeling of refusals as "local" responses to universal scientific rationality. Both the practice of levirate and the Catholic Church, in their very different ways, were as global as seroanthropology was. Indeed, even Santal and Tibetan cultures were far more widely dispersed than the specific localities where they were encountered. The "local" is a historically produced set of relations rather than a self-evident geographical truth. It serves multiple political and epistemic functions and hence requires great analytic caution.[91]

The move beyond the frames of "superstition" and "resistance" and toward one of "refusals" impels us to recognize a polyphony of refusals. Any attempt to find a strong common thread in these rich and distinctive refusals risks resurrecting one or the other reductionist binary.

INTERCHAPTER

Letter 7

Najrul Islam
Kampala, Uganda
20 April 1934

Dear Hemenda,
I just finished reading your letter and felt I had to write back immediately.

What a remarkable experience your friend Mr. Gupta had! It is true that during my travels among the wilder parts of our world, I have often met with both the supercilious colonialist and the mysterious power of religions we do not comprehend, but never have I heard of a clash so remarkable as that between these two. I, too, wish I could understand what the mysterious Santali chant meant.

You mention a chasm between Mr. Sen and the future-humans. On either side of this chasm are two different ways of seeing the world, understanding what it means to be human. You are right, of course. To my way of seeing, however, it eventually comes down to two very different ways of looking at our pasts. Mr. Sen is a modern man, but yet retains a strong sense of his Bengaliness. A sense, I feel, that cannot develop unless one thinks of the past as the firm ground upon which to stand as we reach into the present and beyond it. For the Chief Scientist, however, the past is a shackle. The future can only be welcomed by completely rejecting everything that the past held valuable. The past, for them, is simply a curiosity that can be exhibited in museums, but not something that shapes their lives in any way. The exhibition of the past in museums gives them a foil, against which to mark themselves off; a milestone from which to mark their distance and difference.

Perhaps nothing about Mr. Sen's narrative is more disorienting for me than his description of the hatcheries. In the course of my travels and adventures I have had many opportunities to be surprised, to confront the unknown and the unexpected, to look the exotic in the face. But nothing, I tell you, Hemenda, had ever prepared me to learn that our future descendants might be creating hatcheries to produce human babies like those new businessmen who are starting up poultry businesses all over Bengal.

You must have noticed, Hemenda, from Mr. Sen's diary that Chandrasen, the scientist who created the barrel-dwelling race, concluded that human beings had too many congenital imperfections and got it into his head that the best way to eradicate these and produce a uniformly advanced race would be to stop people from having babies altogether. Can you imagine anything less human than this! Apparently, all of them, men and women, lay eggs, the scientists then test the eggs, and hatch them in the hatcheries. No, Hemenda, I cannot think of this bunch as being even remotely human. If you ask me, they seem to be more chicken than human.

As if that was not bad enough, they then sell the children to would-be parents for a price. A price moreover that varies with the age of the child. How utterly devoid of any semblance of humanity can these people be! Can you blame me if they seem more chicken than human to me? It is like paying to buy chicken by weight in the bazaar.

They claim that by these measures they have eradicated disease, but I say that with it they have also eradicated diversity, difference, distinctiveness, and above all humanity. I keep thinking of our beloved Rabindranath's urging us to strive for unity within diversity, not to create a sameness that abhors anything that looks different.

What right, I want to ask, did Chandrasen have to decide what was good and what was bad, not only for his own compatriots but for yet unborn generations as well?

Mr. Sen tells us that at least some of these future-humans, such as Kamala, thirsted for beauty, for love, for passion, and of course, for children. What right did Chandrasen have to decide that she could not have any of these? This is tyranny, Hemenda. Tyranny, born out of one man's lurid dreams of a perfect future, that came to be realized through science. Again, I am tempted to say with Rabindranath that a mind all logic is much like a knife all blade. It merely wounds the hand that holds it rather than helping it to chop vegetables, sharpen pencils, or do any of those many little useful things that knives do.

The more I think about it, Hemenda, I am convinced that this barrel-dwelling race is not human at all. We may be descended from the same ancestors, but their path has diverged from ours. In seeking to transcend their humanity, they have fallen far below it. Obsessed with the future and with building a brave new world, they forgot the fertility of the old soil that produced them in the first place.

I think theirs is a cautionary tale for us. A warning not to lose sight of our past and our present in our obsession with the future. A warning not to sacrifice love, beauty, and sensory pleasures to some abstract notion of perfection. A warning to temper the hubris of our overheated minds, with the wisdom in our bones. You should write about them, Hemenda.

I know thousands read your books. If you were to put this cautionary tale into your words, I am convinced that many would read it and, perhaps, at least some of them would realize that we are in fact humans, not chickens.

Think about my proposal, Hemenda. I do hope you will write that cautionary tale.

In the meantime, *Shubho noboborsho.*

Your brother,
Najrul

CHAPTER 7

Racing the Future

> Every human problem must be considered from the standpoint of time. Ideally, the present will always contribute to the building of the future.
>
> —Frantz Fanon, *Black Skin, White Masks*

One prominent characteristic that distinguished colonial race science from race science after empire was its temporal orientation. The race science practiced by colonial ethnographers like Sir H. H. Risley looked backward in time. It sought to understand the social divisions of the colonial present with reference to a past of Aryan conquest. By contrast, the overwhelming attitude in the racial science that developed from the late colonial and into the early postcolonial period looked forward in time. Even when researchers spoke of origins, they did so with a view to building a future nation.

The shared orientation toward the future did not mean that everyone agreed on what kind of future they wanted. Sometimes they did not even agree on fundamental questions, such as how races had interacted in the past, how this could be known, or indeed in what spheres of life or by whom this knowledge was to be deployed. Their commitment to building a future nation, however, transcended the extensive disagreements over both the form of the future and how to decide upon it.

As in the previous chapter, I will avoid imposing any misleading coherence to this futurity. To do so would not only be erroneous, but would also miss vital fault lines between people, practices, and politics. Instead, I will delve into four distinct texts written by different authors at different times in the first half of the twentieth century and articulating distinctive futurities. What authorizes their discussion here is that each of the authors was a stalwart in twentieth-century Indian race science. Taken together, they signpost the political complexity of racialized Indian futurisms.

B. N. Seal's Biometric Nationalism

The First Universal Races Congress was held in July 1911 in London. Its declared objective was "to discuss, in the light of science and modern conscience, the general relations subsisting between the peoples of the West and those of the East, between so-called white and so-called coloured peoples, with a view to encouraging between them a fuller understanding, the most friendly feelings, and a heartier co-operation."[1]

The first paper, with which the conference opened, was presented by Dr. Brajendranath Seal. At the time Seal was the principal of the Maharajah of Cooch Behar College. The following year he would join the Philosophy Department of Calcutta University. In 1921 he would become the vice chancellor of Mysore University. Perhaps more pertinent for our discussion, it was Seal's influence that inspired two of the foremost Indian raciologists of the mid-twentieth century to take up the question of race. B. S. Guha, who went on to found the Anthropological Survey of India, was a student of Seal's at Calcutta University and mentioned Seal as one of his two "principal teachers" on his application for admission into the doctoral program at Harvard.[2] P. C. Mahalanobis, the founder of the Indian Statistical Institute and another major figure in race research, shared a close personal relationship with Seal through, among other things, their membership in the reformist Brahmo religion. Indeed, Mahalanobis explicitly mentioned Seal and Seal's address at the Universal Races Congress as the reasons for his having taken up the question of race.[3] Notwithstanding its early date, therefore, the speech's distant echoes drifted well into the postcolonial period.

Seal called race a "dynamical entity" and placed it within an evolutionary framework. Instead of modern races being derived from stable, preexisting races, he argued that races constantly evolved and changed. According to him, "genetical Anthropology" would study "Races and Racial Types as developing entities, tracing the formation of physical stocks or types as radicles, their growth and transmutation into ethnic cultural units (clans, tribes, peoples), and finally, the course of their evolution into historical nationalities."[4] It was only through the study of "genetic conditions and causes of biological, social and psychological forces" shaping the "races of man," Seal argued, that we can "guide and control the future evolution of Humanity by conscious selection in intelligent adaptation to the system and procedure of Nature."[5]

Seal thus laid down a clear futurist outlook on race, and explicitly allied it to a teleological schema whose highest point was the production of "historical nationalities." The concern here was not simply with the evolutionary future of individuals but rather with the evolutionary destiny of individuals within a hierarchy of collectivities that began with clans and ended with nations. I have argued elsewhere that Seal's paper, and his influence more generally, was one of the foundational elements in a new form of nationalism that emerged in twentieth-century India, that is, biometric nationalism.[6] Biometric nationalism was a heterogeneous body of techniques assembled around the idea that political questions about national belonging were susceptible to technocratic clarification through the use of biometric measurements. The actual political content of such nationalism varied, with Mahalanobis, for instance, pulling further to the left of the political spectrum and Guha pulling to the right.[7] But the basic ideas that a nation's contours could be clearly worked out by biometric measurements and that such insights should play some part in national futures were widely shared.

The "starting point" of any evolutionary analysis of race, Seal posited, had to be "Proto-Man." This "generalized form," which expressed a fair degree of "phylogenetic variability," included the gorilla, Pithecanthropus, Neanderthal, and Cro-Magnon. Each of these types had progressively greater cerebral capacities. Such expansion of cerebral capacity, Seal said, was an "index of rapid psychic variation." Moreover, the "psychic (and social) changes" had "zoogenic value." Among such psychosocial characters, Seal included "sexual selection, gregarious impulses, instincts of species-preservation, mutual aid and sympathy." These would lead to characteristics like "foresight, control and coordination." These key psychosocial characteristics were what distinguished man from animals. Seal's formulations resonate with what Erika Milam has termed "rational evolution," an early twentieth-century paradigm that limited the "rational choice" involved in Darwinian sexual selection only to humans.[8] But, crucially, the "range, variability and plasticity" of the rational characteristics varied by race. As races evolved, these psychosocial characters changed.[9]

"Primitive people" or "natural races," Seal continued, often exhibited "abnormal or pathological phenomen[a]" such as "trance phenomenon, black magic, cannibalism, revolting puberty rites, orgies, sexual perversions and inversions etc." But structurally similar traits might even be seen in "civilized man" in the form of "excessive sensuality and, many superstitions." Both arose from "similar excesses or defects of the same normal impulses." The key impulse here, according to Seal, was "sociality" or "sympathy with the horde." Since this was "more adaptive and life-maintaining," evolution has pushed it to ever "expanding circles" and contracted antisocial impulses, while allowing for personal freedom.[10] Among the opposing impulses that had negative evolutionary value, he included cannibalism, infanticide, group marriage, Lewis Henry Morgan's concept of consanguineous marriage, promiscuity, and black magic.[11]

This emphasis on the "social impulse" allowed Seal to move from the "natural races" to the next stage of racial evolution, which he called "cultural race." At this level an "intimate interdependence" developed between "grades of material culture" and "grades of ethnic culture."[12] This in turn led to the development of more complex social collectives, such as clans, tribes, and so on. The problem, according to Seal, with this phase of racial evolution was that "the units are not sufficiently differentiated and the whole is not sufficiently coherent." He explained that at this phase, the central authority of the state could not deal with individuals as individuals but had to do so through "gens or clans." Unlike our contemporary view of the state as an external force that controls individuals, Seal argued that only a strong centralized state allowed individuation to flourish by undermining other, nonstate collectivities. Thus, individuation remained "only rudimentary" when the state remained weak. The "cultural race" was marked by a weak state, insufficient individuation, and strong nonstate collectives like clans, tribes, family, and so on.[13]

The final stage in Seal's schema was the "national race." At this stage "differentiation of the individual and central coherence [went] together." Individuals became "differentiated" from family groups, clans, tribes, and such, and the central authority of the state annulled "all intermediate jurisdictions" in order to deal directly with the individuals.[14] The individual is freed from all customs and affiliations that condition their "social responses" and begins to express "indefinite variability," albeit in conformity with the "biological law of adaptation and survival of the fittest." Natural selection now becomes a "conscious organized rational selection" determined by "ideal satisfactions or ends."[15]

It is worth noting that Seal's formulation of the telos of race as ending in a state where individuals stand stripped of all other ties but those that bind them to a nation-state, perhaps not accidentally, seems to anticipate one of the key ways in which the contemporary Indian state conceptualizes the social. Writing about the recent efforts to push out a single, biometric identification card, or Aadhaar, as the exclusive basis for identifying individualized citizens, Lawrence Cohen has called this a project to produce a "political subject outside of biography."[16]

Remarkably, this is exactly the endpoint of Seal's telos, outlined more than a century ago.

This "political subject outside of biography" is clearly visible in Seal's definition of a "nation." A nation, he asserted, "is a conscious social personality, exercising rational choice as determined by a scheme of ideal ends or values, and having an organ, the State, for announcing and executing its will. Law is nothing but the standing Will of the national Personality."[17] Here is the crux of Seal's vision of the future. There are three key elements to it: an individual entirely atomized and shorn of all customary familial or social ties; a state that is the embodiment of a homogeneous national Will unchallenged by any internal limits to its power; and rationally organized natural selection that will be implemented by the all-powerful national state and rationally obeyed by the atomized individuals.

Almost as an afterthought, Seal acknowledged that nationalism was only a "halting stage on the onward march of Humanity" and that a Universal Humanity would have to mediate any conflict between National Personalities, by having recourse to Justice. International Jurisprudence would allow the distinct nations to exist alongside each other while recognizing each National Personality's right to develop according to its unique ideals.[18] It is clear that, notwithstanding his espousal of a higher, universal humanity, Seal never visualized a future without nations. He foresaw a future without families and tribes, but not without nations. The political subject could be stripped of biography, but not of a biologized nationhood.

Seal developed his ideas through an eclectic and selective reading of a large variety of contemporary scholars. The list of authorities whose researches he cited in his talk included Havelock Ellis, Hermann Klaatsch, Robert Koch, Friedrich Ratzel, Sir G. Archdall Reid, Karl Pearson, Edwin Klebs, Paul Topinard, Augustin Weisbach, Anton Kerner von Marilaun, August Weismann, and many others. The breadth of his readings also meant that, within his teleological framework, he reconciled intellectual positions that were explicitly opposed to each other. For instance, Weismann and Pearson are often seen as hard biological determinists, and Seal certainly took some of the most deterministic aspects of his work from them. Yet, one of the people whom Seal drew upon at greatest length was Franz Boas, who was much more open to biological plasticity.

Herbert Lewis has argued that Boas held a distinctive view of science that was close to that of American pragmatists like William James, John Dewey, and George Herbert Mead. Nourished on the German historicist traditions of Alexander von Humboldt and Johann Gottfried von Herder, the American pragmatists and Boasian anthropology emphasized antifoundationalism, pluralism and diversity, contingency and chance, the individuality of phenomena over totalities, and the importance of the individual. This view of science was explicitly opposed to the "physicalist" conceptions in models of physical science. It did not seek deterministic, universally applicable, and predictive laws, the way the physical sciences did.[19]

It seems ironic that Seal drew copiously on Boas's researches about the plasticity of cranial measurements, which Boas had used to demonstrate the importance of the environment, together with such determinist positions as those of Weismann, in order to postulate modular, teleological, and universal laws of racial and cultural evolution. A similar irony might be glimpsed in Seal's quoting both Archdall Reid and Karl Pearson in the same sentence. While the latter was a hard eugenicist advocating the breeding out of the poor to create a stronger nation, the former thought Pearson was wrong to disregard the importance of nurture. Indeed, Reid went so far as to suggest that Pearson was getting "perilously close to charlatanism."[20]

Yet what seems ironic at first might actually not have been so, after all. Mark Anderson has recently shown how Boas's plasticity had very definite limits about who could and who could not be assimilated. In Anderson's reading, Boas, far from dislodging scientific racism, reinforced the equation between whiteness and America.[21] Kamala Visweswaran has likewise argued that Boas's much celebrated antiracism is premised on a partial and superficial view of all that he had to say on the matter. In fact, Boas was singularly influential in reallocating the study of race away from the social sciences and toward the domain of biology, thereby helping legitimate it as a matter of scientific study.[22]

My purpose here is not to evaluate the "true" legacy of Boas, or any of the other authors Seal drew on, for America or American anthropology, but to explore how these more ominous readings of Boas and others were already available to Seal in early twentieth-century India. Indeed, engaging with Seal shows how readings that have been overlooked for nearly a century in American intellectual circles were clearly visible to someone across the globe even during Boas's own lifetime.

Seal's views also remind us that race is not only about ensuring power to American or European whites. Indeed, whiteness itself becomes a far more contested and conflicted concept when viewed from India, or elsewhere in Asia.[23] Race and whiteness were both reworked and repurposed to serve the aspirations of anticolonial elites seeking to create racially homogenized nation-states.[24] In its entanglements with anticolonial nationalism, race became an idiom through which a colonized elite sought empowerment. But this empowerment was not necessarily envisioned as a general empowerment of all peoples. It crafted and nourished its own hierarchies.

Crucially, the anticolonial nationalist framework bequeathed to racial thought a strong sense of futurity. Since the national state that anticolonial nationalism aspired to was in the future, racial development, accordingly, had to be projected onto a future-oriented timeline. This is where it broke with colonial ethnology, which was mainly satisfied with working out the past of races, but relatively uncommitted to their future development.

One conundrum that Seal did not touch upon in his future-oriented racial project of nationalism, but one that was already beginning to come into view in Indian political life, was how a nation would deal with competing, rival claims of other overlapping nations. In Seal's teleology, the nation is eventually a "conscious social personality" that rationally sets its own ideal of a "national personality" and then achieves it, largely through the active marginalization of other, nonstate sodalities. But if two potential nations were to develop divergent idealized "national personalities" or to take form in two distinctly "conscious social personalities" that laid claim to the same preexisting social groups—as indeed happened in India, along religious and linguistic lines—how was the disagreement to be sorted out?

Likewise, Seal was not explicit about how exactly the relationship between the nation-state, which he privileged, and the prenational cultural groupings was to be managed. As we have seen, Seal clearly wanted the state to stand in direct relationship to the individual. He also said that the evolution of the "national race" out of the preexisting, prenational cultural races was a matter of conscious and deliberate choice rather than of purely undirected processes. This would imply the forcible suppression of these preexisting groupings by the instruments of the nation-state.

Biometric nationalism properly so-called, which provided much of the overall context for seroanthropology and emerged from the 1920s onward, sought to supply bases for the conscious choices that would lead to the eventual emergence of the "national race." In the early 1920s, for instance, Mahalanobis made his first foray into raciology by trying to determine whether mixed-race Anglo-Indians were closer to higher- or lower-caste Hindus. In the early 1930s, when Ambedkar's struggle for Dalit empowerment was becoming more politically challenging for mainstream nationalism, Mahalanobis sought to build a new camera that, together with the new statistical tools he was developing after the

Anglo-Indian study, would help work out more accurately the mutual relatedness of different castes and religions.[25] By the 1940s, as the colony entered the final years of colonial rule and the Partition between India and Pakistan gradually became politically more viable, Mahalanobis encouraged and enabled younger scholars like D. N. Majumdar to undertake a major survey of racial measurements to work out the degrees of relatedness between neighboring groups, such as Bengali Hindus and Bengali Muslims.[26]

None of this went unchallenged. Others, including B. S. Guha, differed on how to determine the mutual relatedness of groups and what conclusions to draw. Yet, they all agreed that there was a technocratic answer to these vexed questions of political representation and national belonging.[27]

Mahalanobis and Guha, through their personal patronage and the institutions they founded (ISI and ASI), provided the direct link between Seal's views and the postcolonial researchers we have met in the earlier chapters. While recalling this continuity, we must also acknowledge that not all elements of Seal's elaborate schema flourished in later decades. The threat of violence, for instance, was fairly obvious, even if tacit, in Seal's teleological schema for the stripping of the biography, in Cohen's sense, of the bionationalized political subject. This, mercifully, was not as explicitly expressed in mid-century biometric nationalisms. Similarly, Seal also clearly distinguished between "good" and "bad" social instincts and practices, and linked them to the evolutionary teleology of the national race. While some of these practices, such as consanguineous marriage, did indeed become favored objects of scorn for later seroanthropologists, such as L. D. Sanghvi, none of these later scientists advocated forcible suppression of such practices by the nation-state. This is where we must acknowledge the limits of Seal's influence. Most later scientists, at least during the period covered in this book, were much more circumspect about how much force they wished the state to wield in shaping racial evolution.

S. S. Sarkar's National Eugenics

Sasanka Sekhar Sarkar delivered his Presidential Address to the Anthropology Section of the Indian Science Congress in 1951 at Bangalore, exactly forty years after Seal's influential address in London. Much water had flowed down the Ganga since Seal had traveled to London. India and Pakistan had emerged in 1947 as two independent nation-states, and the study of human difference was already a well-established academic discipline in many Indian universities. Indeed, as we have seen in the foregoing chapters, Sarkar had been part of the story of those changes. He was the very first person to earn a doctoral degree in anthropology from an Indian university. In 1947 he led the Human Biology Section of the ASI. Later he joined Calcutta University's anthropology department as a charismatic faculty member who trained several physical anthropologists and geneticists of the next generation.[28] His address, therefore, provides a fitting point of comparison with Seal's.

Sarkar began by reminding his audience that "one of the chief aims of science lies in its usefulness in the service of mankind."[29] It is instructive to compare Sarkar's opening comment to that of Seal. Seal had begun by stating that "if modern civilization is distinguished from all other civilizations by its scientific basis, the problems that this civilization presents must be solved by the methods of Science."[30] For Seal, "science" and "modernity" were twin characteristics of the teleological progress of "Universal Humanity." Science laid out the only viable path of Progress. It was its own justification. By contrast Sarkar sought the justification for science in its avowed goal of serving mankind.

This comment was not fortuitous. His students, writing about him after his untimely death, recalled how he had repeatedly and categorically stated that human biologists must contribute toward human welfare. He asserted that the "final aim" of human biological studies must be to "make an individual healthy in mind and body and thus build a healthy nation."[31] Nirmal Kumar Bose, Sarkar's mentor and superior, wrote that "one of the most important aspects of Sarkar's work was that his problems did not arise, merely from what he had read in books and journals, but from his living contact with the Nation, i.e., the people of India, whose life has been overburdened by problems of poverty and inequality and of physical destitution for centuries past."[32] Clearly, unlike Seal, for Sarkar the social purpose and utility of science was most urgent and meaningful.

It was this purpose and utility that he underlined in his Presidential Address. We learn from his students that the address was particularly well received and that it "paved the way for a rethinking in Anthropology which was long overdue."[33] How he imagined this all-important purpose of human biological research is therefore of key importance not merely for understanding Sarkar's own goals, but also for understanding at least one influential current in human biological research in the postcolonial era.

Though he treated a wide variety of subjects and mentioned several specific areas of potential inquiry, in his address Sarkar named his general goal as the establishment of "national eugenics."[34] He asked: "Our population is happy with the fact that the Indian population is increasing. But what type of people are increasing?" Turning to Bengal, the region he himself hailed from, Sarkar reflected on the period in Bengali history that produced intellectuals like Seal: "Bengal gave rise to a galaxy of distinguished men towards the beginning and middle of the 19th century (ca. 1820–1870) whose contributions have enriched the country in so many ways." The nation, Sarkar felt, was still "banking on the achievements of these great men not only in the continuity of their germplasm but also in their master contributions to the entire make up of our culture." Unfortunately, however, "some of the lines of the great men have already ceased to exist," while the descendants of the others are "not capable of being as great as their ancestors, since new gene combinations are always taking place." Thus, "for the well-being of the nation we must always have a continuous chain of intelligentsia."[35]

None of these themes—national eugenics, the anxiety over the "differential fertility" of the different classes, the pinning of hopes to the "germplasm" of Bengali intellectuals of the past century—was new. Sarkar himself noted that he had developed some of these ideas a decade earlier in the bulletin of the Indian Eugenics Society (IES).[36] The text Sarkar referred to was in fact a lengthy introduction to the IES that was appended to its rules, regulations, and goals (fig. 7.1).

The goal of the IES was "to propagate the principles of human genetics and racial hygiene and their practical application for the betterment of the Indian population with a view to enhancing its surviving capacity in the struggle for existence." In furtherance of this eventual goal, it outlined three subsidiary objectives: (1) to "promote scientific research in the fields of Racial and Social Biology"; (2) to "spread the knowledge of racial biology and hygiene, as well as practical rules of conduct flowing therefrom among the people"; and (3) to work together with other scientific bodies "in the field of Eugenics," such as the International Human Heredity Committee, the Eugenics Society (London), the International Commission of Eugenics (United States of America), and so on.[37]

The IES itself had emerged organically. Around 1931, a group of research workers who all had their offices in the Indian Museum at Calcutta had begun meeting informally and discussing their mutual interest in "race, history, demography etc." in the intervals of their research work. B. S. Guha was one of the people in this group. The others were Sarkar; noted physicist and director of the Bose Institute, Debendra Mohan Bose; and the anthropologist, B. K. Chatterjee. Subsequently, Guha also brought in R. B. Seymour Sewell and the famous physicist, Sir Jagadish Chandra Bose (who also happened to be Debendra's uncle).[38]

Sarah Hodges has described a number of other eugenic societies

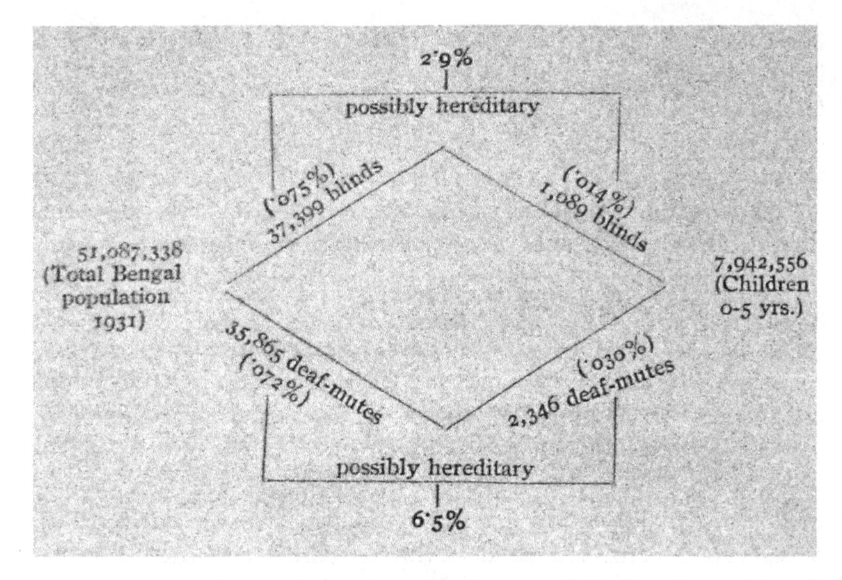

Fig. 7.1a. Diagram showing the number of deaf-mutes and blinds in the total Bengal population in 1931. Source: Sarkar, *Indian Eugenics Society*, 4.

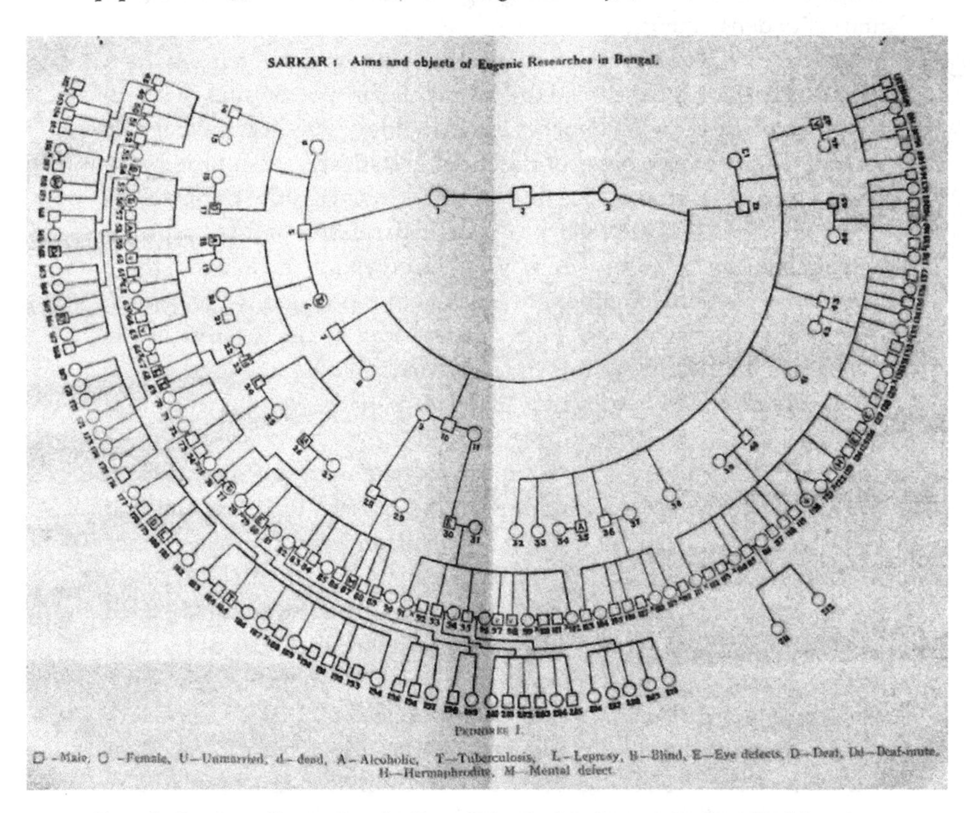

Fig. 7.1b. Family pedigree of an elite Bengali family showing various allegedly deleterious inheritable traits. Source: Sarkar, *Indian Eugenics Society*, n.p.

in interwar India, including those in Lahore (founded 1921), Sholapur (1929), and Bombay (1930). But she does not notice the group in Calcutta. This omission leads her to conclude that, "unlike eugenicists elsewhere, eugenicists in India were unconcerned with understanding the specific workings of heredity."[39] Far from neglecting the mechanisms of heredity, the Calcutta group, made up as it was by scientists, was overwhelmingly concerned with the emerging mechanisms of Mendelian genetics. In fact, Sarkar had commenced his introduction to the society by stating that "with the advancement of research in human heredity it has become possible to recognize a large number of morphological characters, malformations, anomalies and diseases that are hereditarily transmitted from the parent to the offspring."[40] This new nexus between heredity research and eugenics provided the general framework for the IES and distinguished it from the largely philosophical eugenics of Seal's era. Moreover, these ideas, through Sarkar and others, became deeply embedded in the research priorities of a number of postcolonial Indian research institutions, including the ASI, the ISI, and a range of university departments.

In both the 1941 introduction and the address delivered exactly a decade later, Sarkar developed the proposals for specific sites of scientific research that might connect human biology, heredity, and eugenic futures. These diverse areas of potential investigation can broadly be classed into three groups. The first group included a number of studies connected to marriage practices of particular castes and tribes, their potential biogenetic value, and ways of improving them. Second, there was a concern with a number of congenital, or potentially congenital, illnesses or "abnormalities." Finally, there was a keen interest in the "pedigrees" of great men. It is indeed remarkable how many of these concerns remained broadly unchanged over the decade between 1941 and 1951.

With regard to marriage, Sarkar took issue with the Age of Consent Committee (1928–29) and the Child Marriage Restraint Act that followed in 1929.[41] Sarkar called it "bad custom" and "bad law," since it was not based on sound scientific research. He lamented that with the resources put at the disposal of the committee, it would have been possible to undertake much serious research, but that little of this was done.[42] Sarkar argued that there was no sound scientific data upon which to conclude that early marriages were deleterious for the health of the woman or the children she bore. He approvingly cited Augustus Pitt Rivers that "the conviction of its harmfulness is in all probability a superstition arising from the same cause as the demand so passionately advocated in England by sexually dissatisfied women and sexually starved men that the female 'age of consent' should again be postponed beyond the age fixed by existing law."[43]

When Sarkar discussed the age of consent once more in 1951, he again cited the same passage from Pitt Rivers. Now, however, he backed up his comments with research data collected by his own PhD students, Miss Tulika Sen and Mrs. Parukutty Baruah. While Sen collected her data in Sarkar's own ancestral village in West Bengal, Baruah, who belonged to the Nair community in Kerala, collected data among that group.[44] Though Sarkar cited the research of pioneering animal geneticist Francis Albert Eley Crew to establish that menarchal age was determined genetically, he also said that its expression varied with reference to social and environmental factors. On the basis of the data collected by Baruah and Sen, he argued not only that the menarchal age of Bengali girls was going downward, but also that, while the three Bengali castes studied were fairly close in terms of their menarchal age, there was significant difference with the Nair girls.

Sarkar's object was not necessarily to challenge the need for the judicial regulation of an age for sex or marriage. Rather, he was opposed to the scientific ways in which the appropriate age had been determined. Ishita Pande's brilliant recent study has challenged historians to historicize "age" and "childhood" within the medico-judicial discourses. She outlines how chronological "age" and the "age of puberty," more generally, just as much as "gender," were produced by a series of historically specific techniques of observation and inscription, such as the autopsy.[45] What we have in Sarkar's interventions is a rival claim to reconstruct the "age of puberty" using a different set of scientific tools. Indeed, Sarkar reflected upon the difficulties of gathering reliable age data from the low-caste Bagdis, since "these people rarely remember[ed] their dates of birth." But he then outlined how anthropological observation and statistical analysis could correct for this "unreliability."[46]

Like his discussion of the age of marriage, Sarkar's discussion of heritable diseases also became more specific over time. In 1941 he had included a four-page list of various so-called abnormalities that were then considered heritable, along with a list of the basic mechanisms of genetic inheritance. Thus, for hemophilia, he had mentioned that it was "recessive, sex linked." For thrombopenia, he simply wrote "dominant," while for hereditary muscular atrophy, he stated "commonly recessive, dominant known."[47]

In 1951 he linked the relatively superficial descriptions of a decade ago with more concrete research programs. He mentioned that Dr. Jaharlal Ghosh had published one of the first detailed family trees of a hemophiliac family in Bengal. Sarkar had followed up and mapped the family in greater detail. He also stated with some frustration that by the time he took up the family for research, several of the daughters had reached marriageable age, and their parents were trying to find grooms for them. He wistfully lamented that "if only the fathers"—it is not clear if he meant the girl's father or the father of the potential groom, or both—had been aware of the "grave dangers of such unions," and wondered "how many of them have by now married and spread this grave disease further into other families."[48]

Sarkar advocated urgent inquiries into heritable diseases that would combine large-scale hospital statistics with family studies. Statistics and genetically informed family studies were the keys to stop genetic diseases from being passed on and thus to improve the health of the nation and the individual. He also felt that, in 1951, the time was "ripe for compulsory medical examination before all marriages" and the maintenance of a centralized database with hospital statistics about heritable diseases.[49] Furthermore, he advocated for a centralized body akin to the Copenhagen Institute of Human Genetics that would be tasked with creating a master list of all the "defectives" or "dysgenic elements" in the country.[50] The list of diseases he thought might be tackled thus was a long one.

Finally, there was the issue of great men and their germplasms. Here Sarkar's views had evolved in a new direction. In 1941 he had mentioned as an example of such work Egon Freiherr von Eickstedt's tabulation of a pedigree for the Bengali polymath and Nobel Laureate, Rabindranath Tagore (fig. 7.2).[51] In 1951, he connected this kind of research into pedigrees of talent to a more generalized work on caste and race. He pointed out that, out of the sixteen Bengali presidents of the Indian National Congress party between 1885 and 1950, nine had belonged to the Kayastha caste and that eight out of these nine were from the three elite Kulin subcastes (Ghosh, Bose, and Mitra).[52]

Dismissing talk about "mixed races" and the "non-committal term[,] ethnic group" as misguided, and echoing Seal, Sarkar asserted that though races were dynamic rather than stable, they did exist. Citing Cyril Darlington in support, Sarkar argued that races were "formed by a mating group differentiated by the 'unified selective response' at three levels—the environmental, the genetic, and the jointly genetic, environmental, cultural, levels."[53] Races arose through groups being reproductively isolated by either geographic or social barriers.[54] It was this dynamic view of races arising through reproductive isolation that made Sarkar connect his studies of castes, such as those on the Dakshin Rarhi Kayasthas, to his search for superior "germplasm."

Sarkar's Presidential Address is important in many ways. First, it

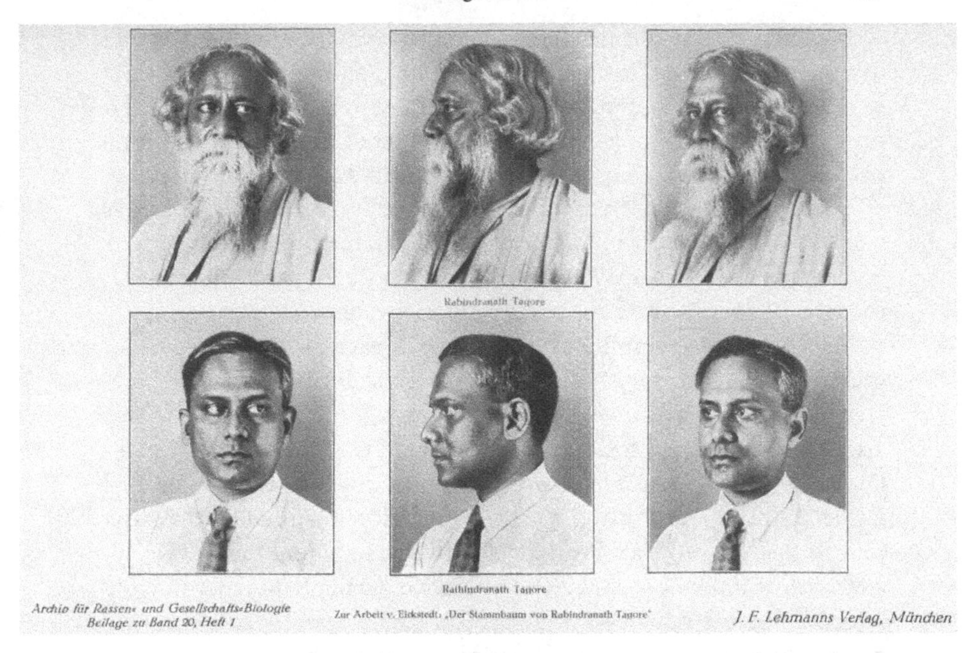

Fig. 7.2. The racial profiles of the poet Rabindranath Tagore and his son Rathindranath Tagore. Source: von Eickstedt, "Der Stammbaum von Rabindranath Tagore," n.p.

shows just how one of the most influential lines of racial thinking had evolved since the time of Seal. Second, it demonstrates how this thinking directly linked up to the history of eugenics and, contrary to existing histories of Indian eugenics, led to a concrete scientific research program. Third, it demonstrates how eugenics survived the end of World War II and continued to flourish in India. In fact, in discussing research into fingerprints, Sarkar lamented the closure of the Kaiser Wilhelm Institute in Berlin by the Allied authorities after the war. (In this regard, it was perhaps not coincidental that he had himself held a research position at the institute in 1938–39.)[55]

Finally, Sarkar's work demonstrates the inapplicability to India of the general chronological schema used for the history of eugenics in Anglo-America. Daniel Kevles argues that the older, classical or "mainline" eugenics was replaced through the 1930s with a new "reform eugenics." This latter differed from the former in many ways. It was more medicalized and less overtly prejudiced, for starters. But it was also focused on social reform as a necessary context within which genetics could operate. Its methods were also somewhat distinct in that it refused to accept the kind of hearsay about long-dead ancestors that Francis Galton and others used to rely on as valid data.[56]

Clearly, Sarkar's eugenic ideas defied this neat partition. In fact, one of the key figures of reform eugenics, Lionel Penrose, as Sarkar's PhD examiner, explicitly stated his disagreement with Sarkar's position. Explaining the grounds of such disagreement while commenting on one of the multiple research projects that constituted Sarkar's doctoral thesis, Penrose wrote that "in the choice of methods in the thesis . . . he has evidently been strongly influenced by the Director and other members of the staff of the Kaiser Wilhelm Institute, whose recommendations . . . though not acceptable to me personally have had wide currency."[57]

The comment exemplifies how Sarkar, in particular, and the wider group of Indian eugenicists drew rather eclectically on anglophone and German scholarship to constitute their own form of eugenics. They also clearly straddled the temporal divide between the older "mainline" eugenics and the later "reform" eugenics and selectively combined elements from both. Indeed, Sarkar's 1951 address ended by invoking Francis Galton and Karl Pearson, the two preeminent names associated with mainline eugenics. The operative principle for these highly selective choices remained premised on what researchers like Sarkar perceived to be necessary for a "national eugenics," though interestingly his conception of the "nation" slipped easily between India and Bengal, and an especially upper-caste Bengali nation. In that sense, his project is perhaps best described as a "*bhadralok* eugenics."[58]

Irawati Karve's Mongrel Nationalism

"I emphasize that we are all thorough mongrels who have been thrown together and must learn to lessen our prejudices and live together," thundered Irawati Karve a little over a decade after Sarkar had dismissed talk of "mixed races" as misguided and inaccurate.[59] The occasion for her speech was a summer school on Indian history organized by the University Grants Commission of India in Kodaikanal during the summer of 1963. The speech was published in a volume that, perhaps fittingly, was dedicated to Prime Minister Jawaharlal Nehru.

Karve was "India's first woman anthropologist" and a remarkable person in many ways. Though unfortunately and unfairly marginalized in some standard histories of Indian anthropology, she is justly celebrated, especially by feminist scholars, as a pioneer and a revered disciplinary ancestor.[60] Her views on race, however, have been a source of some embarrassment to those who look to her for inspiration. Nandini Sundar's insightful biographical essay on Karve has explained Karve's engagement with race in two ways. First, she has argued that Karve's interest in anthropometry, blood groups, and such was an "outdated" leftover from her early years. Second, Sundar has argued that while Karve certainly evinced clear sympathies with Hindu nationalism early in her career, by the mid-1960s her views had changed. In fact, Sundar points precisely to the speech quoted above as evidence of the change. Furthermore, she comments, correctly in my view, that "few secular people [in India] today would quarrel" with the sentiments expressed in Karve's 1963 paper.[61]

While I agree with both these contentions, I argue that such views, far from being a repudiation of the earlier, exclusionary forms of racial thinking, reinvigorated the link between race and Hindu/Indian nationalism. If indeed "few secular people" in India find her views from the 1960s objectionable, this demonstrates how insidious and pervasive certain forms of racial thinking remain even in seemingly politically progressive circles.

To understand the way Karve's statements about all Indians being mongrels could itself engender a racialized nationalism, it is instructive to look at the nexus between race and nationalism in Latin America. Latin Americanists have long described the ways in which claims to *mestizaje* can function as sources of racial pride and exclusion. Richard Graham points out that "the mestizo and the mulatto played an important part in the thinking of both racists and antiracists in Mexico, Brazil and Cuba."[62] The idea of *la raza cósmica* (the cosmic race), developed by philosopher-politician José Vasconcelos in the 1920s, placed the evolving cult of *mestizaje* in a new idiom. As Alan Knight points out in his study of race in Mexico, *mestizaje,* though frequently presented as a "racial fact," was effectively a social designation that was an "achieved—as well as ascribed status."[63] Like all social identities and statuses, therefore, it performed, and continues to perform, particular types of political and social work for particular groups. What *la raza cósmica* did was create a tighter weave linking *mestizaje* and nationalism.[64] This celebrated mixed identity always stood in complex relationships to other identities, such as that of the Indian." Radical *indigenistas* sometimes appropriated the logic of hybridity and at other times repudiated it in favor of an unmixed, "pure Indian" identity.

The situation was not unique to Mexico. In Brazil, for example, Gilberto Freyre's books, such as *Casa grande e senzala* (*The Masters and the Slaves*), also from the interwar period, provided a distinctively Brazilian glorification of *mestizaje*.[65] But mixedness, especially when twinned with the construction of national identities, frequently operated to privilege certain groups and their identities, and undermine others. Many Latin American cults of *mestizaje,* including those in Brazil and Cuba, became conspicuously entangled with aspirations for racial "whitening," that is, a process of mixture whereby the nonwhite could gradually be assimilated, submerged, and neutralized through racial mixture. Marilyn Grace Miller points out how Vasconcelos's *raza cósmica,* too, was deeply tinged with *blanqueamiento* and eugenic ideals.[66] Likewise, celebratory slogans of "racial democracy" in Brazil and Colombia sought to paper over the ways in which *mestizaje* functioned to exclude Black, indigenous, and working-class communities.[67]

Miller's perceptive comment that the problematic political underpinnings of Vasconcelos's *raza cósmica* were "typical of modernist nation-building efforts that tend to assimilate and thus cancel difference and dissent"[68] might with equal justice be applied to Karve's invocations of mongrelization. Indeed, Latin America, which shares many of the features of "caste" societies, offers many more insights for understanding the history of race in South Asia than the more neatly dichotomized racial milieus of Europe and the Anglo settler colonies of the United States, Canada, Australia, and so forth.

Not only did a cult of *mestizaje* operate in Latin America to resuscitate hierarchies of specific mestizo-national identities with the Indian, the Black, or the Asian at their national margins, but it also resonated with mid-century developments in genetics and human biology research. Vanderlei Sebastião de Souza and Ricardo Ventura Santos have pointed out how in Brazil in the 1940s and 1950s, as genetic research was promoted through the Rockefeller Foundation, new research priorities focusing on "racial mixture" also began to flourish. "Geneticists from different parts of the country, who had trained with *Drosophila* [fruit flies] in the techniques and theories of experimental evolutionary biology, began to reframe the history of the biological formation of the Brazilian population, with an emphasis on questions of racial mixing."[69] After World War II, foreign and Brazilian researchers built on this base and agreed that the country provided "exceptionally propitious conditions for studying racial crossing."[70] Just as in India, initially such studies were seroanthropological in nature. One of the Brazilian pioneers of such blood group–based studies of racial mixture was Fritz Ottensooser, a German-born Jewish physician who had come to Brazil to escape Nazism. Later, others like Pedro Henrique Saldanha carried Ottensooser's work forward.[71]

Two things become clear from this all too brief and highly selective perusal of Latin American historiography. First, mongrelization is not a simple opposite of hierarchized, exclusionary racial thinking. Emphases on "diversity" and "mixedness," together with diatribes against uniformity and invocations of political inclusivity, can all still sustain racialized forms of exclusion and subordination. Second, far from seroanthropological studies of race being antithetical to discourses of mongrelization, the two were in fact closely allied in postwar human biology. In fact, studies of "race mixture" directly allowed for an intensification of research into race.

If we revisit Karve's 1963 talk with these two insights in mind, we see her views in very different light. We notice, for instance, that notwithstanding the espousals of mongrelness, Karve had given to the mongrel race a very specific biological profile. "The predominant racial factor seems to [be] what is termed Europoid . . . with mixtures with proto-mongoloid, Australoid, negrito, mongoloid, and perhaps negrid [*sic*]."[72] She added two clarifications to this broad outline: first, that "Europoid" did not mean that the people had actually come from Europe; and second, that she "had not been able to find a truly negrid form except as regards the body proportions."[73]

Not only does this patently racialized understanding of mongrelness seem to fit well with what we have learned from the Latin American uses of *mestizaje*, but it also fits into influential global developments. Consider, for instance, the two UNESCO Statements on Race, published in the early 1950s, which are often held up as marking the end of scientific racism. Jenny Reardon has shown that neither these statements, nor indeed the developments since then, really undermined race as such. What they did do was create a new "population concept of race," which was still about race, but emphatically not about purity. Admixture now was seen to be constitutive of race.[74] This not only allowed quite a few aspects of the earlier race science to persist but also served to intensify research into race.[75] Sebastián Gil-Riaño has argued that this "amplified" race science worked in tandem with the Cold War "development" industry as race research became integrated into developmental planning.[76]

Interestingly, from the little we know of Karve's teaching in the later period ofher career, this very UNESCO Statement formed an important component in her courses. Former students recalled that her master's-level course on "Social Biology" was essentially divided into three parts: a section on the genetic and adaptive basis of various diseases, a section on the UNESCO Statement, and a section on population movements.[77] This was exactly in line with the cutting edge of the international mainstream of human biological research, where unchanging racial typologies were evolving into forms of racial thinking that emphasized racial mixture while also linking it up to medicalized and developmentalist frameworks for operationalizing human difference.

Looking closer at the racial makeup of Karve's mongrel race is also revealing. Particularly illuminating are the two specific points that she chose to further clarify, namely, the geographic origin of the "Europoid" and the unavailability of the "Negrid." The fact that she had been unable to find any "truly Negrid" individual, of course, reinforced the biological proximity of the Indian mongrel to the European and its distance from the African. In fact, she went further still and suggested that whatever Negrid elements were observable in "body proportions," as seen in ancient skeletal remains from northern Gujarat, might in reality have been "Nelotic [*sic*; Nilotic?]" rather than Negrid. She did acknowledge that all castes on the western coast of India, including Brahmins, occasionally have "thick everted lips and tightly curled hair," but said that this Negrid element was of "recent origin."[78]

Her clarification that "Europoid" did not mean people who came from Europe is even more interesting since its political resonance is not as obvious. Only a little later in the talk, Karve explained that "whatever prejudices were there, due to the caste system and 'race' they did not enter into consciousness." This was the case all the way through precolonial and "British times." It was only with the rise of "Hitler's Germany" that the word "Arya" was given a misleading racial meaning. This word then came to be opposed to the category "Dravida." Most problematically, for Karve, the emphasis on Aryanism had led to the "Dravidian nationalistic feelings" and resulted in the emergence of "something which can be termed 'Adidravidianism.'"[79]

The "Adi-Dravida" identity was not simply a misplaced and inchoate expression of "Dravidian nationalistic feelings" that "could be termed" Adi-Dravidianism. The origins of Adi-Dravidianism go all the way back to the 1890s as part of the political assertion of lower-caste groups in the Madras Presidency. By 1918 many of these early political groups were explicitly seeking bureaucratic recognition of the term "Adi-Dravida." The prefix "Adi-" meant "initial" or "primordial" and was positioned in clear opposition to the Brahmin domination and the Sanskrit-centered culture that legitimized that domination. Around the mid-1920s the philosopher, educator, and Dalit political leader, E. V. Ramasamy "Periyar" gave a much sharper, theoretical basis to the "Adi-Dravida" identity by relating it to "self-respect" and equality. By 1944 this had crystallized into a powerful political movement headed by the Dravidar Kazhagam, a party that sought to limit the domination of both Brahmins and the northern Indian cultures, especially through the imposition of the Hindi language. Periyar, like Dr. B. R. Ambedkar, also sought to align the Dalit identity with Buddhism rather than Hinduism. Christophe Jaffrelot calls the Adi-Dravidian movement the "ethnicization of caste."[80]

Whatever label we might give to the way the Adi-Dravidian movement framed and mobilized caste, two things are clear. First, Karve sought to trivialize it by suggesting that it was a mere inchoate movement to which she was applying the term "Adidravidianism," when in fact it had already existed for nearly half a century and mobilized extensively around that term. Second, her "mongrel" thesis clearly sought to undermine the tenability of such an Adi-Dravida identity that was grounded in claims about autochthony. Indeed, the thrust of Karve's argument against the Adi-Dravida is very similar to the ways in which dominant groups in Latin America often deploy arguments about *mestizaje* to deny the claims of marginal groups that mobilize around claims of indigeneity, Blackness, and so on.

That Karve's "mongrel" nationalism was opposed to the Adi-Dravida movement becomes clearer still when we see her return to the subject toward the end of her talk. She stated that "the events of the last ten years show that the peaceful and fruitful co-existence of two communities can be and is often marred by fanaticism of a small minority among both. The Dravida Kadagam [*sic*] demanded a separate Dravidistan on the grounds of separateness of language and culture. The Muslims fearing to remain a permanent minority in India demanded and got a Pakistan."[81] By framing it thus, Karve not only linked the Adi-Dravida movement and its standard-bearer, the Dravida Kazhagam, to the formation of Pakistan, but also depicted both movements as the outcome of the "fanaticism" of "small minorities."

Clearly, as Miller said of Vasconcelos's *raza cósmica*, Karve's "mongrel" nationalism sought to "assimilate" by canceling "difference" and "dissent." Not only was the mongrel nation, biologically speaking, a Europoid mixture with small doses of Australoid and Mongoloid elements, and very little, late, and highly localized Negrid genes, but it was also a mixture within which there was no place for any difference or dissent that could be politically mobilized. It was essentially a mongrelism that smothered all difference within a national majority.

It is true that Karve also explicitly criticized Hindu extremists and stated that "the society we want to build is not a Hindu society or a Hindi society but a multi-cultural, multi-lingual society which has knowledge of one another's culture and has decided to live and work together in comradeship."[82] Yet, this version of multiculturalism was premised on an idiom of mongrelness that could be used flexibly to defuse any politically assertive articulations of difference and dissent.

This is also where Karve's thought seems to be much closer to the way she had framed her explicit Hindu nationalism of earlier decades. In 1947, the year India became independent, for example, Karve had published a lengthy essay on racial conflict, intended as a consciously future-oriented piece that sought to learn from the past and provide a blueprint for the future. Karve wrote that "the cultural unity of India is thus deeper, because much of what makes the culture has become folk-ways deeply ingrained into the lives of all strata of society. Hinduism is not a religion as understood by the Westerners. It has no church, it has almost no dogma, it is a creation of historical growth and culture-contact of centuries. It comprises descriptions of gods, devotional literature, philosophical treatises. From the worship of a thousand spirits it goes to a pantheistic ideal, where the individual identifies himself with the ultimate reality."[83] Clearly, this emphasis on deep cultural unity grounded in Hindu pantheism prefigures the way she imagined a multicultural society two decades later.

Moreover, in 1947 she had clearly stated that cultural unity did not stand independent of biological mongrelization. She referred to "stories of Kshatriyas and Brahmins marrying Nishada wives, and powerful kings seeking alliance of the primitive chieftains. So in the conflict which ensued, contact was limited and certainly not destructive . . . Hinduism in its contact with primitive people gave much and received and made its own many elements of primitive thought."[84] Interestingly, this claim that racial difference and prejudice was entirely a consequence of European colonialism and that before that time racial mixing had happened relatively freely is one that we also find in Latin America. Juan Comas, a Spanish Mexican anthropologist, published a booklet titled *Racial Myths* (1951) through UNESCO, where he argued, in a framework very similar to Karve's, that "'racial prejudice' did not exist before the Iberian conquest of the Americas in the fifteenth century."[85]

The only significant difference between the 1947 and the 1963 texts is that in the former, Karve explicitly attacked both Christianity and Islam for their alleged monotheistic lack of regard for diversity, while in the latter text she resorted to a different strategy. In 1963, instead of attacking the monotheism of Christianity and Islam, she pointed out that they too in India were mongrelized. Among Muslims, there were those who claimed Arab descent, others emphasized their Rajput heritage, and so on. As a result, she alleged that many who had opted to go to Pakistan in 1947 went simply because they thought they would get jobs and later returned to India upon discovering that as dark-complexioned Muslims they were treated as inferiors.[86] The implication, rather appallingly, seems to be that there is no color prejudice in India! Among the Christians too, she pointed out, the caste identity of one's "pre-Christian Hindu ancestors" continued to determine endogamous groupings. Moreover, "among these people," she wrote, "colour of the skin plays quite a role in social hierarchy."[87] None of this was meant to disparage these communities. She pointed out these characteristics not to single out Indian Muslims or Indian Christians, but to completely neutralize their distinctiveness, to render them almost utterly indistinguishable from the larger Hindu majority within which they lived. In place of the vitriolic attacks on the intolerance of monotheistic faiths, she now denied any distinctive identity to these religious communities.

Thiago Pinto Barbosa has shown that Karve, particularly in her later writings, increasingly deployed the adjectival form "racial," but mostly eschewed the noun "race." This, Barbosa perceptively points out, allowed her to constantly "enact and prove" the existence of different races, without actually calling them such.[88] Such linguistic circumlocution operated to bolster the insidiously racialized espousals of admixture.

The performative iterations of the "racial" also allowed Karve to develop her own racialized model of multiculturalism: a model imagined along the lines of a pantheistic, caste society. In such a model difference was neutralized and reduced to discrete endogamous units, akin to caste. Cultural and biological exchange between these units were neither mandated nor strictly forbidden. As she had pointed out about caste in 1947, "Hindu society is a growth of a very long historical process. It has never been a completely homogeneous society. It has allowed racial and cultural elements to live side by side until they fuse in the ripeness of time." This process had only stopped "fifteen hundred years ago."[89] Mongrelization and multiculturalism, for Karve, were part of an assimilative project that produced a racialized national body with little room for radical, politicized difference of the sort seen in the Adi-Dravida or Pakistan movements.

It is this version of multiculturalism—which presents itself as appreciative of diversity and yet quintessentially Hindu; as mixed and yet essentially "Europoid" and not "Negrid"; as tolerant and yet unwilling to concede any ground to political dissent or difference—that is so difficult to repudiate even by the "few secular people" whom Sundar invokes.

L. D. Sanghvi's Professional Opportunities

The "time is ripe for having a new look at the problem of origin in India," declared L. D. Sanghvi in 1969. The "existing racial classifications of the people of India are more like the theoretical frameworks of the fourfold castes," he mused, "and we know the actual situation is very different." This "new look," Sanghvi insisted, could no longer be based on the "old type" of approaches examining cultural and skeletal records.

Rather, they had to be grounded in new types of "analytical studies of the existing populations of India."[90]

Sanghvi's essay, "Perspectives for the Study of Racial Origins in India," was explicitly aimed at promoting new types of PhD research. He stated baldly that in the past researchers in human biology had chosen to do their PhD work on "some endogamous group, chosen for convenience of fieldwork," and remained satisfied with simply mapping their morphological and genetic constitutions. This approach, however, was no longer useful. More focused studies of racial origins were now necessary, and this was what aspiring PhD students ought to pursue.[91] For Sanghvi, therefore, the future of race research was clearly about the kind of PhD projects that were viable and desirable.

We should not be surprised to find that Sanghvi also had other stakes in such research. As we saw in chapter 4, for instance, Sanghvi was deeply committed to a eugenic approach to improve national health by tackling the genetic burden of disease. The 1969 essay, however, demonstrates that notwithstanding such grander commitments, as a professional scientist he mainly thought of race research in terms of the relatively mundane teleology of academic careers.

For Seal, race research was largely a philosophical question tied to the determinations of national and universal futures. For Sarkar, it was part of an incipient eugenics movement that was furthered, at least initially, at the interstices of scientific careers through snatched conversations between co-workers at the Indian Museum. For Karve, the academic pursuit of race still remained entwined with the ideological attractions of Hindu nationalism. As Sundar points out, the disciplinary professionalization of such research in Karve's time remained weak.[92] In contrast to these three researchers, for Sanghvi, race was largely a matter of scientific research, and its future was intimately tied to the professional needs of scientists who pursued such research.

In an essay published three years earlier, in 1966, Sanghvi had argued similarly that India was "unique" in the value that it offered for research into human biological variations. He pointed out that it possessed "almost all racial groups . . . under a social situation that [wa]s favorable for genetic studies." Moreover, alluding to castes, he argued that "some of these groups have undergone a high degree of environmental and occupational specialization over tens and hundreds of generations," and had also "practiced inbreeding at a very high level over two or more millennia." All of this, in turn, raised "interesting theoretical questions . . . and practical problems."[93]

Sanghvi was extremely well placed within the international networks of genetic research. As we have seen earlier, he worked with the foremost geneticists of the time, such as L. C. Dunn, Theodosius Dobzhansky, L. C. Strong, and Lionel Penrose, and at some of the top laboratories in the United States and Britain, before holding important positions at the WHO. Perhaps most significantly, Sanghvi served as the scientific secretary to the high-powered United Nations Scientific Committee on the Effects of Atomic Radiation (UNSCEAR), between 1960 and 1961.[94] As we saw in chapter 1, UNSCEAR was particularly important in reformatting human genetics research for the postwar world. All this had allowed Sanghvi excellent opportunities for promoting India as a key site for genetic research, primarily on account of its caste system.

What we notice in both his 1966 essay "Genetic Adaptation in Man" and the 1969 essay "Perspectives for the Study of Racial Origins in India" is a consistent pattern that follows on that earlier work. For Sanghvi, as a career scientist ensconced within the most elite professional networks of his discipline, race was preeminently a disciplinary object. He thought of its future in terms of the kinds of disciplinary questions it could help answer and the sorts of PhD projects it might sustain. This was no doubt helped by the fact that human genetics research, especially through its association with the United Nations and research into atomic radiation, enjoyed a privileged position within the Cold War research environment.

Yet, it is remarkable that, notwithstanding his greater academic prestige and perhaps a stronger sense of professionalization, Sanghvi relied on and referred to the work of Karve in particular. In his 1969 essay he stated, for instance, that the new line of more "analytic studies" of human biology that he was promoting had strong parallels with the "stimulating hypothesis" postulated by Karve that "a caste is a social fusion or a collection of genetically different population groups" and not the "biological fission" or subdivision of homogeneous ancestral groups.[95]

In fact, neither the invocation of Karve's research nor the similarity between her and Sanghvi's framings of racial mixture was coincidental. Sanghvi and Karve had been collaborating with each other in one way or another since 1952. Recalling their collaboration after Karve's death, Sanghvi mentioned that they had developed "common intellectual interests in the study of human populations in India since 1952."[96] They carried out some minor collaborative research on populations in Maharashtra in the mid-1950s. In 1958, however, they began to imagine a more elaborate collaboration. During an informal chat that year, they came to the conclusion that the "time was ripe for a major, multi-disciplinary project on a single cultural linguistic region." Initially, Karve suggested Karnataka, but Sanghvi insisted on Tamil Nadu instead. In keeping with his more professional focus, he argued that the "problems were more stimulating" in Tamil Nadu.[97] This was most likely linked to Sanghvi's longstanding interest in consanguineous marriages, which were said to be more common within South Indian kinship systems, and their overall genetic consequences.[98]

The existence of this project also casts a more putatively political light on Karve's disparaging of the Adi-Dravida movement, which was largely headquartered in Tamil Nadu. According to Sanghvi, he had noted in his efforts to convince Karve about the project that researching human biological difference in Tamil Nadu had the added advantage of being likely to provide "clues to the development of Dravidian culture in South India" and also of illuminating the "controversial role of Harappan culture in influencing the developments in the South." Little wonder, then, that she agreed.[99]

Accordingly, over the next seven years Karve and Sanghvi jointly undertook a large-scale project to study human variation in Tamil Nadu. They identified fifteen different groups and carried out extensive fieldwork on them. Deccan College, Pune, where Karve was based, had the primary responsibility for the project, but the genetical and statistical analyses were carried out by Sanghvi's team at the Human Variation Unit in the Cancer Research Institute in Bombay. The project formally commenced in 1959, and the bulk of the work was concluded in 1969. In August 1970, while the final report was under preparation, Karve died. The project stalled after that, but was later revived and jointly published in 1981 by the ASI and the Indian Society of Human Genetics, Pune.[100] When Sanghvi wrote his 1969 essay outlining a new line of inquiry into the racial origins of Indian populations, therefore, he had just concluded the Tamil study. Indeed, he had alluded in that essay itself to an unnamed but recently concluded inquiry conducted jointly with Deccan College.[101]

The history of this collaboration is revealing. First, it reinforces the point I have made earlier in this chapter that Karve was not pursuing outdated or old-fashioned research programs. It only appears as such if we ignore the history of human biological research more broadly and reduce anthropology's polymorphous disciplinary existence narrowly and exclusively to social or cultural anthropology. Her research into human variations was actually entirely in keeping with the emerging international consensus around questions of race. Second, it demonstrates how race research transcended narrow disciplinary boundaries and was pursued by researchers in different kinds of institutions and with distinctive disciplinary backgrounds. Third, it shows once again that acknowledging the "dynamic nature" of race or the "mongrel" nature of the national body did not signal the end of the scientific idea of race.

Rather, it presaged the reinvention and intensification of race through new forms of genetic research. Finally, the collaboration attests most conspicuously to the ways in which an older, ideologically redolent investment in questions about the racial origins of the Indian nation gradually but seamlessly morphed into a newer professional imaginary that focused on "interesting theoretical questions and practical problems" and forestaged potential research careers for aspiring PhD students.

The future of the nation and the nation-state now became even more tangible, as its scientific pursuit promised satisfying professional futures for aspiring young researchers. Ultimately, part of the success of race science in India lay in its ability to provide stable careers to a small but growing group of men and women trained in biology, genetics, and statistics.

INTERCHAPTER

Letter 8

Hemendrakumar Ray
21, Pathuriaghata Bylane
Calcutta, India
4 July 1934

My dear Najrul,
You must forgive me the unconscionable delay in writing back. I had traveled out of Calcutta for a couple of months. I could not take the city and its bustle anymore and needed some quiet to gather my thoughts. Everyone is always speaking and no one is listening. Not to others, not to themselves. Even the cinema, where I used to go sometimes merely because I enjoyed the quiet and darkness inside the halls, has now begun to speak.

In your case, too, it is the quiet of the jungle around you, I feel, Najrul, that gives you the clarity and depth that I so enjoy in your letters. The unease about the blind pursuit of an abstract idea of a perfect future, to the detriment of the fullness of the present that makes you uncomfortable, troubles me no less. And yet, so few of those around us seem to worry about it. With our national awakening has come a welcome spirit of hope and enthusiasm about the future, a willingness to leave the past behind and forge ahead. But at what cost?

Let me confide in you a small secret. Years ago, when I was still an aspiring writer, I had embarked upon a small personal experiment. To write, I thought, I must first listen. Not just to those around me but, perhaps more important, to those who I thought were least like me. I began to go out late at night and roam the city. I made it a point to go to those areas and mix with those people who were furthest from my respectable, bhadralok *world. Gradually, this became a habit. Over the years, I got to know* goondas, *murderers, prostitutes, drunkards, you name it. These were people whom we consider vicious, corrupt, and almost inhuman. Indeed, there were at least two occasions when*

I myself came perilously close to being implicated in murders I had witnessed. But as I got to know these fiends in the flesh, befriended them, talked to them, and listened to them, I realized one thing: God has not created anyone out of pure evil. Even the most vicious and heartless murderers always have sides that are deeply humane and emotional. I still remember one man, a notorious upcountry fellow, a gun for hire who had killed many, who always made it a point to stop on the street to help any old or disabled beggar he saw with money and kind words, even as the more respectable denizens of the city pushed past.

I had written up these experiences and published them as a small book under a pseudonym in the hope that it might make us less sure of our humanity and more unsure of the inhumanity of those unlike us. Alas, Najrul, it remains a matter of deep regret for me that most people read that book as mere salacious gossip, a collection of lurid scandals. You tell me to write a cautionary tale. But my experience makes me worry.

Mr. Sen's diary is so singular, so remarkable, that it would not be difficult to render it, with minor emendations, into a book for public dissemination. But will it remain a cautionary tale? Or will it be read as mere cheap entertainment?

It is not that I do not want to entertain through my writing. In fact, I have always disliked the overly self-conscious writings of the self-appointed avant-garde precisely because they forget the need to entertain. I have consciously avoided the so-called "realism" of some of our modern authors precisely because they dig a moat between the author's duty to describe the world and his job to entertain. But entertainment for its own sake and for the sake of enriching the almighty publisher cannot be the sole ends for which I write. My hope has always been that description and entertainment can be aligned in the pursuit of a higher moral purpose. A purpose that is deeply enchanting, yet lightly worn.

Moral suasion, I am convinced, my dear brother, can only be effective when it is devoid of pedantry and ponderous affectations. It must work by touching the reader's soul, by kindling in it a sense of beauty and fulfillment through humility and wonder. That, for me, is the true power of literature. Without it, I fear, literature too, like science, will descend all too easily to being a knife all blade that can only make the hand bleed.

Can I rewrite Mr. Sen's diary in such a way? Perhaps. Perhaps not. Perhaps, like my Calcutta by Night, *it too shall be read simply as a cheap entertainment and fail to achieve its moral purpose. But I am certainly willing to give it a try. Both because you, my brother, urge me to do so, and also because I agree with the reasons you lay out. If I end up writing it, I think I will call it* Amanushik Manush, The Inhuman Humans. *Don't you think that'll be a fitting title for the future-humans?*

Whether I succeed in telling the cautionary tale well or not, however, I want to reassure you that humanity will not necessarily be obliterated so easily by these dreams of the perfect future. There are deeper forces than what my feeble pen might be able to muster up that nourish our present and connect us to the past. The hatcheries that worry you so are not a new idea. Human hubris has long dreamt of such things. But here we are, approaching the middle of the twentieth century, and there are at least some of us—you and I, for instance—who are still revolted by it.

As a reassuring testament to both the longevity of such ideas and the opposition to them, let me end my letter with a few lines from that recondite balladeer of our city from almost a century ago, Rupchand Pakshi,

> *Jute mills, wheat mills, oil mills, cloth mills and brick-grinding mills,*
> *Water pumps, stone-crushing machines, machines can make roads like Aiyravat in a single day*
> *I bow to the feet of machines, they have connected city and village*
> . . .
> *Machines are doing everything, soon they will be making babies too*
> *No one will be childless in the world, the machine-made child will inherit their wealth* after they die.

Like Pakshi, we worry, Najrul. But I hope like us, too, people will continue to worry a century after we are gone and with that worrying, we will keep the hatcheries forever in the future, but never in our present.

Affectionately,
Hemenda

Conclusion

Having arrived at the end of the book, I feel I should clarify the basis of my critique of seroanthropology. Let me begin by reiterating up front that I am not a discursive solipsist who would argue that all differences lie in the eye of the beholder or are "merely" discursive constructs. Neither do I believe that seroanthropology or the population genetics that has descended from it are simple stereotyping devices. Differences between two human beings are real, and scientific research is not a mere exercise in discursive stereotyping. I well know, for instance, that I am quite a bit taller, quite a bit darker, and have different-colored eyes than my grandmother. These are obvious, measurable differences, and no matter what epistemic presumptions you carry, you will notice these differences right away. Shown photographs of the two of us, and without being told of our relationship, many people might not even think we are in any way related.

Yet, we are related. That relationship is also, obviously, much closer than whatever relationship I might have with some random stranger who might share my eye color, skin tone, or height. The point I want to make, therefore, is this: seroanthropology was not about individual differences. It was about using differences to sort people into groups.

Depending on what differences one chooses, one can sort people into different types of groups. Making up these groups means privileging certain shared features and underplaying other features that are shared with a different set of individuals. More important, nobody perceives an entire group, leave alone its differences, as a whole, directly, in the way one notices differences between two individuals. The production of a group, any group, and its distinction from other groups is, therefore, always a contingent matter, dependent upon assumptions, protocols of observation, and methods used.

All this is obvious, and I doubt any serious scientists would object to it. They would insist, however, that their groupings are simply biological groups and that their sorting into groups has nothing to do with social perceptions. This is where I will differ. I do not think there was any technique in seroanthropology's toolkit to get at any kind of "pure" biological relatedness without going through the socialized and political narratives of relatedness. As I have shown in the chapters of this book, all kinds of assumptions about endogamy, group identity, community histories, and so on, as well as concrete political relationships that mediated contact between the scientist and their subject, shaped the kind of biological samples that were collected and analyzed—not to mention what was made of those samples. Yet, unfortunately, the nonbiological data that fed into seroanthropological research formed the least rigorous part of most of the research projects I have studied.

Even without raising the fundamental question of whether it is even possible to produce scientific truths about human relatedness that entirely eschew the contingencies of hermeneutics, group interests, and historical situatedness, we might at least expect that scientists using nonbiological data would be as rigorous in evaluating this form of data as they are with their biological samples. Yet, at least in India, this has not been the case. The analysis of sociological and historical data that informed and underwrote so many of the seroanthropological researches was frequently dated, misleading, or even simply erroneous. Even now, the kinds of assumptions that underwrite hugely influential and resource-rich projects, such as the Indian Genome Variation Database, are entirely out of sync with social-scientific or humanistic research into caste, conjugation, and communalism.

To return to my late grandmother, what connected me to her above everything else were the stories she used to tell me as a child. What constitutes family groups, communities, and even larger national groups, for most people, are the stories they share, especially the stories about their shared pasts. Seroanthropologists wanted to explore human relatedness and yet for the most part ignored the rich world of stories. When they did take them seriously, they did so with cavalier disregard for narrative complexity, performance, and context, treating them instead like a mine to be plundered for bits of decontextualized information that could be selectively plugged into their own "snapshot biohistories."

More troubling still was that scientists claimed and, at least in the eyes of the state that funded them, had a lot more resources and authority to determine community identities, pasts, and relatedness than nonscientists such as storytellers who also concerned themselves with human relatedness. Whether their seroanthropological stories actually translated into government policies or not is doubtful, but they certainly received the funds and the benediction of the state to tell their snapshotted biohistories. The stories of the uncanny and the inhuman that seemed to push against and beyond the increasingly biologized figure of the human enjoyed neither the resources nor the authority that the late colonial and postcolonial states offered seroanthropologists. They lived on in cheap pulp fictions such as those that my grandmother would often read to me.

It is this divergence and hubris that brings me back to Frantz Fanon. Scientists told their own stories but did not pay much attention to other stories being told around them. Yet, crucially, it was the scientists' stories that received the blessings of the state. Their seroanthropological stories were therefore a tool by which to create increasingly abstract maps of social groupings without ever fully participating in the intricate weave of narratives and lifeworlds that made up the social groups they studied. In short, seroanthropology was a technology of alienation, the basic figure through which Fanon mapped the operations of race.

Manyfoldedness of Alienation

In this book you have already encountered two distinct narratives. One of these has developed through the foregoing chapters. Another has flowed through the interchapters. The first have narrated the emergence, evolution, and efflorescence of a cluster of scientific investments in race that I have gathered under the loose designation of "seroanthropology." The second have built up a narrative that draws sustenance from Hemendrakumar Ray's novel *Amanushik Manush* [Inhuman Humans] along with a host of his short stories. Ray's literary output was prolific, and he frequently explored the myriad contemporary intersections of race and science. Most of the seroanthropologists and Ray lived in the same city, namely, Calcutta, and spoke the same language, namely, Bengali. It is more than likely that some of their paths might occasionally have crossed. And, yet, these two narratives seem to flow along entirely distinct channels. The ghosts, curses, dystopian futures, and critiques of race that surface in Ray's writings never trouble the self-assured scientific pronouncements of the seroanthropologists. Is this not alienation?

In my choice to mark the chapters neatly away from the interchapters, I have tried to mirror the alienation produced by race science in the structure of my own critique. Yet, by placing them side by side, I have also tried to show that, in spite of the alienation, these alternative forms of engagement with human difference existed cheek by jowl in the same social world.

Ray's popularity, if nothing else, suggests that there was a much larger cultural appetite for these themes in the very city and among the very classes from which the seroanthropologists emerged. The terse, objective language of scientific publications, however, never permitted itself to acknowledge such unruly ideas. I am not suggesting that scientific publications should have cited literature in the larger sense. All that I am arguing is that Ray's themes were not idiosyncratic. They tapped into more widely dispersed discussions about race, which the seroanthropologists were bound to have known about, but never had any thought of engaging with: critiques of race, primitivism, or the totalizing authority of science.

Whether one finds this lack of engagement surprising or entirely expected is another matter, but it certainly indicates a certain alienation of the seroanthropologists, or at least their scientific production, from other, vibrant parts of their sociocultural milieu.

Banu Subramaniam has recently reminded us that "'genres' are themselves epistemological choices of writing."[1] No aspect of life, including race, comes prelabeled as "literature" or "science." Writing about it in one genre rather than another entails particular epistemic choices. One aspect of this choice in mid-twentieth-century India was certainly about the degree to which the discussion was related to older, deeper stories, themes, and concerns available in the author's own milieu. S. S. Sarkar, one of the central figures in our history of seronathropology, for instance, waited until the last year of his life to publish a small book, written in Bengali, compiling the folktales he had collected over the years. Two aspects of the book illuminate the point I am trying to make. First, in introducing the book, Sarkar mentioned how he had originally wanted to research these stories, but then had had to put them aside because professional duties propelled him in another direction.[2] Second, he lamented how children no longer grew up hearing such tales, before recounting at length a simple tale that his grandmother used to tell him before he learned to read.[3] This slippage from fieldwork to his own childhood memories was exactly what his professional duties had disallowed. This is the alienation Fanon alerts us to: the alienation that prevents a scientist like Sarkar from accessing his memories of childhood, his image of his grandmother, and the stories he so desired to hear.

In his better known scientific publications, dominated by seronathropological numbers, there was little room for such stories or the shared mnemonic structures they evoked. In fact, even late in life when Sarkar finally published the little book, he explicitly pitched it as a book for children and marked it off clearly from his work as a researcher. This separation of the "fictional" and the "factual" was thus, in effect, also an epistemic choice premised on alienation. It was a choice made more significant by the fact that only a generation before Sarkar's these generic choices had been less rigidly demarcated. As Charu Singh has recently shown, a "sensibility for science" in South Asia was originally produced by creatively drawing upon literary genres, such as the detective mystery and fictional dialogues.[4] With the professionalization of science and scientific writing, however, such generic choices clearly narrowed and narratives diverged.

The alienation and the epistemic choice grounded in it also foreclosed certain alternative imaginaries. Again, as Subramaniam eloquently puts it, the "fictional and the fantastical . . . unleash narrative resources for other imaginaries and enable thought experiments."[5] As I have tried to show in the interchapters, the imaginaries embedded in Ray's fiction articulated a distinctive moral universe and generated, at least hypothetically, alternative futures that reflected critically on some of the scientific pursuits of the present.

Alienation through race science, however, was far from being a simple, monolithic experience of social estrangement that affected the researcher alone. Indeed, we have witnessed repeatedly in the foregoing chapters the "manyfolded" character of such alienation. We have seen, for instance, in Sanghvi in particular, an utter alienation from the lived reality of caste-as-*jati*. His repeated invocations of perfect caste endogamy as a social fact, we have noticed, had already been challenged decades earlier by Dr. Ambedkar. His own experience as a South Asian man in the mid-twentieth century would likely have contradicted such cardboard stereotypes about caste. Yet, he kept insisting upon it. We have seen likewise how Irawati Karve, despite noting the ongoing prevalence of levirate, for instance, continued to work with Sanghvi and reinforce ideas about caste endogamy. These modes of alienation show that the researchers were alienated not only from the narrative universes they had grown up with, as Sarkar was, but also from sources of knowledge and insight that would have made them rethink the categorical logic of race.

If belief in perfect caste endogamy was one of the foundational pillars of seroanthropology, numeralization and statistics were another.[6] While the level and complexity of the statistical tools used varied widely—from the famous statistician C. R. Rao's work on some of D. N. Majumdar's seroanthropological data, on the one hand, to rudimentary chi square calculations by ASI researchers, on the other—numerical tables were an unmissable part of every seroanthropological publication. Numbers are the consummate "technologies of distance."[7] All the complexity of in-field interaction between the researcher and the researched vanishes behind a wall of faceless numbers. While some of the earlier publications from the 1930s and early 1940s, such as Macfarlane's study of the Cochin Jews, immersed the tables within a larger narrative that was rich in names and lore, later publications were barely a few pages in length and focused almost exclusively on the numbers.

Indeed, the very constitution of seroanthropological research objects was alienating. We have noticed how, in seeking to racialize the bitter taste of phenylthiocarbamide, researchers progressively reduced a polymorphous, multisensory affective universe to the simple absorption process linked to a single, impoverished genetic marker. After first trying to localize and pin down a discrete quantifiable taste, the researchers eventually rendered even this minimum gustatory response redundant by invoking ageusia and other tasting "disorders." Whereas in the former instances, alienation had taken the form of alienation from groups of people or bodies of opinion, in this instance alienation comes from an estrangement of the sensible human body from the sensuous affective universe within which it is immersed.

A more tragic form of alienation is to be glimpsed in the racialization of sickle cell disease. The clinical experience of the disease is almost entirely absent from these studies. Despite the intersection with medicine and the involvement of medical doctors, such as Dr. J. B. Chatterjea, the studies are almost entirely bereft of clinical information. The patient and their suffering has been utterly eclipsed by a pursuit of racialized "risk."

This pursuit of risk also resonates with the temporal alienations we witnessed in chapters 2 and 7. While seroanthropologists were keenly cognizant of snapshot biohistories of the communities they studied, and variously invested in distinctive futurities, they seldom dwelled on the present by itself. The communities they studied were often among the most marginal in postcolonial India, and frequently subjected to multiple forms of privation. Indeed, sometimes the seroanthropologists opportunistically benefited from these exploitative structures, but never did they directly comment on these circumstances or actively engage with these inequities.

Social alienation emerges most conspicuously through the friction we witness in the seropraxis of the researchers, as well as in their almost complete ignorance of the grounds on which people refused their research. The physical alienation of drops of blood, not only of others but also regularly of themselves and their own team members, as well as the complete nonengagement with the intellectual, cosmological, and practical reasons why people declined to participate in the research, evinces an indifference that testifies to the degree to which their science authorized a studied lack of interest in social relations.

Alienation, too often, is understood as a homogeneous and monolithic experience. Yet, it can take several forms and be articulated in myriad distinctive ways. These articulations mutually reinforce each other. The temporal alienation organized around the binary of "superstitious traditionals" and "enlightened moderns," for instance, intensifies the social and cosmological alienations. The alienation through numbers similarly resonates with the alienation from affective engagement with the world. Alienation, therefore, remains both singular and internally diversified. This is what, as we have seen in chapter 5, Annemarie Mol calls "manyfoldedness," rather than "plurality." Only by capturing this manyfoldedness of alienation can we grapple with the ways in which race produces alienation.

The Allure of Race

The "black man's alienation," Fanon had insisted, was "not an individual question." The question was "sociogenic" and hence demanded a "sociodiagnostic."[8] Where the long tradition of alienation critique had tended to posit alienation as a tragedy of the individual self, Fanon insisted that disalienation could only follow by recognizing the fundamentally sociogenic nature of alienation itself. Thus, elsewhere, he had clarified that "the sane human being is a social human being; or else, that the measure of the sane human being, psychologically speaking, will be his or her more or less perfect integration into the socius." He called these the "social constants of personality."[9] Commenting on Fanon's insistence that the alienation of individual selves had to be seen within a historical and social context of racialized colonialism, Gilles Deleuze and Félix Guattari, in their characteristically irreverent prose, wrote that "it is strange that we had to wait for the dreams of colonized peoples in order to see that, on the vertices of the pseudo [oedipal] triangle, mommy was dancing with the missionary, daddy was being fucked by the tax collector, while the self was being beaten by a white man."[10]

The question that has motivated this book, however, is what happens when the white man leaves? Does he take his masks with him?

Across this book, we have seen that race neither died with Hitler nor ebbed with the waning of colonialism. It mutated, was nationalized, and, in fact, intensified after World War II. We have recognized with much greater clarity the truth of David Arnold's observation that it is a fallacy to think of "race as a relatively homogeneous set of ideas and practices, driven by material greed and social anxieties in the West, and capable of delivering social power and political authority to whites across the globe." Rather, we must learn to see race as "a far more nebulous and often self-contradictory concept, and rather than being the voice of white authority alone, [it] could form part of an interactive process by which ideas of race were internalized and reworked by the subjects of European racial discourse and practice, in search of their own empowerment."[11]

Furthermore, this kind of racialization was not always merely about anticolonial empowerment. As Elise Burton has argued for the Middle East, national investment in race did not remain satisfied by the construction of homogeneous national subjects. They too, like their colonial predecessors, sought to classify people into hierarchized taxonomies.[12] It is precisely here that Fanon, once more, is illuminating. He had written eloquently that "the Negro enslaved by his inferiority, the white man enslaved by his superiority alike behave in accordance with a neurotic orientation."[13] This astute observation that race is a form of structural violence that objectifies both the group in power and the group deprived of it, opens up f a way or us to think about postcolonial racial violence without engaging in the banal pursuit of villains and victims.

Drawing on a very different genealogy through Dr. Martin Luther King Jr., Paul Gilroy has recently made a similar point. Reemphasizing the reflexive and bidirectional violence of "race-thinking," Gilroy pithily writes that "race-thinking has the capacity to make its beneficiaries inhuman even as it deprives its victims of their humanity."[14] Though Gilroy acknowledges his debt to Fanon repeatedly, the object of his critique is different.

Rather than the racialized colonialism of the mid-twentieth century, the target of Gilroy's critique was the new "allure of race": an "allure of automatic, pre-political uniformity" at the dawn of the twenty-first century.[15] This allure was expressed in a New Racism, where an "idea of natural difference" had been annexed to "claims to mutual exclusivity" and mutually exclusive "national cultures." Gilroy calls this a "bioculturalism" that inverted and intensified earlier forms of race-thinking. This kind of bioculturalism relied on a new synthesis of nature, culture, biology, and history.[16] It has authorized the embrace of essentialist race-thinking by groups that were themselves once subjected to racial logics. "Ethnic absolutism," "primordialism," and "lazy essentialisms" thus became entrenched through a careful blending of new advances in genetic technologies, "corporate multiculturalism," and the inertia of earlier anticolonial political traditions that had deployed race.[17]

Gilroy traces one of the key political genealogies of this New Racism through what he describes as the "brittle-traveling nationalisms firmly rooted in African American circumstances."[18] A much better resourced and explicitly articulated version of this kind of "ethnic absolutism," together with an explicit and vigorous uptake of scientific tools, happened in majoritarian postcolonial nations like India.

Alison Bashford's insightful study of Radhakamal Mukerjee has already given us a small glimpse of the racialized anticolonialism germinating in Indian scientific circles in the interwar years.[19] What this book has shown is that seroanthropology provided a more vibrant, institutionalized, and distinct disciplinary structure for what in Mukherjee's hands had been an interdisciplinary and boundary-pushing project. In other words, seroanthropology, as it came to be institutionalized in India, established such racialized anticolonialism and its attendant biometric nationalism as a kind of unexceptional "normal science." Racialized anthropology and nationalism came to enjoy enormous cultural legitimacy as a result. It was little wonder that the ASI, in time, became the world's largest employer of anthropologists. As one appreciative recent historian has pointed out, not only did "the question of nationalism" remain supremely important to a number of late colonial and postcolonial Indian anthropologists, but "no Indian anthropologist or sociologist [of the time] did oppose nationalism."[20]

While Indian nationalism today has evolved into a largely majoritarian, intolerant configuration, in the late colonial and early postcolonial periods the majoritarian elements in it were much more intimately intertwined with an anticolonial strand. The dominant influence of this entangled Indian nationalism on Indian anthropologists has rendered it difficult to narrate their history within any simple dramaturgy that clearly distinguishes between victims and villains. Sarkar, for instance, was both a leading member of the Indian Eugenics Society and someone deeply concerned with the uplift of marginal groups. Likewise, Guha was utterly convinced of the racial basis of Indian castes, but was scrupulously free of communal sentiment. Their alienation through race science did not make them racists.

Biometric nationalism provided the political form, while seroanthropology provided the practical tools for the unfolding of a national raciology that long predated Gilroy's New Racism but anticipated much of its "allure" for a cladistic "automatic, pre-political uniformity." Even the "nano-politics," which Gilroy so insightfully describes, emerged, I would argue, much earlier and in majoritarian postcolonial nation-states like India, Iran, and Turkey. Gilroy's "nano-politics" is a political form that succeeds biopolitics and takes the "skin, bones and even blood" as the "primary referents of racial discourse," forsaking the "integrity of the body" and moving "inside the threshold of the skin" with much greater momentum than earlier raciologies.[21] Evidently, seroanthropology, rather than the late-twentieth-century biotechnological disciplines descended from it, was the discipline where nano-politics as such began to emerge.

Indeed, here Fanon himself is instructive once more. As Fanon sarcastically pointed out,

> in the first chapter of the history that the others have compiled for me, the foundation of cannibalism has been made eminently plain in order that I may not lose sight of it. My chromosomes were supposed to have a few thicker or thinner genes representing cannibalism. In addition to the *sex-linked*, the scholars had now discovered the *racial-linked*. What a shameful science!
>
> But I understand this "psychological mechanism." For it is a matter of common knowledge that the mechanism is only psychological."[22]

While rejecting the essentialism that national seroanthropologists would enthusiastically take up, Fanon is unequivocal in already recognizing the intimations of a nano-politics of race in mid-century genetics.

While Fanon's attention to embodiment is reasonably widely acknowledged, his explicit reference to genetic raciology is seldom remarked on. In fact, most English editions of *Black Skin, White Masks* do not include Fanon's references to genetics in the index. Even on the rare occasions when scholars have rightly found his comments presciently anticipating the "reassertion of racism in contemporary genetics," such observations have once more been subsumed within explorations of Fanon's ideas about consciousness.[23] Fanon's writing, however, defies such closure. Rather, his repeated references to genes and chromosomes as linked to racial alienation clearly look ahead toward the nano-politics of New Racism.

Criticizing Mayotte Capécia's novelistic depiction of black women desiring to marry white men, Fanon wrote, for instance, "Meanwhile, André has departed to carry the white message to other Mayottes under other skies: delightful little genes with blue eyes, bicycling the whole length of the chromosome corridor."[24] Likewise, in discussing the African American folkloric figure of Br'er Rabbit, Fanon once again approvingly cited the American Trotskyist Bernard Wolfe's discussion of how arguments about "African genes" were mobilized in America to establish the alleged animality of African Americans.[25] In criticizing Carl Jung's notions of heredity, too, Fanon criticized the tendency to invoke genes.[26] These references to genetics, though not exhaustive, all occur at key moments in *Black Skin, White Masks*. Elsewhere he is even more explicit. While trying to argue for the need to develop locally relevant projective psychological tests in Algeria, he wrote that "to say that the Muslim is unable to invent, by invoking a particular genetic constitution, one subsumed within the more general framework of some primitivism, seems a difficult position to defend in our view."[27] These repeated, explicit, and critical comments suggest that Fanon recognized the growing importance of genetics as a new and distinct form of racialization that was in the process of intensifying older forms of alienation.

This New Racism was able to slip with ease into national frameworks because the national was never a mere discourse, derivative or otherwise.[28] It was engendered in and through practices of nationalization of the body.[29] Nationalization, as I have argued elsewhere, needs to be a seen as an active process by which the nation's "bodies, beings, practices, customs, spaces, substances and sites" were actualized.[30] This entailed an actual and representational reworking of preexisting bodies, practices, and such within a new national framework. Medicine had provided one of the earliest and most productive sites for such reworkings in India since the latter decades of the nineteenth century. Bodies were particularly salient to the overall process of nationalization.

The emergence of biometric nationalism in the early decades of the twentieth century revamped the nationalization of bodies by rendering them measurable in new ways. New tools and techniques, along with new mathematical models of a dynamic, racialized idea of national races, emerged particularly through the influence of scholars like Sir Brajendranath Seal and Ramaprasad Chanda.[31] But biometric nationalism really flourished with the emergence and growth of seroanthropology from the 1920s onward. To that extent we might also see in biometric nationalism a precursor, or even perhaps an early instantiation, of nano-politics.

This nano-politics replaced other forms of biometric differentiations inscribed on the surfaces of the body—forms whose organization Gilroy would perhaps call "dermo-politics"[32]— with a new and acute reorganization of biometric nationalism that went much deeper, underneath the visible surfaces of the body.

In 1949 the ASI, for instance, published the first of its planned vernacular texts on Indian anthropology with a view to public dissemination of anthropological knowledge. The book, authored by B. S. Guha, was a Bengali rendition of his arguments about the racial constituents of the Indian population. In it, Guha explained how "common people [*sadharan lok*]" erroneously thought that every child was merely a "tiny replica [*khudra pratimurti*]" of their parents. In reality, he continued, the parents were merely holders of specific "biological heritage [*jaiba uttaradhikar*]" as "trust funds [*nyasta dhan*]," which they in turn passed on. This "trust fund" took the form of "thin, long, stick-like things" called chromosomes. There were, Guha explained, forty-eight chromosomes in every human being, and these in turn were made up of a type of "even smaller ingredients [*khudratara upadan*]" called "genes." These genes were the "ultimate factor [*maulik karan*]" of "heredity [*bangsagati*]," the "unit of life [*jeebsattar byasti*]," and all physical and mental traits were the result of these genes [*daihik o manoshik lakshan ihaderi srishti*].[33] Explicitly, the force of the description seeks to demonstrate the fallacy of imagining a child's identity in terms of its parents' visible features, leave alone their immediate social milieu. Instead, everything that the child is, both physically and mentally, derives from long-forgotten ancestors who have bequeathed a "biological heritage." All this, however, is presented in the service of the nation. The pamphlet is thus titled *Bharater Jati Parichay* [India's Racial Identity].

Even the linguistic solidarities, which in the early postcolonial years had often provided a powerful counterweight to the Indian state's centralizing and homogenizing tendencies, were dismissed by Guha. Rejecting the then highly emotive issue of mother tongues, he said people could always learn a new language, but they could not change their biological inheritance. Hence, in his view, the latter was a much more reliable and primordial basis of identity.[34]

In this admittedly propagandist and official publication aimed at changing how the "common people" viewed their identities, the ASI clearly articulated a genetically based, racialized form of embodied alienation that sought to nationalize Indian bodies. The rest of the pamphlet described how the Indian peoples were in fact made up of six, originally distinct, ancestral races that had mixed to various degrees to give rise to the extant castes and tribes. These ancestral races were the Negrobut (Negrito), the Adi-ostral-rup (Proto-Australoid), the Mongol-rup (Mongoloid), the Bhu-madhya (Mediterranean), the Pashchatya hrasha-kapal (Western Bracycephals) and finally, the Nordiya (Nordic) race.[35] While these races had all combined to create a heterogeneous Indian population, the pamphlet made clear that these racial identities did not in any way line up with any linguistic or regional identities.[36]

This early form of nano-politics, by diving deep into the body and revealing a new architecture of heredity, evidently operated to alienate people from the various putatively accessible bases of identity, whether that was everyday familial kinship, linguistic communities, or even forms of visible physical resemblance. What mattered, as the only true and authentic basis of identity, were the genes and chromosomes that nobody but the seroanthropologists could see in their laboratories. In a way, seroanthropology was what began actualizing Sir B. N. Seal's dream of creating the political subject stripped of biography whose image is now institutionalized in the Aadhaar card.

Alienation as the "Relation of Relationlessness"

Fanon adopted alienation as the key lens through which to explore the violence of race. In his hands, the concept of alienation became particularly effective in accessing shared social and historical forms of estrangement, rather than being a window into individualized angst and anomie. Yet, Fanon was not the first to have come upon alienation. Alienation critique had already enjoyed a diverse and rich career in European philosophical and political writings before Fanon adopted it. From Jean-Jacques Rousseau to Karl Marx, several powerful critics had used alienation, in their own distinctive ways, as an important part of their critical repertoire.

Notwithstanding this hoary genealogy, of late alienation has fallen somewhat out of favor among social critics. The reason is not difficult to discern. With the growing suspicion of essentialist thinking, social critics have developed a strong distaste for the seemingly unavoidable essentialism that creeps into any mobilization of alienation as a critical concept. Whether in its romantic Rousseau-esque garb of alienation from an essential human nature, the Marxist version of alienation from an essentialized and frequently biologized notion of use values, or indeed the Heideggerian account of alienation from an authentic individuality, it is difficult to shake off the musty smell of stale essentialisms that cling to alienation.[37]

Recycling the notion of alienation, Rahel Jaeggi has offered a new, de-essentialized account of the concept that, in many ways, is also remarkably congruent with Fanon's use of the term. Instead of seeing alienation as an estrangement from a preexisting, essentialized being or state, Jaeggi argues that alienation is in fact a "relation of relationlessness."[38] While avoiding the spurious search for an authentic essence, Jaeggi's proposition, like Fanon's, offers us a way to think of alienation in fundamentally social and political terms.

Alienation, Jaeggi clarifies, "means indifference and internal division, but also powerlessness and relationlessness with respect to oneself and to a world experienced as indifferent and alien. Alienation is the inability to establish a relation to other human beings, to things, to social institutions and thereby also—so the fundamental intuition of the theory of alienation—to oneself."[39] The emphasis on simultaneous failure to relate to certain aspects of one's own self as well as to the world one inhabits seems to capture the broad contours of what we have encountered in this book as the "manyfoldedness of alienation." The temporal alienation from the present, the alienation from the remembered modes of socialization, including the stories one had heard, the alienation through numbers from the social worlds of those among whom one did fieldwork, the alienation from all the affectively loaded, lived bases of identity: all can be organized as forms of alienation either from oneself or from the social world one inhabits.

What is truly productive about Jaeggi's formulation, though, is that this alienation is not conceptualized as a lack of social relationships. Rather, it is itself a specific type of social and self-relationship: a relationship of unrelatedness—and a relationship that distorts. Explaining this notion of a distorted relationship, she points out that any "experience is miseducative that has the effect of arresting or distorting the growth of further experience."[40] To wit, distorted relationships follow from an impeding of further experience. This is indeed what we see in the practice of Indian race science. The alienation follows from the development of a particular type of relationship between the researcher and the research subjects that forecloses certain types of further experience. The researcher, for instance, does not inquire about the social lives, economic woes, or cosmological beliefs of their subjects, and becomes almost obsessively focused on determining deeper bases of identity that are almost entirely beyond putative social access. Again, Sarkar's deliberate turning away from the stories he longed to hear, stories that reminded him of his childhood and connected him to those he studied, due to his "professional duty" to search for gene frequencies, is exemplary of how further experiential growth is explicitly stunted, and a relationship of unrelatedness established.

Moreover, Jaeggi's de-essentialized framework for thinking about alienation also calls attention to the constitutive function of social roles and their political character. There is no "automatic, pre-political uniformity" from which one is exiled via alienation. Rather, alienation proceeds from a particular and distorting social role that deliberately breaks the Tardian chain of mimicry that binds together the "social" in the first place. By refusing to hear or remember stories that transcend, subvert, or question the cladistic world of neatly endogamous gene pools, the seroanthropologist deliberately breaks a series of chains of mimicry. The nationalized precursor of "corporate multiculturalism" produces alienation by a conscious and deliberate unraveling of earlier social bonds, themselves overdetermined by acts of learned mimicry.

Ziauddin Sardar, explicating Fanon's references to genetics and chromosomes, has proposed that the comments constitute a critique of both the human and the social sciences in general. Anthropology, Sardar continues, "was developed specifically to describe, manage and contain the black man."[41] Though Sardar's attempt to metaphorize and universalize a singular white/Black dichotomy seems to run counter to Fanon's own attention to the specificities of historically constituted Blackness, his indictment of the human and social sciences is extremely insightful.[42] Becoming a (sero)anthropologist entailed inhabiting a social role that alienated by recoding old modes of relatedness and inaugurating new relationships of unrelatedness. This latter, distorting relationship was what allowed the seroanthopologist to "describe, manage and control" both themselves and the objects of their study. Put another way, those who donned the *white coats* were called upon to describe, manage, and control those who had *brown skins*.

The alienation of the *white coats* from the *brown skins* is as much a self-alienation of the researchers, who are cut off from aspects of their social being, as it is a social alienation that produces an estrangement between the researcher and their research subjects. This was not simply the self-hating race-thinking that inspired those with *black skins* to wear *white masks*. In fact, in a way, this new race science did not provide any explicit grounds for love or hate. It simply painted a portrait of the "human being as an essentially irrelevant transitory medium for the dynamic agency of their . . . genes."[43] The brownness of one's skin evoked neither shame nor joy as such, for it was a mere code for a set of genes and by itself entirely unrelated to the worlds of stories and struggles in which it was placed.[44]

Critical Utopianism and a Brown Planetary Humanism

One of Gilroy's most astute observations is that Fanon himself has been perverted and instrumentalized by New Racism. He finds that "a simplistic version of racial phenomenology mistakenly attributed to Fanon" has been aligned with a "'race'-entrenching pragmatism."[45] It is part of a wider phenomenon where the quest for disalienation has ironically itself become a way to produce prepolitical, mutually exclusive "national encampments." With such "national encampments," identities "cease to be an ongoing process of self-making and social interaction," becoming instead a "thing possessed and displayed." Every identity becomes akin to a fortified national camp.[46] Faced with such encampments, the project of disalienation must be radically rethought.

At the moment of Gilroy's writing, at the dawn of the millennium, the Human Genome Project's assertions of human biological unity led him to believe that the emergent contradictions of New Racism were producing a "crisis of raciology" that might eventually undermine race-thinking. The two decades since then have shown how such perceptions of crisis have, once more, intensified genomic "national encampments."[47] Developments internal to the world of biotechnology, therefore, have certainly not thus far opened up any pathways to disalienation.

Gilroy's larger route map to disalienation, however, is still viable. Combining the projects of Fanon and Martin Luther King Jr., Gilroy proposes a road to disalienation that leads through a "radically nonracial humanism." Such a radically nonracial humanism would be "counter-anthropological" in its primary concern with "forms of human dignity that race-thinking strips away." This "radically nonracial humanism" is counteranthropological because it categorically rejects bodily essences as the basis of its humanity. Instead, it locates humanist identification and empathy in the "recurrence of pain, disease, humiliation and loss of dignity, grief and the care for those one loves."[48]

Such a "radically nonracist humanism" is also "planetary" in its scope. Unlike the commoditized, financialized, and national-securitized scale of the "global," the scale of the "planetary" that informs Gilroy derives from the Bandung Conference of 1955 and its expansive, "translocal, transcultural" non-aligned impulse.[49] The resources for such a radically nonracist planetary humanism must be gathered from a "principled cross-cultural approach to the history and literature of extreme situations in which the boundaries of what it means to be human were being negotiated and tested minute by minute, day by day."[50] Such a humanism is thereby constantly forestalled from reinserting any singular, privileged, transhistorical, and prepolitical figure of "the human" as its stable referent, and especially eschews the classical liberal humanist standard of the European male, bourgeois subject.[51]

This shimmering, unstable, and political figure of the human recognized in moments of shared suffering is, I would argue, a Brown figure. This is not the genetically grounded, individualized Brownness that a seroanthropologist or one of their intellectual heirs might describe. Indeed, this Brown figure is not a singular, individualized subject at all, but "an impulse to go beyond the singular . . . and individualized subjectivities." While this Brownness is implicated in a shared social world that is simultaneously about "human" suffering and persisting in the face of suffering, it does not belong to a biologized national camp. It is a sharing that is engendered in "contact" and "nothing like continuousness." José Esteban Muñoz called it the "brown commons."[52]

Such an account of Brownness flickers ephemerally in the space just prior to its (perhaps) inevitable capture by national encampments, their geneticists, and their activists. Yet, in its fleeting, transient forms, it might open up for us a space of "critical utopianism": a utopianism that does not aim to "enact" but to "touch"; to affect; to feel.[53] The fundamental instability constitutive of Brownness, by clearly interrupting the hegemonic "Black-white relational chain," I argue, becomes particularly available to operate generatively as a utopian moment "between camps."[54]

This Brown planetary humanism, so long as it can elude capture by national encampments that biologize and nationalize it, is nourished by the "prophylactic powers of memory to work against future evils."[55] Memories, of course, are overdetermined by the social: they are simultaneously located in social interactions, but also thereby constitute the social, at least in the Tardian sense of a loose chain of reciprocal mimicries.

It is memory that brings me back again to the grandmothers: both Sarkar's and mine. It was the stories themselves as much as the acts of their storytelling that allowed us to "overhear" other relationships and other pasts.[56] It was the memory of those stories and storytellings that led Sarkar, however late in his life, to reflect on a larger, shared human heritage that transcended the abstract seroanthropological identities he so assiduously created in his professional publications. Completely inverting the modernist progress narrative that his eugenic work so celebrated, Sarkar remembered that "folklore [*lokgatha*] is the highest development [*sarbochcha bikash*] of man's affectability [*rasanubhuti*] and affective inspiration [*rasanuprerana*]. Social behaviors [*achar*], judgment/values [*bichar*], industries [*Shilpa*], the arts [*kala*], everything has been passed on orally [through folklore]." He lamented that grandmothers "nowadays [*ajkal*]" no longer tell such stories. Perhaps, he mused, "the storehouse of stories that modern grandmothers [*adhunika didima-thakuma*] had has been emptied." It was not just the modern grandmothers, however. He asserted that he could no longer hear stories that could affect people like the bygone stories of his grandmother did [*rasala galpa ar shuni na*].[57]

Not only did Sarkar repudiate his progressivist teleology in these recollections, he also asserted that these stories were "not simply parables to entertain children—they were [rather] the unbloomed language [*asphuta bhasa*] in which that stratum of human society which could not express their mental states like poets [*kabi*] and intellectuals [*chintasheel*] expressed themselves. These were the treasures [*sampad*] from which a nation's [*jatir*] moral history [*naitik itihas*] could be written."[58] Once again contradicting his programmatic statements and statistical calculations of Bengali intellectual stalwarts and their caste backgrounds, here Sarkar was unambiguously asserting that the true genius of the nation lay accumulated in folktales told by the unlettered and the unintellectual.

These remarkable inversions that led the avowed and lifelong eugenicist to lament modernity and its diminishing store of stories were all occasioned by the memory of his grandmother, the storyteller.

Of course, genre also mattered. The genre that affected the aging eugenicist so deeply and nudged him to repudiate, at least for the moment, his faith in both progress and the role of the national intellectual was the folktale, with its enchanted world of fairies, goblins, goddesses, frog princes, and much more. It was, in short, a world that refused the biologized figure of the human: an uncanny world of the superhuman and the transhuman.

The fairy tale, as Walter Benjamin reminds us, "to this day is the first tutor of children because it was once the first tutor of mankind . . . The first true storyteller is, and will continue to be, the teller of fairy tales."[59] It was this world of stories, magical and moral, that interrupted the telos of progress, of cladistics, and of measurable human differences. The stories of the tribal Indians (such as the Santals and the Khasis) as well as foreigners (such as the Polynesians and the western Amazonians), all groups seroanthropologically clearly distinguishable from Sarkar's own self, thus found a place right next to his own beloved grandmother's stories in the little collection he published. It was this uncanny and more-than-human world that offered the graying eugenicist a glimpse of a Brown planetary humanism. It is also this uncanny and possibly magical world of stories that draw upon folklore and more that has given me, in this book, a space to pursue a plurigeneric experiment in critical utopianism.

NOTES

Introduction

1. Arnold, "'An Ancient Race Outworn': Malaria and Race in Colonial Bengal, 1860–1930."

2. Chandra, "Whiteness on the Margins of Native Patriarchy: Race, Caste, Sexuality, and the Agenda of Transnational Studies."

3. Trautmann, *Aryans and British India*; Ballantyne, *Orientalism and Race: Aryanism in the British Empire*; Mukharji, "The Bengali Pharaoh."

4. Pande, *Medicine, Race and Liberalism in British Bengal*, 10, 21–43.

5. Anderson, *Cultivation of Whiteness*.

6. Anderson, *Cultivation of Whiteness*; Abu El-Haj, *The Genealogical Science: The Search for Jewish Origins and the Politics of Epistemology*; Burton, *Genetic Crossroads: The Middle East and the Science of Human Heredity*; Clouser, "Chinese Whiteness: The Discourse of Race in Modern and Contemporary Chinese Culture"; Rogaski, *Hygienic Modernity: Meanings of Health and Disease in Treaty-Port China*, 240–44; Dikötter, *The Discourse of Race in Modern China*; El Shakry, *The Great Social Laboratory: Subjects of Knowledge in Colonial and Postcolonial Egypt*.

7. Bashford, "Anticolonial Climates: Physiology, Ecology, and Global Population, 1920–1950," 604.

8. Balibar and Wallerstein, *Race, Nation, Class: Ambiguous Identities*; Dikötter, *The Discourse of Race in Modern China*; Ergin, *"Is the Turk a White Man?" Race and Modernity in the Making of Turkish Identity*.

9. Bayly, "Caste and 'Race' in the Colonial Ethnography of India"; Bates, "Race, Caste and Tribe in Central India: The Early Origins of Indian Anthropometry"; Sinha, *Colonial Masculinity*; Béteille, "Race and Caste;" Roy, "Race and Recruitment in the Indian Army: 1800–1918"; Rai, "From Colonial 'Mongoloid' to Neoliberal 'Northeastern': Theorising 'Race', Racialization and Racism in Contemporary India." And so on.

10. This claim to a single, defining "scientific method," was itself a nineteenth-century invention. Cowles, *The Scientific Method: An Evolution of Thinking from Darwin to Dewey*.

11. Gordin, *The Pseudoscience Wars: Immanuel Velikovsky and the Birth of the Modern Fringe*, 1.

12. See also Gordin, *On the Fringe: Where Science Meets Pseudoscience*.

13. Stepan, *The Hour of Eugenics: Race, Gender and Nation in Latin America*, xvi.

14. Campbell, *Race and Empire: Eugenics in Colonial Kenya*, 8.

15. Roberts, *Fatal Invention: How Science, Politics, and Big Business Re-Create Race in the Twenty First Century*, xi.

16. Gannett, "Racism and Human Genome Diversity Research: The Ethical Limits of 'Population Thinking.'"

17. Reardon, *Race to the Finish: Identity and Governance in an Age of Genomics*, 17.

18. Reardon, *Race to the Finish: Identity and Governance in an Age of Genomics*, 23–26.

19. Reardon, *Race to the Finish: Identity and Governance in an Age of Genomics*, 26.

20. Reardon, *Race to the Finish: Identity and Governance in an Age of Genomics*, 29.

21. Reardon, *Race to the Finish: Identity and Governance in an Age of Genomics*, 31.

22. Reardon, *Race to the Finish: Identity and Governance in an Age of Genomics*, 3.

23. Reardon, *Race to the Finish: Identity and Governance in an Age of Genomics*, 2.

24. Abu El-Haj, *The Genealogical Science: The Search for Jewish Origins and the Politics of Epistemology*; Burton, *Genetic Crossroads: The Middle East and the Science of Human Heredity*; Hyun, "Blood Purity and Scientific Independence: Blood Science and Postcolonial Struggles in Korea, 1926–1975"; Bashford, "Anticolonial Climates."

25. Abu El-Haj, *The Genealogical Science: The Search for Jewish Origins and the Politics of Epistemology*, 137.

26. Burton, *Genetic Crossroads: The Middle East and the Science of Human Heredity*, 115–18.

27. On the transition from gentlemanly science to industrial science, see Shapin, *The Scientific Life: A Moral History of a Late Modern Vocation*.

28. Anderson, "Racial Hybridity, Physical Anthropology, and Human Biology in the Colonial Laboratories of the United States"; Lipphardt, "Isolates and Crosses in Human Population Genetics; or, A Contextualization of German Race Science"; Roque, "The Blood That Remains: Card Collections from the Colonial Anthropological Missions"; Rees, "Doing 'Deep Big History': Race, Landscape and the Humanity of H. J. Fleure (1877–1969)."

29. Duster, "A Post-Genomic Surprise: The Molecular Reinscription of Race in Science, Law and Medicine"; Fujimura, Duster, and Rajagopalan, "Introduction: Race, Genetics, and Disease: Questions of Evidence, Matters of Consequence"; Benjamin, *Race after Technology: Abolitionist Tools for the New Jim Code*.

30. McMahon, "The History of Transdisciplinary Race Classification: Methods, Politics and Institutions, 1840s–1940s."

31. El Shakry, *The Great Social Laboratory: Subjects of Knowledge in Colonial and Postcolonial Egypt*, 55–86.

32. Savary, "Vernacular Eugenics? Santati-Sastra in Popular Hindi Advisory Literature (1900–1940)." See also Savary, *Evolution, Race and Public Sphere in India: Vernacular Concepts and Sciences (1860–1930)*.

33. Majumder, "People of India: Biological Diversity and Affinities" (article).

34. Widmer, "Making Blood 'Melanesian': Fieldwork and Isolating Techniques in Genetic Epidemiology."

35. Gissis, "When Is 'Race' a Race? 1946–2003."

36. Lipphardt and Niewöhner, "Producing Difference in an Age of Biosociality: Biohistorical Narratives, Standardisation and Resistance as Translations."

37. Reardon, *Race to the Finish: Identity and Governance in an Age of Genomics*, 17.

38. Wade et al., "Introduction: Genomics, Race Mixture, and Nation in Latin America."

39. Anderson and Lindee, "Pacific Biologies: How Humans Become Genetic."

40. Gannett, "Racism and Human Genome Diversity Research: The Ethical Limits of 'Population Thinking'"; Reardon, *Race to the Finish: Identity and Governance in an Age of Genomics.*

41. Lindee and Ventura Santos, "The Biological Anthropology of Living Human Populations."

42. El Shakry, *The Great Social Laboratory: Subjects of Knowledge in Colonial and Postcolonial Egypt*, 70.

43. Bangham, *Blood Relations: Transfusion and the Making of Human Genetics.*

44. Burton, *Genetic Crossroads: The Middle East and the Science of Human Heredity.*

45. Abu El-Haj, *The Genealogical Science: The Search for Jewish Origins and the Politics of Epistemology*; Anderson, *Cultivation of Whiteness.*

46. Seth, *Difference and Disease: Medicine, Race, and the Eighteenth-Century British Empire*, 170.

47. Rakshit, *Bio-Anthropological Research in India: Proceedings of the Seminar in Physical Anthropology and Allied Disciplines.*

48. Bhattacharjee and Rakshit, "Sero-Anthropology of the Indian Mongoloids."

49. Seth, *Difference and Disease: Medicine, Race, and the Eighteenth-Century British Empire*, 11.

50. Quack, *Disenchanting India: Organized Rationalism and Criticism of Religion in India.*

51. Nanda, *Prophets Facing Backwards: Postmodern Critiques of Science and Hindu Nationalism in India*, 391.

52. Nanda, *Prophets Facing Backwards: Postmodern Critiques of Science and Hindu Nationalism in India*, 307.

53. Schiebinger, *Plants and Empire: Colonial Bioprospecting in the Atlantic World*, 3. See also Proctor and Schiebinger, *Agnotology: The Making and Unmaking of Ignorance.*

54. Trautmann, "Discovering Aryan and Dravidian: A Tale of Two Cities"; Trautmann, *Languages and Nations: The Dravidian Proof in Colonial Madras.*

55. Trautmann, *Aryans and British India.*

56. Ludden, "Orientalist Empericism: Transformations of Colonial Knowledge."

57. Mantena, *Alibis of Empire: Henry Maine and the Ends of Liberal Imperialism.*

58. Dirks, *Castes of Mind: Colonialism and the Making of Modern India.*

59. Christopher Fuller, *Anthropologist and Imperialist: H. H. Risley and British India, 1873–1911.* See also his "Colonial Anthropology and the Decline of the Raj: Caste, Religion and Political Change in India in the Early Twentieth Century."

60. Singha, *A Despotism of the Law: Crime and Justice in Early Colonial India*, 168–228. Wagner, "Confessions of a Skull: Phrenology and Colonial Knowledge in Early Nineteenth-Century India"; Nigam, "Disciplining and Policing the 'Criminals by Birth,' Part 1: The Making of a Colonial Stereotype—The Criminal Tribes and Castes of North India"; Nigam, "Disciplining and Policing the 'Criminals by Birth,' Part 2: The Development of a Disciplinary System, 1871–1900"; Major, "State and Criminal Tribes in Colonial Punjab: Surveillance, Control and Reclamation of the 'Dangerous Classes'"; Mishra, "Of Poisoners, Tanners and the British Raj: Redefining Chamar Identity in Colonial North India, 1850–90."

61. For arguments about precolonial antecedents for "criminal tribe" stereotypes, see Piliavsky, "The 'Criminal Tribe' in India before the British." For the continued use of colonial-era practices in postcolonial India, see D'Souza, "De-Notified Tribes: Still 'Criminal'?"

62. Bates, "Race, Caste and Tribe in Central India: The Early Origins of Indian Anthropometry"; Bayly, *The New Cambridge History of India IV.3—Caste, Society and Politics in India from the Eighteenth Century to the Modern Age.*

63. Kolsky, *Colonial Justice in British India: White Violence and the Rule of Law*; Fischer-Tiné, "Hierarchies of Punishment in Colonial India: European Convicts and the Racial Dividend, c. 1860–1890"; Singha, *A Despotism of the Law: Crime and Justice in Early Colonial India*, 289–93.

64. Bailkin, "The Boot and the Spleen: When Was Murder Possible in British India?"; Sen, "Confessions of the Unfriendly Spleen: Medicine, Violence, and the Mysterious Organ of Colonial India."

65. Sharafi, "The Imperial Serologist and Punitive Self-Harm: Bloodstains and Legal Pluralism in British India."

66. Sharafi, "Abortion in South Asia, 1860–1947: A Medico-Legal History."

67. Roy, "Race and Recruitment in the Indian Army: 1800–1918"; Rand and Wagner, "Recruiting the 'Martial Races': Identities and Military Service in Colonial India"; Caplan, "Martial Gurkhas: The Persistence of a British Military Discourse on 'Race.'"

68. Buxton, "Imperial Amnesia: Race, Trauma and Indian Troops in the First World War"; Imy, *Faithful Fighters: Identity and Power in the British Indian Army.*

69. Ballhatchet, *Race, Sex, and Class under the Raj: Imperial Attitudes and Policies and Their Critics, 1793–1905*; Wald, *Vice in the Barracks.*

70. Harrison, *Climates & Constitutions*; Harrison, "'The Tender Frame of Man': Disease, Climate, and Racial Difference in India and the West Indies, 1760–1860."

71. Arnold, "Diabetes in the Tropics: Race, Place and Class in India, 1880–1965."

72. Seth, *Difference and Disease: Medicine, Race, and the Eighteenth-Century British Empire*, 280–87.

73. Ernst, "Colonial Policies, Racial Politics and the Development of Psychiatric Institutions in Early Nineteenth-Century British India."

74. Ernst, "Colonial Psychiatry, Magic and Religion: The Case of Mesmerism in British India"; Pande, *Medicine, Race and Liberalism in British Bengal.*

75. Pande, *Medicine, Race and Liberalism in British Bengal*; Mukharji, "Vernacularizing the Body: Informational Egalitarianism, Hindu Divine Design, and Race in Physiology Schoolbooks, Bengal 1859–1877."

76. Sinha, *Colonial Masculinity.*

77. Sinha, "Britishness, Clubbability, and the Colonial Public Sphere: The Genealogy of an Imperial Institution in Colonial India," 497.

78. Mizutani, *The Meaning of White.*

79. Majeed, "Race and Pan-Islam in Iqbal's Thought."

80. Hellman-Rajanayagam, "Is There a Tamil Race?"

81. Chandra, "Whiteness on the Margins of Native Patriarchy: Race, Caste, Sexuality, and the Agenda of Transnational Studies."

82. Fischer-Tiné, "From Brahmacharya to 'Conscious Race Culture': Victorian Discourses of 'Science' and Hindu Traditions in Early Indian Nationalism"; Ramaswamy, "Remains of the Race: Archaeology, Nationalism, and the Yearning for Civilization in the Indus Valley."

83. Béteille, "Race and Caste"; Gupta, "Caste Is Not Race: But, Let's Go to the UN Forum Anyway"; Oomen, "Race and Caste: Anthropological and Sociological Perspectives."

84. Ghoshal, "Race in South Asia: Colonialism, Nationalism and Modern Science."

85. Hodges, "South Asia's Eugenic Past."

86. Ramaswamy, "Remains of the Race: Archaeology, Nationalism, and the Yearning for Civilization in the Indus Valley," 107–8.

87. Parfitt and Egorova, *Genetics, Mass Media and Identity: A Case of Genetic Research on the Lemba*; Egorova and Parfitt, "Genetics, History, and Identity: The Case of the Bene Israel and the Lemba"; Egorova, "De/Geneticizing Caste: Population Genetic Research in South Asia"; Egorova, "Castes of Genes? Representing Human Genetic Diversity in India"; Egorova, "The Substance That Empowers: DNA in South Asia."

88. Ray, "Studying Laboratories: A Sociological Inquiry into Research Practices in Genetics."

89. Rajan, *Biocapital.*

90. Uberoi, Sundar, and Deshpande, *Anthropology in the East: Founders of Indian Sociology and Anthropology.*

91. Dasgupta, "The Journey of an Anthropologist in Chhotanagpur"; Guha, *In Search of Nationalist Trends in Indian Anthropology: Opening a New Discourse.*

92. Sundar, "In the Cause of Anthropology: The Life and Work of Irawati Karve," 373.

93. IGVC, "The Indian Genome Variation Database (IGVdb): A Project Overview," 4.

94. IGVC, "The Indian Genome Variation Database (IGVdb): A Project Overview," 1.

95. IGVC, "The Indian Genome Variation Database (IGVdb): A Project Overview," 2.

96. Majumder, "People of India: Biological Diversity and Affinities" (article), 100. See also Majumder, "People of India: Biological Diversity and Affinities" (chapter).

97. Newman and Principe, *Alchemy Tried in the Fire: Starkley, Boyle, and the Fate of Helmotian Chymistry*. Newman, *Atoms and Alchemy: Chymistry and the Experimental Origins of the Scientific Revolution*; Roos, *The Salt of the Earth: Natural Philosophy, Medicine, and Chymistry in England, 1650–1750.*

98. Prasad, "Burdens of the Scientific Revolution: Euro/West-Centrism, Black Boxed Machines, and the (Post) Colonial Present," 1061.

99. Reardon, *Race to the Finish: Identity and Governance in an Age of Genomics*, 27.

100. Gannett, "Racism and Human Genome Diversity Research: The Ethical Limits of 'Population Thinking'"; Gannett, "The Biological Reification of Race."

101. Lipphardt, "Isolates and Crosses in Human Population Genetics; or, A Contextualization of German Race Science"; Lipphardt, "'Geographical Distribution Patterns of Various Genes.'"

102. Hage, "Recalling Anti-Racism," 124.

103. Subramaniam, *Holy Science: The Biopolitics of Hindu Nationalism*, 36.

104. Guha, "The Prose of Counter-Insurgency."

105. Fanon, *Black Skin, White Masks*, 64.

106. Alam, "The Care of Foreigners: A History of South Asian Physicians in the United States, 1965–2016."

107. "Overhearing" is a form of anonymized speech, embedded in Bengali social and linguistic practices, that brings into view ephemeral worlds of political possibilities. Khan, "Marginal Lives and the Microsociology of Overhearing in the Jamuna Chars."

108. Hartman, "Venus in Two Acts," 2.

109. Nandy, *The Intimate Enemy.*

110. Smith, *Decolonizing Methodologies: Research and Indigenous Peoples*; Mavhunga, *Transient Workspaces: Technologies of Everyday Innovation in Zimbabwe.*

111. Subramaniam, *Holy Science: The Biopolitics of Hindu Nationalism.*

112. Subramaniam, *Holy Science: The Biopolitics of Hindu Nationalism*, 34–40.

113. Nandy, *The Intimate Enemy*, xv.

114. Nanda, *Prophets Facing Backwards: Postmodern Critiques of Science and Hindu Nationalism in India.*

115. Hartman, "Venus in Two Acts," 11.

116. Hartman, "Venus in Two Acts," 11.

117. Hartman, "Venus in Two Acts," 12.

118. Aquil and Chatterjee, *History in the Vernacular*. See also Amin, *Event, Metaphor, Memory, Chauri Chaura 1922–1992*; Amin, *Conquest and Community*.

119. Historical analysis of the genres in which South Asians wrote about science has only just begun. See Singh, "The Shastri and the Air-Pump: Experimental Fiction and Fictions of Experiment for Hindi Readers, 1915–1919."

120. Mukharji, "Hylozoic Anticolonialism: Archaic Modernity, Internationalism, and Electromagnetism in British Bengal, 1909–1940."

121. Vargas et al., "The Debate between Tarde and Durkheim," 767.

122. Interestingly, there were comparable attempts to reconceptualize "the social" around the same time in colonial South Asia as well. See Banerjee, "Between the Political and the Nonpolitical: The Vivekananda Moment and a Critique of the Social in Colonial Bengal, 1890s–1910s."

123. Fanon, "Parallel Hands."

124. Tarde, *Underground Man*.

125. Jakobsen, "A Fragment of Future History: Gabriel Tarde's Archival Utopia."

126. Chakravorty, "'Skeletons of History': Fact and Fiction in Rakhaldas Bandyopadhyay's Sasanka."

127. Ghosh, *Occasional Paper No. 125: The Slave of MS.H.6.*; Ghosh, *In n Antique Land: History in the Guise of a Traveler's Tale.*

128. Chakrabarty, *The Calling of History: Sir Jadunath Sarkar and His Empire of Truth*; Rochona Majumdar, "Feluda on Feluda."

129. Ray, *Amanushik Manush* [Inhuman Humans]. For the original publication details, see Basu, *Bangla Sishushahitya: Granthapanji* [A Bibliography of Bengali Books for Children and Young Adults], 413. Ray's name appears in English both as "Ray" and as "Roy." I will standardize it in these pages as "Ray."

130. Chattopadhyay, "Hemendrakumar Ray and the Birth of Adventure Kalpabigyan."

131. Chattopadhyay, "Hemendrakumar Ray and the Birth of Adventure Kalpabigyan."

132. Chattopadhyay, "Hemendrakumar Ray and the Birth of Adventure Kalpabigyan."

133. Khan, "Marginal Lives and the Microsociology of Overhearing in the Jamuna Chars."

134. Gupta, *Rater Kolkata* [Night Life of Calcutta].

135. Gilroy, *Between Camps: Nations, Cultures and the Allure of Race*, 79, 2.

136. Ray, "Najruler Janmadin Smarane" [Remembering Nazrul's Birthday].

137. Ray, "Najruler Janmadin Smarane" [Remembering Nazrul's Birthday], 279.

Chapter 1

1. Hirschfeld [Hirszfeld] and Hirschfeld [Hirszfeld], "Serological Differences between the Blood of Different Races: The Result of Researches on the Macedonian Front." Predictably, such linear narratives of discovery can be easily complicated. Recent scholarship, for instance, has drawn attention to the Hirszfelds' reliance on older, imperial networks of specialists in tropical medicine. See Mikanowski, "Dr. Hirszfeld's War: Tropical

Medicine and the Invention of Sero-Anthropology on the Macedonian Front." Others have drawn attention to the long European genealogy of works trying to establish racial "purity" within which the Hirszfelds worked. See Cleminson and Roque, "Imagining the 'Biochemical Race': Seroanthropology and Concepts of Racial Purity in Portugal (1900–1950)." Notwithstanding such constructed genealogies, as William Schneider points out, the "key development" that produced the explosion of seroanthropological studies was undoubtedly the publication of the Hirszfeld study. Schneider, "Blood Group Research in Great Britain, France, and the United States Between the World Wars."

2. For a fuller discussion of the discovery, see Ludwik Hirszfeld, *Ludwik Hirszfeld: The Story of One Life.*

3. Bangham, *Blood Relations: Transfusion and the Making of Human Genetics.*

4. On the visual presentation of the Biochemical Race Index, see Gannett et al., "Classical Genetics and the Geography of Genes."

5. Hirschfeld [Hirszfeld] and Hirschfeld [Hirszfeld], "Serological Differences between the Blood of Different Races: The Result of Researches on the Macedonian Front," 679.

6. Schneider, "Blood Group Research in Great Britain, France, and the United States between the World Wars," 90.

7. Burton, *Genetic Crossroads: The Middle East and the Science of Human Heredity*, 76.

8. Hyun, "Blood Purity and Scientific Independence: Blood Science and Postcolonial Struggles in Korea, 1926–1975."

9. Mukharji, "From Serosocial to Sanguinary Identities: Caste, Transnational Race Science and the Shifting Metonymies of Blood Group B, India c. 1918–1960."

10. Roque, "The Blood That Remains: Card Collections from the Colonial Anthropological Missions," 35.

11. Malone and Lahiri, "The Distribution of the Blood-Groups of Certain Races and Castes in India," 964.

12. Arnold, *Science, Technology and Medicine in Colonial India*, 60–61.

13. Hirszfeld, *Ludwik Hirszfeld: The Story of One Life*, 51.

14. Summers, "Cholera and Plague in India: The Bacteriophage Inquiry of 1927–1936," 284.

15. Lahiri, "Observations on the Medico-Legal Applications of Blood Grouping with a Note on Blood Groups in a Polyandrous Family."

16. Mukharji, "Profiling the Profiloscope: Facialization of Race Technologies and the Rise of Biometric Nationalism in Inter-War British India."

17. Annandale, "Introductory Note."

18. Mahalanobis, "A Revision of Risley's Anthropometric Data Relating to the Tribes and Castes of Bengal"; Mahalanobis, "A Revision of Risley's Anthropometric Data Relating to the Chittagong Hill Tribes."

19. Mahalanobis et al., "Anthropometric Survey of the United Provinces, 1941: A Statistical Study"; Majumdar and Rao, "Bengal Anthropometric Survey, 1945: A Statistical Study."

20. Ghoshal, "Race in South Asia: Colonialism, Nationalism and Modern Science."

21. Unfortunately, despite my best efforts I have been unable to find Eileen Macfarlane's papers or indeed much information about her personal life. The difficulties arise partly from the fact that she changed her name several times throughout her life, mainly through multiple marriages but also by using different combinations of her maiden names. The lack of a stable teaching position might also have contributed to the dif-

ficulties. What I have gathered of her life has mainly been by accumulating scattered pieces of information in letters, obituaries, entries in her university yearbook, and even a flier she had printed to advertise herself as a public speaker. Macfarlane, "Letter to J. B. S. Haldane," 12 July 1942; Anon., "[Entry on] Eileen M. Erlanson MacFarlane"; Macfarlane, "Eileen Macfarlane: Lecturer, Traveller, Scientist"; Anon., "[Obituary of] Eileen Whitehead Erlanson MacFarlane"; Lewis, Reznicek, and Rabeler, "Identifications and Typifications of Rosa (Rosaceae) Taxa in North America Described or Used by E. W. Erlanson, 1925–1934."

22. Macfarlane, "Preliminary Note on the Blood Groups of Some Cochin Castes."

23. Macfarlane, "Untitled Communication on Bagdi Blood Groups."

24. Macfarlane and Sarkar, "Blood Groups in India."

25. Macfarlane, "Blood Grouping in the Deccan and the Eastern Ghats," 49.

26. Fuller, "Colonial Anthropology and the Decline of the Raj: Caste, Religion and Political Change in India in the Early Twentieth Century," 481.

27. Sen, "'No Matter How, Jogendranath Had to Be Defeated': The Scheduled Castes Federation and the Making of Partition in Bengal, 1945–1947."

28. Macfarlane, "Blood Grouping in the Deccan and the Eastern Ghats," 45–46.

29. Singh, "Evolution of Dalit Identity: History of Adi Hindu Movement in United Province (1900–1950)."

30. Duncan, "Dalits and the Raj: The Persistence of the Jatavs in the United Provinces," 132.

31. Risley had proposed dividing the British Indian population into six basic racial groups. Malone and Lahiri, "The Distribution of the Blood-Groups of Certain Races and Castes in India," 964.

32. Macfarlane, "Blood Group Distribution in India with Special Reference to Bengal."

33. Malone and Lahiri, "The Distribution of the Blood-Groups of Certain Races and Castes in India," 966–67.

34. Malone and Lahiri, "The Distribution of the Blood-Groups of Certain Races and Castes in India," 967.

35. Malone and Lahiri, "The Distribution of the Blood-Groups of Certain Races and Castes in India," 967.

36. Macfarlane and Sarkar, "Blood Groups in India," 409.

37. On the upper-caste claims to have originated outside India, see Ballantyne, *Orientalism and Race: Aryanism in the British Empire.*

38. Fuller, "Colonial Anthropology and the Decline of the Raj: Caste, Religion and Political Change in India in the Early Twentieth Century," 464.

39. Fuller, "Colonial Anthropology and the Decline of the Raj: Caste, Religion and Political Change in India in the Early Twentieth Century," 486.

40. Sarkar, *Modern India, 1885–1947*, 21.

41. Sen, "'No Matter How, Jogendranath Had to Be Defeated': The Scheduled Castes Federation and the Making of Partition in Bengal, 1945–1947."

42. Duncan, "Dalits and the Raj: The Persistence of the Jatavs in the United Provinces," 121.

43. Appadurai, "Number in the Colonial Imagination"; Dirks, *Castes of Mind: Colonialism and the Making of Modern India*; Peabody, "Cents, Sense, Census: Human Inventories in Late Precolonial and Early Colonial India"; Guha, "The Politics of Identity and Enumeration in India, c. 1600–1990"; Samarendra, "Census in Colonial India and the Birth of Caste"; Piliavsky, "The 'Criminal Tribe' in India before the British."

44. Fuller, "Anthropologists and Viceroys: Colonial Knowledge and Policy Making in India, 1871–1911."

45. For an excellent discussion of Haldane's politics, science, and life in India see Subramaniam, *A Dominant Character: The Radical Science and Restless Politics of JBS Haldane.*

46. Macfarlane, "Letter to J. B. S. Haldane," 12 July 1942.

47. Mukharji, "The Bengali Pharaoh: Upper-Caste Aryanism, Pan-Egyptianism, and the Contested History of Biometric Nationalism in Twentieth-Century Bengal."

48. Breckenridge, *The Biometric State: The Global Politics of Identification and Surveillance in South Africa, 1850 to the Present.*

49. Fanon, "The Trials and Tribulations of National Consciousness," 97.

50. Fanon, "The Trials and Tribulations of National Consciousness," 103.

51. Fanon, "The Trials and Tribulations of National Consciousness," 113.

52. Majumdar, *Race Realities in Cultural Gujarat: Report on the Anthropometric, Serological and Health Survey of Maha Gujarat.*

53. Bandyopadhyay, *Caste, Culture and Hegemony: Social Domination in Colonial Bengal,* 191–239.

54. Mukharji, "Profiling the Profiloscope."

55. Singh, "Introduction," in *The History of the Anthropological Survey of India: Proceedings of a Seminar,* 1.

56. Edney, *Mapping an Empire: The Geographical Construction of British India, 1765–1843.*

57. Bhattacharya and Sarkar, "B. S. Guha: The Founder Director of the Anthropological Survey of India." For the reference to Elwin's proposal, see Singh, "Introduction."

58. Anon., "Curriculum Vitae of Dr. Biraja Sankar Guha (a Sketch)."

59. Thomas, "Reminiscences in the Service of the Anthropological Survey of India," 20–21.

60. Thomas, "Reminiscences in the Service of the Anthropological Survey of India," 33–34.

61. Thomas, "Reminiscences in the Service of the Anthropological Survey of India," 27, 30–31.

62. Basu et al., "Physical Anthropological Research in the Survey: 1948–1990," 73.

63. Thomas, "Reminiscences in the Service of the Anthropological Survey of India," 27.

64. Bhattacharya and Sarkar, "B. S. Guha: The Founder Director of the Anthropological Survey of India," 2–3.

65. Bhattacharya and Sarkar, "B. S. Guha: The Founder Director of the Anthropological Survey of India," 3.

66. Arnold, "Nehruvian Science and Postcolonial India."

67. Basu et al., "Introduction," 4.

68. Madan and Sarana, "Introduction," 4.

69. University Grants Commission, *University Development in India: Basic Facts and Figures, 1964–65,* 104.

70. University Grants Commission, *University Development in India: Basic Facts and Figures, 1964–65,* 83–85.

71. Mukerji, "Majumdar: Scholar and Friend," 8–11.

72. Bashford, "Anticolonial Climates."

73. Anon., "Dhirendra Nath Majumdar: Vita."

74. University Grants Commission, *University Development in India: Basic Facts and Figures, 1964–65,* 102, 104, 107.

75. Undevia et al., "Dr. L.D. Sanghvi."

76. Beatty, "Genetics in the Atomic Age"; Lindee, *American Science and the Survivors at Hiroshima.*

77. Undevia et al., "Dr. L.D. Sanghvi."

78. Béteille, "Race and Descent as Social Categories in India," 448.

79. Béteille, "Race and Descent as Social Categories in India," 449.

80. Béteille, "Race and Descent as Social Categories in India," 461.

81. Béteille, "Race and Descent as Social Categories in India," 448.

82. Sarkar, "Race and Race Movement in India," 371.

83. Sarkar, "Race and Race Movement in India," 371.

84. Dobzhansky, "Commentary," 279–80.

85. Dobzhansky, "Commentary," 279.

86. Sarkar, "Race and Race Movement in India," 372.

87. Bhalla, "Caste as Evolutionary Force in Genetic Change," 99.

88. Bhalla, "Caste as Evolutionary Force in Genetic Change," 100.

89. Mitra, "'Surplus Woman': Female Sexuality and the Concept of Endogamy," 5.

90. M'Lennan, *Primitive Marriage: An Inquiry into the Origins of the Form of Capture in Marriage Ceremonies.*

91. Dodson, *Orientalism, Empire, and National Culture,* 94, 80.

92. Dodson, *Orientalism, Empire, and National Culture.*

93. In ʿAmmar's case, he drew more explicitly upon Ibn Khaldun's precolonial notion of tribal ʾ*asabiyya*. El Shakry, *The Great Social Laboratory: Subjects of Knowledge in Colonial and Postcolonial Egypt.*

94. Guha, "The Role of Social Sciences In Nation Building," 150.

95. Sarkar, "Race and Race Movement in India," 371.

96. Sanghvi, "Inbreeding in India."

97. Beatty, "Genetics in the Atomic Age."

98. Wallace, *Genetic Load: Its Biological and Conceptual Aspects,* 1.

99. Paul, "'Our Load of Mutations' Revisited," 328.

100. Dobzhansky, "Genetic Loads in Natural Populations," 191–94.

101. Chadarevian, "Chromosome Surveys of Human Populations: Between Epidemiology and Anthropology."

102. Paul, "'Our Load of Mutations' Revisited," 332.

103. Dobzhansky, "Of Flies and Men."

104. Morton, Crow, and Muller, "An Estimate of the Mutational Damage in Man from Data on Consanguineous Marriages"; Crow, "Some Possibilities for Measuring Selection Intensities in Man."

105. Sanghvi, "The Concept of Genetic Load: A Critique."

106. Sanghvi, "The Concept of Genetic Load: A Critique," 307.

107. Sanghvi, "Inbreeding in India," 301.

108. Sanghvi, "Genetics of Caste in India."

Chapter 2

1. Macfarlane, "The Racial Affinities of the Jews of Cochin," 13–16.

2. Gilman, *The Jew's Body.*

3. Weitzman, *The Origin of the Jews: The Quest for Roots in a Rootless Age,* 290.

4. Abu El-Haj, *The Genealogical Science: The Search for Jewish Origins and the Politics*

of Epistemology, 17; Burton, *Genetic Crossroads: The Middle East and the Science of Human Heredity*, 116.

5. Malone and Lahiri, "The Distribution of the Blood-Groups of Certain Races and Castes in India."

6. Glass et al., "Genetic Drift in a Religious Isolate."

7. Glass et al., "Genetic Drift in a Religious Isolate,"147–48.

8. Glass et al., "Genetic Drift in a Religious Isolate,"145.

9. Lipphardt, "The Jewish Community of Rome: An Isolated Population? Sampling Procedures and Bio-Historical Narratives in Genetic Analysis in the 1950s," 306–7.

10. Widmer, "Making Blood 'Melanesian': Fieldwork and Isolating Techniques in Genetic Epidemiology," 119.

11. Burton, *Genetic Crossroads: The Middle East and the Science of Human Heredity*, 101–6; Pols and Anderson, "The Mestizos of Kisar: An Insular Racial Laboratory in the Malay Archipelago."

12. Germann, "'Nature's Laboratories of Human Genetics': Alpine Isolates, Hereditary Diseases and Medical Genetic Fieldwork, 1920–1970"; Anderson, "The Anomalous Blonds of the Maghreb: Carleton Coon Invents the African Nordics."

13. Anderson, "From Racial Types to Aboriginal Clines: The Illustrative Career of Joseph B. Birdsell."

14. Lipphardt, "Isolates and Crosses in Human Population Genetics; or, A Contextualization of German Race Science," S76.

15. Glass et al., "Genetic Drift in a Religious Isolate," 147.

16. Burton, *Genetic Crossroads: The Middle East and the Science of Human Heredity*, 73–75, 108–9.

17. Dronamraju, "Mating Systems of the Andhra Pradesh People"; Dutta, "A Note on the Ear Lobe."

18. Hacking, "Making Up People."

19. Hirschfeld [Hirszfeld] and Hirschfeld [Hirszfeld], "Serological Differences between the Blood of Different Races: The Result of Researches on the Macedonian Front," 676.

20. Hirschfeld [Hirszfeld] and Hirschfeld [Hirszfeld], "Serological Differences between the Blood of Different Races: The Result of Researches on the Macedonian Front,"677.

21. Bates, "Race, Caste and Tribe in Central India: The Early Origins of Indian Anthropometry."

22. Malone and Lahiri, "The Distribution of the Blood-Groups of Certain Races and Castes in India," 965.

23. Malone and Lahiri, "The Distribution of the Blood-Groups of Certain Races and Castes in India," 965.

24. Macfarlane, "The Racial Affinities of the Jews of Cochin," 4–5.

25. Macfarlane, "The Racial Affinities of the Jews of Cochin," 4.

26. Macfarlane, "The Racial Affinities of the Jews of Cochin," 5.

27. Macfarlane, "The Racial Affinities of the Jews of Cochin," 16.

28. Macfarlane, "The Racial Affinities of the Jews of Cochin," 17–18.

29. Macfarlane, "Genetics," 3.

30. Sanghvi and Khanolkar, "Data Relating to Seven Genetical Characters in Six Endogamous Groups in Bombay," 62.

31. Lipphardt, "Isolates and Crosses in Human Population Genetics; or, A Contextualization of German Race Science," S78.

32. Gannett, "The Biological Reification of Race."

33. Lipphardt and Niewöhner, "Producing Difference in an Age of Biosociality: Biohistorical Narratives, Standardisation and Resistance as Translations," 48.

34. Sirsat, "Exploring Nature's Secrets."

35. Sirsat, "Effects of Migration on Some Genetical Characters in Six Endogamous Groups in India."

36. Sirsat, "Effects of Migration on Some Genetical Characters in Six Endogamous Groups in India," 152.

37. Bhattacharjee, "Genetic Survey of Rarhi Brahmins and Muslims"; Bhattacharjee, "Distribution of the Blood Groups (A_1 A_2 B O, MNSs, Rh), and the Secretor Factor among the Muslims and the Pandits of Kashmir."

38. Burton, "Red Crescents: Race, Genetics, and Sickle Cell Disease in the Middle East," 251.

39. Bhattacharjee, "The Blood Groups (A1 A2 B O, MNS and Rh) of the Ladakhis," 82.

40. Bhattacharjee, "A Genetical Study of the Santals of the Santal Parganas," 101.

41. Bhattacharjee, "Genetic Survey of Rarhi Brahmins and Muslims."

42. Mahapatra and Das, "Taste Threshold for Phenylthiocarbamide in Some Endogamous Groups of Assam."

43. Bansal, "A Study of ABO Blood Groups of the People of Ladakh."

44. Majumdar, "Muslim Blood Groups with Particular Reference to the U.P."

45. Kumar and Ghosh, "ABO Blood Groups and Sickle-Cell Trait Investigations in Madhya Pradesh: Ujjain and Dewas Districts," 55.

46. Molina, "Amerindians, Europeans, Makiritare, Mestizos, Puerto Rican, and Quechua: Categorical Heterogeneity in Latin American Human Biology," 657.

47. Molina, "Amerindians, Europeans, Makiritare, Mestizos, Puerto Rican, and Quechua: Categorical Heterogeneity in Latin American Human Biology," 657.

48. Molina, "Amerindians, Europeans, Makiritare, Mestizos, Puerto Rican, and Quechua: Categorical Heterogeneity in Latin American Human Biology," 667–72.

49. Skaria, "Shades of Wildness: Tribe, Caste, and Gender in Western India"; Ghosh, "A Market for Aboriginality: Primitivism and Race Classification in the Indentured Labour Market of Colonial India."

50. Macfarlane, "The Racial Affinities of the Jews of Cochin," 14–15.

51. Sirsat, "Effects of Migration on Some Genetical Characters in Six Endogamous Groups in India," 150.

52. Vyas et al., "Study of Blood Groups and Other Genetical Characters in Six Endogamous Groups in Western India," 185.

53. Sanghvi, "Comparison of Genetical and Morphological Methods for a Study of Biological Differences," 387.

54. Dobzhansky, "Mendelian Populations and Their Evolution," 405.

55. Gannett, "Theodosius Dobzhansky and the Genetic Race Concept," 259.

56. Dobzhansky, "Mendelian Populations and Their Evolution," 407.

57. Gannett, "Racism and Human Genome Diversity Research: The Ethical Limits of 'Population Thinking,'" S485.

58. Molina, "Amerindians, Europeans, Makiritare, Mestizos, Puerto Rican, and Quechua: Categorical Heterogeneity in Latin American Human Biology," 672–73.

59. Paik, "Amchya Jalmachi Chittarkatha (The Bioscope of Our Lives): Who Is My Ally?"; Paik, "Mangala Bansode and the Social Life of Tamasha: Caste, Sexuality, and Discrimination in Modern Maharashtra."

60. Soneji, *Unfinished Gestures: Devadasis, Memory, and Modernity in South India*; Ramberg, *Given to the Goddess: South Indian Devadasis and the Sexuality of Religion.*

61. Sarkar, "Talking about Scandals: Religion, Law and Love in Late Nineteenth Century Bengal"; Shodhan, "Women in the Maharaj Libel Case."

62. Chatterjee, "Genealogy, History and Law: The Case of Tripura Rajamala"; Chatterjee, "Monastic 'Governmentality': Revisiting 'Community' and 'Communalism' in South Asia."

63. Mitra, "'Surplus Woman': Female Sexuality and the Concept of Endogamy," 23.

64. Mitra, "'Surplus Woman': Female Sexuality and the Concept of Endogamy," 21–22.

65. Rocher and Davis Jr., "The Aurasa Son."

66. Ambedkar, "Castes in India."

67. Sekhar Bandyopadhyay, "Caste, Social Reform and the Dilemmas of Indian Modernity: Reading Acharya Prafulla Chandra Ray," 39.

68. Parikh, Baxi, and Jhala, "Blood Groups, Abnormal Haemoglobins and Other Genetical Characters in Three Gujarati-Speaking Groups," 486.

69. Lipphardt, "The Jewish Community of Rome: An Isolated Population? Sampling Procedures and Bio-Historical Narratives in Genetic Analysis in the 1950s," 308.

70. Lipphardt, "The Jewish Community of Rome: An Isolated Population? Sampling Procedures and Bio-Historical Narratives in Genetic Analysis in the 1950s," 308.

71. Lipphardt, "The Jewish Community of Rome: An Isolated Population? Sampling Procedures and Bio-Historical Narratives in Genetic Analysis in the 1950s," 309.

72. Lindee, "Provenance and the Pedigree: Victor McKusick's Fieldwork with the Old Order Amish."

73. Macfarlane, "The Racial Affinities of the Jews of Cochin."

74. Sirsat, "Effects of Migration on Some Genetical Characters in Six Endogamous Groups in India."

75. Bhattacharjee, "Genetic Survey of Rarhi Brahmins and Muslims"; Bhattacharjee, "Distribution of the Blood Groups (A_1 A_2 B O, MNSs, Rh), and the Secretor Factor among the Muslims and the Pandits of Kashmir."

76. Kumar and Ghosh, "ABO Blood Groups and Sickle-Cell Trait Investigations in Madhya Pradesh: Ujjain and Dewas Districts."

77. Lal, *Hindu America: Revealing the Story of the Romance of the Surya Vanshi Hindus and Depicting the Imprints of Hindu Culture on the Two Americas*; For "weird history" see Melleuish, Sheiko, and Brown, "Pseudo History/Weird History: Nationalism and the Internet."

78. Lipphardt and Niewöhner, "Producing Difference in an Age of Biosociality: Biohistorical Narratives, Standardisation and Resistance as Translations," 52.

79. Abu El Haj, *The Genealogical Science: The Search for Jewish Origins and the Politics of Epistemology*, 238.

80. Lipphardt and Niewöhner, "Producing Difference in an Age of Biosociality: Biohistorical Narratives, Standardisation and Resistance as Translations," 52.

81. Abu El Haj, *The Genealogical Science: The Search for Jewish Origins and the Politics of Epistemology*, 239..

82. Ragab, "Islam Intensified: Snapshot Historiography and the Making of Muslim Identities," 216.

83. Ragab, "Islam Intensified: Snapshot Historiography and the Making of Muslim Identities," 210.

84. Sirsat, "Effects of Migration on Some Genetical Characters in Six Endogamous Groups in India," 150–53. On the historical heterogeneity of the Marathas, see Deshpande, "Caste as Maratha: Social Categories, Colonial Policy and Identity in Early Twentieth-Century Maharashtra."

85. Sirsat, "Effects of Migration on Some Genetical Characters in Six Endogamous Groups in India," 146.

86. Egorova and Parfitt, "Genetics, History, and Identity: The Case of the Bene Israel and the Lemba," 206.

87. Egorova and Parfitt, "Genetics, History, and Identity: The Case of the Bene Israel and the Lemba," 205–7.

88. Egorova and Parfitt, "Genetics, History, and Identity: The Case of the Bene Israel and the Lemba," 207–8.

89. Egorova and Parfitt, "Genetics, History, and Identity: The Case of the Bene Israel and the Lemba," 208–9.

90. Ahmed, *The Bengal Muslims, 1871–1906*.

91. Banerji, *West Bengal District Gazetteers: Hooghly*, 230–31, 219–20.

92. Banerji, *West Bengal District Gazetteers: Hooghly*, 229–30.

93. Ahmed, *The Bengal Muslims, 1871–1906*, 7.

94. Ahmed, *The Bengal Muslims, 1871–1906*, 10.

95. Crawford, *A Brief History of the Hughli District*, 78–79.

96. Macfarlane, "The Racial Affinities of the Jews of Cochin," 2.

97. Segal, "White and Black Jews at Cochin, the Story of a Controversy," 236–37.

98. Macfarlane, "The Racial Affinities of the Jews of Cochin," 9.

99. Segal, "White and Black Jews at Cochin, the Story of a Controversy," 236–37.

100. Anon., "Manifests of Passengers Arriving at St. Albans, VT, District through Canadian Pacific and Atlantic Ports, 1895–1954" (1919).

101. Goldstein, *The Price of Whiteness: Jews, Race, and American Identity*, 124.

102. Goldstein, *The Price of Whiteness: Jews, Race, and American Identity*, 122.

103. Brodkin, *How Jews Became White Folks and What That Says about America*.

104. Snow, "The Civilization of White Men: The Race of the Hindu in United States v. Bhagat Singh Thind," 265.

105. Snow, "The Civilization of White Men: The Race of the Hindu in United States v. Bhagat Singh Thind."

106. Coulson, *Race, Nation, and Refuge: The Rhetoric of Race in Asian American Citizenship Cases*, 51.

107. Macfarlane, "The Racial Affinities of the Jews of Cochin," 2.

108. Macfarlane, "The Racial Affinities of the Jews of Cochin," 18.

109. Reardon and TallBear, "'Your DNA Is Our History': Genomics, Anthropology, and the Construction of Whiteness as Property"; TallBear, *Native American DNA: Tribal Belonging and the False Promise of Genetic Science*; TallBear, "Genomic Articulations of Indigeneity"; Kowal, Radin, and Reardon, "Indigenous Body Parts, Mutating Temporalities, and the Half-Lives of Postcolonial Technoscience"; Kowal, "Orphan DNA: Indigenous Samples, Ethical Biovalue and Postcolonial Science."

110. Baviskar, "Adivasi Encounters."

111. Gould, *Hindu Nationalism and the Language of Politics in Late Colonial India*, 69.

112. Gould, *Hindu Nationalism and the Language of Politics in Late Colonial India*, 267.

113. Rajagopal, "Hindu Nationalism in the US: Changing Configurations of Political Practice," 467.

114. McGonigle, *Genomic Citizenship: The Molecularization of Identity in the Contemporary Middle East*, 31–62.

115. Egorova, "The Substance That Empowers: DNA in South Asia."

116. TallBear, "Genomic Articulations of Indigeneity"; Watt, Kowal, and Cummings, "Traditional Laws Meet Emerging Biotechnologies: The Impact of Genetic Genealogy on Indigenous Land Title in Australia"; Burton, *Genetic Crossroads: The Middle East and the Science of Human Heredity*.

117. Bald, *Bengali Harlem and the Lost Histories of South Asian America*.

118. Robbins and Sohoni, *Jewish Heritage of the Deccan, Mumbai, the Northern Konkan, Pune*.

Chapter 3

1. Seth, "PTC Taste Distribution among the Betel Chewers, Non-Vegetarians and Smokers," 62.

2. Race and Sanger, *Blood Groups in Man*, 370.

3. Vyas et al., "Study of Blood Groups and Other Genetical Characters in Six Endogamous Groups in Western India," 185.

4. Chakraborty, "A Theorem on Race Mixture."

5. Chadarevian, "Chromosome Surveys of Human Populations: Between Epidemiology and Anthropology"; Lipphardt, "'Geographical Distribution Patterns of Various Genes'"; Bangham, "Blood Groups and Human Groups: Collecting and Calibrating Genetic Data after World War Two."

6. Anon., "Scientific News," 14.

7. Fox, "The Relationship between Chemical Constitution and Taste," 115.

8. Sanghvi and Khanolkar, "Data Relating to Seven Genetical Characters in Six Endogamous Groups in Bombay."

9. Sirsat, "Effects of Migration on Some Genetical Characters in Six Endogamous Groups in India."

10. Bhattacharya, "Tasting of P.T.C. among the Anglo-Indians of India," 164.

11. Anderson, *Cultivation of Whiteness*.

12. Mukharji, "Bloodworlds: A Hematology of the 1952 Indo-Australian Genetical Survey of the Chenchus."

13. Simmons et al., "A Genetical Survey in Chenchu, South India: Blood, Taste and Secretion," 502.

14. Mukharji, "Bloodworlds: A Hematology of the 1952 Indo-Australian Genetical Survey of the Chenchus."

15. Khullar, "Taste Sensitivity to Phenylthiocarbamide among Sindhi Children of Delhi"; Sharma, "Taste Sensitivity to Phenylthiocarbamide among Three Mongoloid Populations of the Indian Borders"; Chattopadhyay, "Taste Sensitivity to Phenylthiocarbamide among the Jats."

16. Parmar, "Taste Sensitivity to Phenyl Thio Carbamide (P.T.C.) in Gorkhas of Dhauladhar Range (Himachal Pradesh)."

17. Tripathy, "PTC Taste Sensitivity in Some Orissan Castes"; Mahapatra and Das, "Taste Threshold for Phenylthiocarbamide in Some Endogamous Groups of Assam."

18. Sharma, "Taste Sensitivity to Phenylthiocarbamide among Three Mongoloid Populations of the Indian Borders," 317.

19. Sharma, "Taste Sensitivity to Phenylthiocarbamide among Three Mongoloid Populations of the Indian Borders," 323.

20. Bhattacharya, "Tasting of P.T.C. among the Anglo-Indians of India," 165.

21. Mahapatra and Das, "Taste Threshold for Phenylthiocarbamide in Some Endogamous Groups of Assam."

22. Sirsat, "Effects of Migration on Some Genetical Characters in Six Endogamous Groups in India."

23. Das, "Application of Phenylthiocarbamide Taste Character in the Study of Racial Variation," 69.

24. OED Online, Oxford University Press, www.oed.com.

25. On the fascinating cultural history of early PTC testing in the United States see, Berenstein, "There's No Voting on Matters of Taste: Phenylthiocarbamide and Genetics Education."

26. Blakeslee, "Genetics of Sensory Thresholds: Taste for Phenyl Thio Carbamide," 122.

27. Blakeslee, "Genetics of Sensory Thresholds: Taste for Phenyl Thio Carbamide," 123.

28. Blakeslee, "Genetics of Sensory Thresholds: Taste for Phenyl Thio Carbamide," 124.

29. Falconer, "Sensory Thresholds for Solutions of Phenyl-Thio-Carbamide," 212.

30. Simmons et al., "A Genetical Survey in Chenchu, South India: Blood, Taste and Secretion," 502.

31. Das, "A Contribution to the Heredity of the P.T.C. Taste Character Based on a Study of 845 Sib-Pairs," 336.

32. Miki, Tanaka, and Furuhata, "On the Distribution of the ABO Blood Groups and the Taste Ability for Phenyl-Thio-Carbamide (P.T.C.) of the Lepchas and the Khasis," 79–80.

33. Agarwal, "A Study on ABO Blood Groups, P.T.C. Taste Sensitivity, Sickle Cell Trait and Middle Phalangeal Hairs among the Burmese Immigrants of Andaman Islands."

34. Sharma, "Taste Sensitivity to Phenylthiocarbamide among Three Mongoloid Populations of the Indian Borders."

35. Tyagi, "Taste Sensitivity to Phenyl-Thio-Urea (P.T.C.) among Oraons and Mundas of Ranchi (India)"; Tripathy, "PTC Taste Sensitivity in Some Orissan Castes."

36. Solomon, *Metabolic Living: Food, Fat, and the Absorption of Illness in India*, 51, 176.

37. Dey, *The Indigenous Drugs of India*, 307–8.

38. Roy, *Malarial Subjects*, 360–61.

39. Blakeslee and Salmon, "Odor and Taste Blindness"; Blakeslee, "Genetics of Sensory Thresholds: Taste for Phenyl Thio Carbamide."

40. Falconer, "Sensory Thresholds for Solutions of Phenyl-Thio-Carbamide," 211.

41. Falconer, "Sensory Thresholds for Solutions of Phenyl-Thio-Carbamide," 212.

42. Harris and Kalmus, "The Measurement of Taste Sensitivity to Phenylthiourea (P.T.C.)," 25.

43. Mohr, "Taste Sensitivity to Phenylthiourea in Denmark"; Büchi, "Blood Secretion and Taste among the Pallar—A South Indian Community."

44. Boyd and Boyd, "Sexual and Racial Variations in Ability to Taste Phenyl-Thio-Carbamide with Some Data on the Inheritance."

45. Barnicot, "Taste Deficiency for Phenylthiourea in African Negroes and Chinese."

46. Tyagi, "Taste Sensitivity to Phenyl-Thio-Urea (P.T.C.) among Oraons and Mundas of Ranchi (India)"; Agarwal, "ABO Blood Groups, P.T.C. Taste Sensitivity, Sickle Cell Trait, Middle Phalangeal Hairs, and Colour Blindness in the Coastal Nicobarese

of Great Nicobar"; Khullar, "Taste Sensitivity to Phenylthiocarbamide among Sindhi Children of Delhi"; Srivastava, "Frequency of Non-Tasters among the Danguria Tharu of Uttar Pradesh"; Srivastava, "Further Data on Non-Tasters among the Tharus of Uttar Pradesh"; Parmar, "Taste Sensitivity to Phenyl Thio Carbamide (P.T.C.) in Gorkhas of Dhauladhar Range (Himachal Pradesh)."

47. Das, "Application of Phenylthiocarbamide Taste Character in the Study of Racial Variation."

48. Srivastava, "Measurement of Taste Sensitivity to Phenylthiourel (P.T.C.) in Uttar Pradesh."

49. Bhattacharjee, "Genetic Survey of Rarhi Brahmins and Muslims," 24.

50. Das, "Application of Phenylthiocarbamide Taste Character in the Study of Racial Variation," 66.

51. Chi square values are a single number that expresses the difference between the observed value and the value one would expect if there was no statistically valid relationship uniting the group.

52. Tripathy, "PTC Taste Sensitivity in Some Orissan Castes."

53. Mahapatra and Das, "Taste Threshold for Phenylthiocarbamide in Some Endogamous Groups of Assam."

54. Mukharji, "Profiling the Profiloscope"; Mukharji, "From Serosocial to Sanguinary Identities."

55. Ambedkar, "Castes in India," 84. A similar, though not identical, conclusion was arrived at by the chemist Sir P. C. Ray about the historical variations of the practice of caste endogamy. See Bandyopadhyay, "Caste, Social Reform and the Dilemmas of Indian Modernity," 39.

56. Arya and Rathore, *Dalit Feminist Theory: A Reader*. See also Mitra, "'Surplus Woman': Female Sexuality and the Concept of Endogamy."

57. Ambedkar, "Castes in India," 83.

58. Rao, *The Caste Question: Dalits and the Politics of Modern India*, 125.

59. Paik, *Dalit Women's Education in Modern India: Double Discrimination*.

60. Mitra, "'Surplus Woman': Female Sexuality and the Concept of Endogamy," 22.

61. Boyd and Boyd, "Sexual and Racial Variations in Ability to Taste Phenyl-Thio-Carbamide with Some Data on the Inheritance," 48.

62. Boyd and Boyd, "Sexual and Racial Variations in Ability to Taste Phenyl-Thio-Carbamide with Some Data on the Inheritance," 48.

63. Bhattacharjee, "Genetic Survey of Rarhi Brahmins and Muslims," 24–25.

64. Das and Bhattacharjee, "Blood Groups (A1A2BO), ABH Secretion, Sickle-Cell, P.T.C. Taste and Colour Blindness in the Rajbanshi of Midnapur District, West Bengal," 4.

65. Mahapatra and Das, "Taste Threshold for Phenylthiocarbamide in Some Endogamous Groups of Assam," 26.

66. Bhattacharjee, "Genetic Survey of Rarhi Brahmins and Muslims," 24.

67. Amrhein, Greenland, and McShane, "Scientists Rise Up against Statistical Significance," 306.

68. Amrhein, Greenland, and McShane, "Scientists Rise Up against Statistical Significance," 306.

69. Vyas et al., "Study of Blood Groups and Other Genetical Characters in Six Endogamous Groups in Western India," 193.

70. Vyas et al., "Study of Blood Groups and Other Genetical Characters in Six Endogamous Groups in Western India," 192.

71. Sanghvi and Khanolkar, "Data Relating to Seven Genetical Characters in Six Endogamous Groups in Bombay," 56.

72. Khullar, "Taste Sensitivity to Phenylthiocarbamide among Sindhi Children of Delhi"; Agarwal, "A Study on ABO Blood Groups, P.T.C. Taste Sensitivity, Sickle Cell Trait and Middle Phalangeal Hairs among the Burmese Immigrants of Andaman Islands."

73. Gordon, *Ghostly Matters: Haunting and the Sociological Imagination*, 197.

74. Gordon, *Ghostly Matters: Haunting and the Sociological Imagination*, 8.

75. Gordon, *Ghostly Matters: Haunting and the Sociological Imagination*, 200.

76. Das, "Inheritance of the P.T.C. Taste Characters in Man: An Analysis of 126 Rarhi Brahmin Families of West Bengal.," 210.

77. Das, "Inheritance of the P.T.C. Taste Characters in Man: An Analysis of 126 Rarhi Brahmin Families of West Bengal."

78. Skude, "Sweet Taste Perception for Phenylthiourea (P.T.C.)."

79. Skude, "Complexities of Human Taste Variation"; Skude, "Saliva and Sweet Taste Perception for Phenylthiourea (P.T.C.)"; Skude, "Studies in Sweet Taste Perception for Phenylthiourea (P.T.C.)."

80. Das, "Application of Phenylthiocarbamide Taste Character in the Study of Racial Variation," 68.

81. Sen, *Comprehensive Anglo-Bengali Dictionary*, 108.

82. Sen, *Comprehensive Anglo-Bengali Dictionary*, 13.

83. Mitter, *Bengali and English Dictionary, for the Use of Schools*, 122.

84. Mitter, *Bengali and English Dictionary, for the Use of Schools*, 51.

85. Haughton, *A Dictionary, Bengali and Sanskrit, Explained in English*, 1332.

86. Haughton, *A Dictionary, Bengali and Sanskrit, Explained in English*, 584.

87. Highmore, "Bitter after Taste: Affect, Food, and Social Aesthetics," 119–20.

88. Highmore, "Bitter after Taste: Affect, Food, and Social Aesthetics," 119.

89. Solomon, *Metabolic Living: Food, Fat, and the Absorption of Illness in India*, 15.

90. Highmore, "Bitter after Taste: Affect, Food, and Social Aesthetics."

91. Mukharji, "Historicizing 'Indian Systems of Knowledge': Ayurveda, Exotic Foods, and Contemporary Antihistorical Holisms."

92. Kaviraj, *Dravyagun Darpan*, 44.

93. On Ayurvedic pathogenesis in early modern Bengal, see Mukharji, *Doctoring Traditions*.

94. Kaviraj, *Dravyagun Darpan*, 100.

95. Kaviraj, *Dravyagun Darpan*, 10, 29.

96. Kaviraj, *Dravyagun Darpan*, 94, 68, 64–65.

97. Kaviraj, *Dravyagun Darpan*, 13.

98. Kaviraj, *Dravyagun Darpan*, 17.

99. Kaviraj, *Dravyagun Darpan*, 79.

100. Kaviraj, *Dravyagun Darpan*, 44.

101. Highmore, "Bitter after Taste: Affect, Food, and Social Aesthetics," 119.

102. Sharma, "Blood and P.T.C. Taste Studies in Punjabis and the Effects of Age and Certain Eating Habits on Taste Thresholds."

103. Seth, "PTC Taste Distribution among the Betel Chewers, Non-Vegetarians and Smokers."

104. Tripathy, "PTC Taste Sensitivity in Some Orissan Castes."

105. Seth, "PTC Taste Distribution among the Betel Chewers, Non-Vegetarians and Smokers," 37.

106. Dragsdahl, "The Practices of Indian Vegetarianism in a World of Limited Resources: The Case of Bengaluru."

107. Lugg and Whyte, "Taste Thresholds for Phenylthiocarbamide of Some Population Groups: The Thresholds of Some Civilized Ethnic Groups Living in Malaya"; Lugg, "Taste-Thresholds for Phenylthiocarbamide of Some Population Groups: The Threshold of Two Uncivilized Ethnic Groups Living in Malaya."

108. Das, Mukherjee, and Bhattacharjee, "P.T.C. Taste Threshold Distribution in the Bado Gadaba and the Bareng Paroja of Koraput District in Orissa."

109. Das and Mukherjee, "Phenylthiocarbamide Taste Sensitivity Survey among the Pareng Gadaba, the Ollaro Gadabaand the Konda Paroja of Koraput District, Orissa."

110. Das, "Application of Phenylthiocarbamide Taste Character in the Study of Racial Variation."

111. Matson, "Blood Groups and Ageusia in Indians of Montana and Alberta."

112. Tracy, "Delicious Molecules: Big Food Science, the Chemosenses, and Umami," 91.

113. Tracy, "Delicious Molecules: Big Food Science, the Chemosenses, and Umami."

114. Anderson and Lindee, "Pacific Biologies: How Humans Become Genetic," 487.

115. Roberts, "The Death of the Sensuous Chemist."

116. Roos, *The Salt of the Earth: Natural Philosophy, Medicine, and Chymistry in England, 1650–1750*.

117. Roberts, "The Death of the Sensuous Chemist."

118. Lourdusamy, *Science and National Consciousness in Bengal*. See also Nandy, *Alternative Sciences: Creativity and Authenticity in Two Indian Scientists*.

119. Anon., "Life Sketch of Sudhir Ranjan Das."

120. Anon., "Life Sketch of Sudhir Ranjan Das."

121. Anon., "Life Sketch of Sudhir Ranjan Das."

122. Anon., "Life Sketch of Sudhir Ranjan Das."

123. Dasgupta and Hauspie, "Editor's Note."

124. Kapoor, "The Smell of Caste: Leatherwork and Scientific Knowledge in Colonial India," 992. On the sensorial dimensions of caste see also Sarbadhikary, "The Leftover Touch: Sensing Caste in Modern Urban Lives of a Devotional Instrument."

125. Kapoor, "The Smell of Caste: Leatherwork and Scientific Knowledge in Colonial India," 993.

126. Smith, *How Race Is Made: Slavery, Segregation and the Senses*, 2.

127. Smith, *How Race Is Made: Slavery, Segregation and the Senses*, 4. See also Smith's comments on Benjamin Rush's ideas about the "smell of the Negro" (18).

128. Gordin, *On the Fringe: Where Science Meets Pseudoscience*, vii.

Chapter 4

1. Dacie, "Hermann Lehmann, 8 July 1910–13 July 1985," 418.

2. Dacie, "Hermann Lehmann, 8 July 1910–13 July 1985," 418.

3. Fullwiley, *The Enculturated Gene: Sickle Cell Health Politics and Biological Difference in West Africa*, x.

4. Wailoo, *Drawing Blood: Technology and Disease Identity in Twentieth-Century America*; Chadarevian, "Following Molecules: Hemoglobin between the Clinic and the Laboratory"; Tapper, *In the Blood: Sickle Cell Anemia and the Politics of Race*; Wailoo, *Dying in the City of Blues: Sickle Cell Anemia and the Politics of Race and Health*; Fullwiley, *The Enculturated Gene: Sickle Cell Health Politics and Biological Difference in West*

Africa; Burton, "Red Crescents: Race, Genetics, and Sickle Cell Disease in the Middle East."

5. Burton, "Red Crescents: Race, Genetics, and Sickle Cell Disease in the Middle East."

6. Pandey, *Role of the Modulating Factors on the Phenotype of Sickle Cell Disease.*

7. Wailoo, *Drawing Blood: Technology and Disease Identity in Twentieth-Century America*, 134.

8. Wailoo, *Drawing Blood: Technology and Disease Identity in Twentieth-Century America*, 145–48.

9. Fullwiley, *The Enculturated Gene: Sickle Cell Health Politics and Biological Difference in West Africa*, x.

10. Wailoo, *Drawing Blood: Technology and Disease Identity in Twentieth-Century America*, 141.

11. On asymptomatic carriers, see Leavitt, "'Typhoid Mary' Strikes Back: Bacteriological Theory and Practice in Early Twentieth-Century Public Health."

12. Chadarevian, "Following Molecules: Hemoglobin between the Clinic and the Laboratory," 160.

13. Burton, "Red Crescents: Race, Genetics, and Sickle Cell Disease in the Middle East," 256.

14. Chadarevian, "Following Molecules: Hemoglobin between the Clinic and the Laboratory," 167.

15. Wailoo, *Drawing Blood: Technology and Disease Identity in Twentieth-Century America*, 154–59.

16. Wailoo, *Drawing Blood: Technology and Disease Identity in Twentieth-Century America*, 158.

17. Dacie, "Hermann Lehmann, 8 July 1910–13 July 1985," 415; Chadarevian, "Following Molecules: Hemoglobin between the Clinic and the Laboratory," 168.

18. Chadarevian, "Following Molecules: Hemoglobin between the Clinic and the Laboratory," 169.

19. Chakrabarti, *Bacteriology in British India: Laboratory Medicine and the Tropics.*

20. Dacie, "Hermann Lehmann, 8 July 1910–13 July 1985."

21. Lehmann and Cutbush, "Sickle-Cell Trait in Southern India."

22. Chadarevian, "Following Molecules: Hemoglobin between the Clinic and the Laboratory," 172.

23. Burton, "Red Crescents: Race, Genetics, and Sickle Cell Disease in the Middle East."

24. Burton, "Red Crescents: Race, Genetics, and Sickle Cell Disease in the Middle East," 255.

25. Burton, "Red Crescents: Race, Genetics, and Sickle Cell Disease in the Middle East," 253–59.

26. Dunlop and Mozumder, "The Occurrence of Sickle Cell Anaemia among a Group of Tea Garden Labourers in Upper Assam."

27. Dunlop and Mozumder, "The Occurrence of Sickle Cell Anaemia among a Group of Tea Garden Labourers in Upper Assam," 391.

28. Ghosh, "A Market for Aboriginality: Primitivism and Race Classification in the Indentured Labour Market of Colonial India"; Sharma, *Empire's Garden: Assam and the Making of India.*

29. Dunlop and Mozumder, "The Occurrence of Sickle Cell Anaemia among a Group of Tea Garden Labourers in Upper Assam," 391.

30. Büchi, "Is Sickling a Weddid Trait?"

31. Mukharji, "Bloodworlds: A Hematology of the 1952 Indo-Australian Genetical Survey of the Chenchus."

32. Huxley, "On the Geographical Distribution of the Chief Modifications of Mankind," 404–5.

33. Haddon, *The Races of Man and Their Distribution*, 7.

34. Giuffrida-Ruggeri, *The First Outlines of Systematic Anthropology of Asia*, 53.

35. Giuffrida-Ruggeri, *The First Outlines of Systematic Anthropology of Asia*, 54–56.

36. Bhasin and Walter, *Genetics of Castes and Tribes of India*, 27–29.

37. On the Sarasins and their work on the Veddas, see Schär, "From Batticaloa via Basel to Berlin: Transimperial Science in Ceylon and Beyond around 1900."

38. Sysling, *Racial Science and Human Diversity in Colonial Indonesia.*

39. Guha, "Lower Strata, Older Races, and Aboriginal Peoples: Racial Anthropology and Mythical History Past and Present"; Sysling, *Racial Science and Human Diversity in Colonial Indonesia*, 103.

40. Büchi, "Is Sickling a Weddid Trait?" 27.

41. Büchi, "Is Sickling a Weddid Trait?" 28.

42. Luzzato, "In Memoriam: P. K. Sukumaran."

43. Lehmann and Sukumaran, "Examination of 146 South Indian Aboriginals for Haemoglobin Variants."

44. Sukumaran, Sanghvi, and Vyas, "Sickle-Cell Trait in Some Tribes of Western India."

45. Shukla and Solanki, "Sickle-Cell Trait in Central India"; Shukla, Solanki, and Parande, "Sickle Cell Disease in India."

46. Shukla and Solanki, "Sickle-Cell Trait in Central India," 297.

47. Shukla and Solanki, "Sickle-Cell Trait in Central India," 298.

48. Shukla and Solanki, "Sickle-Cell Trait in Central India," 297.

49. Deshpande, "Caste as Maratha: Social Categories, Colonial Policy and Identity in Early Twentieth-Century Maharashtra," 8.

50. Shukla and Solanki, "Sickle-Cell Trait in Central India," 297.

51. Negi, "The Incidence of Sickle-Cell Trait in Two Bastar Tribes, I"; Negi, "The Incidence of Sickle-Cell Trait in Bastar, II"; Negi, "New Incidence of Sickle-Cell Trait in Bastar, III."

52. Negi, "New Incidence of Sickle-Cell Trait in Bastar, III," 173.

53. Negi, "The Incidence of Sickle-Cell Trait in Bastar, II"; Negi, "New Incidence of Sickle-Cell Trait in Bastar, III."

54. Fullwiley, *The Enculturated Gene: Sickle Cell Health Politics and Biological Difference in West Africa*, xi–xii.

55. Fullwiley, *The Enculturated Gene: Sickle Cell Health Politics and Biological Difference in West Africa*, xi–xii.

56. Oka and Kusimba, "Siddi as Mercenary or as African Success Story on the West Coast of India"; McLeod, "Marriage and Identity among the Siddis of Janjira and Sachin"; Obeng, "Religion and Empire: Belief and Identity among African Indians of Karnataka, South India"; Robbins and McLeod, *African Elites in India: Habshi Amarat.*

57. Delafresnaye, "Foreword."

58. Basu, "J. B. Chatterjea: An Obituary."

59. Burton, "Red Crescents: Race, Genetics, and Sickle Cell Disease in the Middle East," 254–55.

60. Basu, "J. B. Chatterjea: An Obituary."

61. Chatterjea, "Haemoglobinopathy in India," 336.

62. Mukherji, "Cooley's Anaemia (Erythroblastic or Mediterranean Anaemia)."

63. Chatterjea, "Haemoglobinopathy in India."

64. Chadarevian, "Following Molecules: Hemoglobin between the Clinic and the Laboratory," 160–61.

65. Chatterjea, "Haemoglobinopathy in India," 334.

66. Chatterjea, "Haemoglobinopathy in India," 333.

67. Chatterjea, "Haemoglobinopathy in India," 337.

68. Sanghvi, Sukumaran, and Lehmann, "Haemoglobin J Trait in Two Indian Women: Associated with Thalassaemia in One."

69. Sukumaran et al., "Haemoglobin L in Bombay: Findings in Three Gujarati Speaking Lohana Families," 204.

70. Roy and Roy Chaudhuri, "Sickle-Cell Trait in the Tribal Population in Madhya Pradesh and Orissa (India)."

71. Swarup Mitra and Dutta, "Jyoti Bhusan Chatterjea," 158.

72. Anon., "Council of the Indian Anthropological Society, 1968–69."

73. Cantor, "The Frustrations of Families: Henry Lynch, Heredity, and Cancer Control, 1962–1975"; World Health Organization, "Research on Genetics in Psychiatry: Report of a WHO Scientific Group."

74. Raper, "Unusual Haemoglobin Variant in a Gujerati Indian."

75. Sukumaran et al., "Haemoglobin L in Bombay: Findings in Three Gujarati Speaking Lohana Families"; Chatterjea, "Haemoglobinopathies, Glucose-6-Phosphate Dehydrogenase Deficiency and Allied Problems in the Indian Subcontinent."

76. Vella and de V. Hart, "Sickle-Cell Anaemia in an Indian Family in Malaya."

77. Dunlop and Mozumder, "The Occurrence of Sickle Cell Anaemia among a Group of Tea Garden Labourers in Upper Assam."

78. Chatterjea, "Haemoglobinopathies, Glucose-6-Phosphate Dehydrogenase Deficiency and Allied Problems in the Indian Subcontinent."

79. Botha and van Zyl, "Abnormal Haemoglobins in Cape Town."

80. Sinha, "Totaram Sanadhya's Fiji Mein Mere Ekkis Varsh: A History of Empire and Nation in a Minor Key."

81. Berk and Bull, "A Case of Sickle Cell Anaemia in an Indian Woman."

82. Mital et al., "A Focus of Sickle Cell Gene near Bombay." See also Parekh, "Hereditary Haemolytic Anaemias."

83. Chatterjea, "Haemoglobinopathies, Glucose-6-Phosphate Dehydrogenase Deficiency and Allied Problems in the Indian Subcontinent."

84. Banerjee, "Debt, Tome and Extravagance: Money and the Making of 'Primitives' in Colonial Bengal."

85. Chatterjea, "Haemoglobinopathy in India," 335–36.

86. Ager and Lehmann, "Haemoglobin K in an East Indian and His Family," 1449.

87. Wailoo, "Who Am I? Genes and the Problem of Historical Identity," 17.

88. Abu El-Haj, "The Genetic Reinscription of Race."

89. Wailoo, *Drawing Blood: Technology and Disease Identity in Twentieth-Century America*, 131.

90. Sanghvi, "Mystery That Is Heredity," 7.

91. Sanghvi, "Mystery That Is Heredity (Part II)," 23.

92. Sanghvi, "Mystery That Is Heredity (Part II)," 24.

93. Sanghvi, "Mystery That Is Heredity (Part II)," 24.

94. Sanghvi, "Mystery That Is Heredity (Part II)," 25.

95. Sanghvi, "Mystery That Is Heredity (Part II)," 25.

96. Sanghvi, "Mystery That Is Heredity (Part II)," 29–30.

97. Sanghvi, "Mystery That Is Heredity (Part II)," 26.

98. Desai, "Place of Eugenics in a Democracy."

99. TOI News Service, "Minister Explains Bill to Rationalise Abortion Law."

100. Anon., "Sterilisation of the Unfit Bill."

101. Hodges, "South Asia's Eugenic Past," 234.

102. Hodges, "South Asia's Eugenic Past," 232–33.

103. Hodges, "South Asia's Eugenic Past," 231–32.

104. Savary, *Evolution, Race and Public Sphere in India: Vernacular Concepts and Sciences (1860–1930)*.

105. Raina, "Family Planning with Special Reference to Medical Aspects," 155.

106. Raina, "Family Planning with Special Reference to Medical Aspects," 155.

107. There was of course a hoary colonial tradition linking race and malaria. It would be worth exploring how such racial thinking evolved in postcolonial malariology. On colonial malariology and malaria-related discourse, see Arnold, "'An Ancient Race Outworn': Malaria and Race in Colonial Bengal, 1860–1930"; Roy, *Malarial Subjects*; Samanta, *Living with Epidemics in Colonial Bengal*.

108. Raina, "Family Planning with Special Reference to Medical Aspects," 156.

109. Raina, "Family Planning with Special Reference to Medical Aspects," 156.

110. Raina, "Family Planning with Special Reference to Medical Aspects," 157.

111. Aronowitz, *Risky Medicine: Our Quest to Cure Fear and Uncertainty*, 4–5.

112. It is worth clarifying, however, that I have mainly been interested in sero-anthropological researchers and authors. I have not looked at actual clinical practice and whether race and risk featured in the same way there.

Chapter 5

1. Golinski, *Making Natural Knowledge: Constructivism and the History of Science*.

2. Radin, *Life on Ice: A History of New Uses for Cold Blood*; Bangham, *Blood Relations: Transfusion and the Making of Human Genetics*. Ricardo Roque, "The Blood That Remains: Card Collections from the Colonial Anthropological Missions," *BJHS Themes* 4 (2009): 29–53.

3. Burton, *Genetic Crossroads: The Middle East and the Science of Human Heredity*, 106.

4. Burton, *Genetic Crossroads: The Middle East and the Science of Human Heredity*, 101–27.

5. Copeman and Banerjee, *Hematologies: The Political Life of Blood in India*, 117.

6. Copeman and Banerjee, *Hematologies: The Political Life of Blood in India*, 86–126.

7. Copeman and Banerjee, *Hematologies: The Political Life of Blood in India*, 46–86.

8. Mumtaz, Bowen, and Mumtaz, "Meanings of Blood, Bleeding and Blood Donations in Pakistan: Implications for National vs Global Safe Blood Supply Policies."

9. Cohen, "The Other Kidney: Biopolitics beyond Recognition," 15.

10. Bangham, *Blood Relations: Transfusion and the Making of Human Genetics*, 14.

11. Kowal, Greenwood, and McWhirter, "All in the Blood: A Review of Aboriginal Australians' Cultural Beliefs about Blood and Implications for Biospecimen Research," 348.

12. Mol, *The Body Multiple: Ontology in Medical Practice*, viii.

13. Mol, *The Body Multiple: Ontology in Medical Practice*, 5.

14. Mol, *The Body Multiple: Ontology in Medical Practice*, 6.

15. Bangham, *Blood Relations: Transfusion and the Making of Human Genetics*, 13.

16. Majumdar, "Blood Groups of Tribes and Castes of the U.P. with Special Reference to the Korwas," 87–88.

17. Macfarlane, "Blood Grouping in the Deccan and the Eastern Ghats," 39.

18. Macfarlane, "Blood Grouping in the Deccan and the Eastern Ghats," 39.

19. Burton, *Genetic Crossroads: The Middle East and the Science of Human Heredity*, 101.

20. Owen, *Bihar and Orissa in 1921*, 114.

21. Anon., "How Government Dealt with Influenza Epidemic."

22. The Portuguese seroanthropologist António de Almeida is known to have used a very similar strategy, harking back to colonial-era practices, to collect samples in East Timor in the 1950s and 1960s. See Roque, "Bleeding Languages: Blood Types and Linguistic Groups in the Timor Anthropological Mission." For an earlier Southeast Asian example, see Sysling, *Racial Science and Human Diversity in Colonial Indonesia*, 48, 56.

23. Malone and Lahiri, "The Distribution of the Blood-Groups of Certain Races and Castes in India," 964–65.

24. Chatterjee, *A Time for Tea: Women, Labor, and Post/Colonial Politics*, 75.

25. Sharma, *Empire's Garden: Assam and the Making of India*; Bhattacharya, *Contagion and Enclaves: Tropical Medicine in Colonial India*.

26. Sarkar, "An Analysis of Indian Blood Group Data with Special Reference to the Oraons."

27. Macfarlane, "Blood Grouping in the Deccan and the Eastern Ghats," 40.

28. Sarkar, "Blood Grouping Investigations in India with Special Reference to Santal Perganas, Bihar," 90.

29. Banerjee, "Debt, Tome and Extravagance: Money and the Making of 'Primitives' in Colonial Bengal," 431.

30. Banerjee, "Debt, Tome and Extravagance: Money and the Making of 'Primitives' in Colonial Bengal," 432.

31. Guha, *Elementary Aspects of Peasant Insurgency in Colonial India*.

32. Sarkar, *The Jarawa*, 47–48.

33. Gates, "Blood Groups from the Andamans," 55; my emphasis.

34. Sarkar, "Blood Groups from the Andaman and Nicobar Islands."

35. Figure 5.1 is a traumatic and ethically complicated image. I was hesitant to use it here because it might amplify the exploitation already inflicted upon this boy and his family. Upon reflection, however, I decided that since the image is already available in print in the original publications, any additional harm caused by its reproduction here is outweighed by the potential of this image to restore the humanity so brutally snatched from this little boy in the name of science. His stare, I feel, challenges us to recognize him as a fellow human being in ways that are far stronger than what my words can do.

36. Sarkar, "The Jarawa of the Andaman Islands," 672.

37. Sarkar, "The Jarawa of the Andaman Islands", 676–77.

38. Bangham, *Blood Relations: Transfusion and the Making of Human Genetics*, 13.

39. Bangham, "Writing, Printing, Speaking: Rhesus Blood-Group Genetics and Nomenclatures in the Mid-Twentieth Century," 338.

40. Schiff and Boyd, *Blood Grouping Technic: A Manual for Clinicians, Serologists, Anthropologists, and Students of Legal and Military Medicine*, 3–5.

41. Bangham, *Blood Relations: Transfusion and the Making of Human Genetics*, 14.

42. Bangham, *Blood Relations: Transfusion and the Making of Human Genetics*, 61.

43. Bangham, *Blood Relations: Transfusion and the Making of Human Genetics*, 69.

44. Directorate General of Health Services, *Tri-Annual Reports of the Directorate General of Health Services, 1954–56*, 66.

45. Directorate General of Health Services, *Tri-Annual Reports of the Directorate General of Health Services, 1954–56*, xiii.

46. Central Bureau of Health Intelligence, *Annual Report of the Directorate General of Health Services 1960*, 266.

47. Central Bureau of Health Intelligence, *Annual Report of the Directorate General of Health Services 1960*, 271.

48. Majumdar, "Blood Groups among the Makranis of Western Khandesh," 128.

49. Majumdar and Kishen, "Blood Group Distribution in the United Provinces: Report on the Serological Survery of the United Provinces," 10.

50. Das et al., "Blood Groups (ABO, M-N and Rh), ABH Secretion, Sickle-Cell, P.T.C. Taste, and Colour Blindness in the Mahar of Nagpur," 346.

51. Das, Mukherjee, and Bhattacharjee, "Survey of the Blood Groups and PTC Taste among the Rajbanshi Caste of West Bengal (ABO, MNS, Rh, Duffy and Diego)"; Chaudhuri et al., "Study of Blood Groups and Haemoglobin Variants among the Santal Tribe in Midnapore District of West Bengal, India"; Bhattacharya, "A Study of ABO, RH-HR and MN Blood Groups of the Anglo-Indians of India."

52. Central Bureau of Health Intelligence, *Annual Report of the Directorate General of Health Services 1957*, 151.

53. Sarkar, "Blood Grouping Investigations in India with Special Reference to Santal Perganas, Bihar," 90.

54. Boyd, "Genetics and the Human Race," 1058.

55. Macfarlane, "The Racial Affinities of the Jews of Cochin," 14.

56. Majumdar and Kishen, "Blood Group Distribution in the United Provinces: Report on the Serological Survery of the United Provinces," 9.

57. Sarkar, "Analysis of Indian Blood Group Data with Special Reference to the Oraons," 2.

58. Majumdar, "Blood Groups of Tribes and Castes of the U.P. with Special Reference to the Korwas," 93.

59. Sarkar and Sen, "Further Blood Group Investigations in Santal Parganas," 8.

60. Bose, "Blood Groups of the Tribes of Travancore."

61. Srivastava, "Blood Groups in the Tharus of Uttar Pradesh and Their Bearing on the Ethnic and Genetic Relationships," 3.

62. Sarkar, "Analysis of Indian Blood Group Data with Special Reference to the Oraons," 2.

63. McKennan, "The Physical Anthropology of Two Alaskan Athapaskan Groups," 44.

64. McKennan, "The Physical Anthropology of Two Alaskan Athapaskan Groups," 45.

65. Schiff and Boyd, *Blood Grouping Technic: A Manual for Clinicians, Serologists, Anthropologists, and Students of Legal and Military Medicine*, 23.

66. Kuklick, "After Ishmael: The Fieldwork Tradition and Its Future," 62–63.

67. Martin, "The Potentiality of Ethnography and the Limits of Affect Theory."

68. Sysling, "Measurement, Self-Tracking and the History of Science: An Introduction."

69. Büchi, "Blood Groups of Tibetans," 72.

70. Chakraborty, "ABO Blood Groups of the Zeliang Naga," 311.

71. Singh, "History of Development of Physical Anthropology in India," 222.

72. Sibum, "Reworking the Mechanical Value of Heat: Instruments of Precision and Gestures of Accuracy in Early Victorian England"; Raj, "When Human Travellers Become Instruments: The Indo-British Exploration of Central Asia in the Nineteenth Century"; Roberts, "The Death of the Sensuous Chemist"; Mukharji, *Doctoring Traditions.*

73. Schiff and Boyd, *Blood Grouping Technic: A Manual for Clinicians, Serologists, Anthropologists, and Students of Legal and Military Medicine*, 24.

74. Schiff and Boyd, *Blood Grouping Technic: A Manual for Clinicians, Serologists, Anthropologists, and Students of Legal and Military Medicine*, 24.

75. Schiff and Boyd, *Blood Grouping Technic: A Manual for Clinicians, Serologists, Anthropologists, and Students of Legal and Military Medicine*, 26.

76. Schiff and Boyd, *Blood Grouping Technic: A Manual for Clinicians, Serologists, Anthropologists, and Students of Legal and Military Medicine*, 22.

77. Bangham, *Blood Relations: Transfusion and the Making of Human Genetics*, 136.

78. Daston, "Objectivity and the Escape from Perspective."

79. Daston and Galison, *Objectivity*; Porter, *Trust in Numbers: The Pursuit of Objectivity in Science and Public Life.*

80. Golinski, *Making Natural Knowledge*, 6.

81. Heidegger and Lovitt, "The Question concerning Technology."

82. Heidegger and Lovitt, "The Question concerning Technology."

83. Bowler and Pickstone, *The Cambridge History of Science Volume 6: The Modern Biological and Earth*; Strasser and Chadarevian, "The Comparative and the Exmplary: Revisiting the Early History of Molecular Biology."

84. Sivasundaram, "Imperial Transgressions: The Animal and Human in the Idea of Race," 157.

85. Sivasundaram, "Imperial Transgressions: The Animal and Human in the Idea of Race," 169.

86. Bangham, *Blood Relations: Transfusion and the Making of Human Genetics*, 86.

87. Pickstone, *Ways of Knowing: A New History of Science, Technology and Medicine*, 176–78.

88. Bangham, *Blood Relations: Transfusion and the Making of Human Genetics*, 8.

89. Bangham, *Blood Relations: Transfusion and the Making of Human Genetics*, 124.

90. Schiff and Boyd, *Blood Grouping Technic: A Manual for Clinicians, Serologists, Anthropologists, and Students of Legal and Military Medicine*, 14.

91. Büchi, "Frequency of ABO-Blood Groups and Secretor Factor in Bengal," 50.

92. Trofa et al., "Dr. Kiyoshi Shiga: Discoverer of the Dysentery Bacillus."

93. O'Brien and Holmes, "Shiga and Shiga-like Toxins."

94. Dabassa and Bacha, "The Prevalence and Antibiogram of Salmonella and Shigella Isolated from Abattoir, Jimma Town, South West Ethiopia."

95. Shahan and Huffman, *Diseases of Sheep and Goats: Farmer's Bulletin No. 1943*, 11–12.

96. Ministry of Food and Agriculture, Government of India, *Report of the Ad-Hoc Committee on Slaughter-Houses and Meat Inspection Practices*, 9–11.

97. Bose, *The Hindoos as They Are: A Description of the Manners, Customs and Inner Life of Hindoo Society*, 147–49.

98. Gupta, "The Domestication of a Goddess: Carana-Tirtha Kalighat, the Mahapitha of Kali," 71.

99. Samanta, "The 'Self-Animal' and Divine Digestion: Goat Sacrifice to the Goddess Kali in Bengal."

100. Ministry of Food and Agriculture, Government of India, *Report of the Ad-Hoc Committee on Slaughter-Houses and Meat Inspection Practices*, 1.

101. Yang, "Sacred Symbol and Sacred Space in Rural India: Community Mobilization in the 'Anti-Cow Killing' Riot of 1893"; Adcock, "Sacred Cows and Secular History: Cow Protection Debates in Colonial North India."

102. Das et al., "Blood Groups (ABO, M-N and Rh), ABH Secretion, Sickle-Cell, P.T.C. Taste, and Colour Blindness in the Mahar of Nagpur"; Das, Sastry, and Mukherjee, "Blood Groups (A_1A_2BO, M-N, Rh) and ABH Secretion in the Pareng Gadaba, the Ollaro Gadaba and the Konda Paroja of Koraput District in Orissa."

103. Kumar and Sastry, "A Genetic Survey among the Riang: A Mongoloid Tribe of Tripura (North East India)."

104. Bhattacharjee, "Distribution of the Blood Groups (A_1 A_2 B O, MNSs, Rh), and the Secretor Factor among the Muslims and the Pandits of Kashmir"; Bhattacharjee and Kumar, "A Blood Group Genetic Survey in the Dudh Kharias of the Ranchi District (Bihar, India)."

105. Kabat, *Blood Group Substances: Their Chemistry and Immunochemistry*, 187–88.

106. Sharafi, "The Imperial Serologist and Punitive Self-Harm: Bloodstains and Legal Pluralism in British India," 66.

107. Maluf, "History of Blood Transfusion."

108. Schneider, *The History of Blood Transfusion in Sub-Saharan Africa*; Schneider and Drucker, "Blood Transfusions in the Early History of AIDS in Sub-Saharan Africa."

109. Sunseri, "Blood Trials: Transfusions, Injections, and Experiments in Africa, 1890–1920," 296.

110. Sunseri, "Blood Trials: Transfusions, Injections, and Experiments in Africa, 1890–1920," 296.

111. Sunseri, "Blood Trials: Transfusions, Injections, and Experiments in Africa, 1890–1920," 297.

112. Sunseri, "Blood Trials: Transfusions, Injections, and Experiments in Africa, 1890–1920," 309–10.

113. Sunseri, "Blood Trials: Transfusions, Injections, and Experiments in Africa, 1890–1920," 314.

114. Harrison, "'The Tender Frame of Man': Disease, Climate, and Racial Difference in India and the West Indies, 1760–1860."

115. Sunseri, "Blood Trials: Transfusions, Injections, and Experiments in Africa, 1890–1920," 314.

116. Schiff and Boyd, *Blood Grouping Technic: A Manual for Clinicians, Serologists, Anthropologists, and Students of Legal and Military Medicine*, 12.

117. Mukharji, "Cat and Mouse: Animal Technologies, Trans-Imperial Networks and Public Health from Below, British India, c. 1907–1918"; Pemberton, "Canine Technologies, Model Patients: The Historical Production of Hemophiliac Dogs in American Biomedicine."

118. Mol, *The Body Multiple: Ontology in Medical Practice*, 6.

119. Anderson, "Objectivity and Its Discontents," 571.

120. Mol, *The Body Multiple: Ontology in Medical Practice*, 84.

121. Radin, *Life on Ice*.

122. Radin, *Life on Ice*, 2.

123. Lindee and Ventura Santos, "The Biological Anthropology of Living Human Populations."

Chapter 6

1. Macfarlane, "Eastern Himalayan Blood-Groups," 127.

2. Majumdar, "Blood Groups of the Doms," 154.

3. Simmons et al., "A Genetical Survey in Chenchu, South India: Blood, Taste and Secretion," 500.

4. Kumar, "Blood Group and Secretor Frequency among the Galong," 55.

5. Bharati, "The Use of 'Superstition' as an Anti-Traditional Device in Urban Hinduism."

6. Bharati, "The Use of 'Superstition' as an Anti-Traditional Device in Urban Hinduism."

7. Quack, *Disenchanting India: Organized Rationalism and Criticism of Religion in India*; Ranganathan, "Healing Temples, the Anti-Superstition Discourse and Global Mental Health: Some Questions from Mahanubhav Temples in India"; Binder, "Magic Is Science: Atheist Conjuring and the Exposure of Superstition in South India."

8. Bharati, "The Use of 'Superstition' as an Anti-Traditional Device in Urban Hinduism," 41.

9. Amrith, *Decolonizing International Health: India and Southeast Asia, 1930–65*, 135–37.

10. Visvanathan, *A Carnival of Science: Essays on Science, Technology and Development*; Santos, *The End of the Cognitive Empire: The Coming of Age of the Epistemologies of the South.*

11. See, for instance, Brimnes, "Variolation, Vaccination and Popular Resistance in Early Colonial South India"; Amrith, *Decolonizing International Health.*

12. McGranahan, "Theorizing Refusal: An Introduction," 319.

13. Simpson, "On Ethnographic Refusal: Indigeneity, 'Voice' and Colonial Citizenship."

14. Arnold, *Colonizing the Body: State Medicine and Epidemic Disease in Nineteenth-Century India*, 218–20; White, *Speaking with Vampires: Rumor and History in Colonial Africa.*

15. Büchi, "Blood Groups of Tibetans," 71.

16. Gerke, *Long Lives and Untimely Deaths: Life-Span Concepts and Longevity Practices among Tibetans in the Darjeeling Hills, India*, 7.

17. Gerke, *Long Lives and Untimely Deaths: Life-Span Concepts and Longevity Practices among Tibetans in the Darjeeling Hills, India*, 137–65.

18. Gerke, *Long Lives and Untimely Deaths: Life-Span Concepts and Longevity Practices among Tibetans in the Darjeeling Hills, India*, 138, 144.

19. Gerke, *Long Lives and Untimely Deaths: Life-Span Concepts and Longevity Practices among Tibetans in the Darjeeling Hills, India*, 144.

20. Gerke, *Long Lives and Untimely Deaths: Life-Span Concepts and Longevity Practices among Tibetans in the Darjeeling Hills, India*, 145.

21. Gerke, *Long Lives and Untimely Deaths: Life-Span Concepts and Longevity Practices among Tibetans in the Darjeeling Hills, India*, 145.

22. Majumdar, "The Tharus and the Blood Groups"; Sarkar and Sen, "Further Blood Group Investigations in Santal Parganas"; Kumar, "Blood Group and Secretor Frequency among the Galong."

23. Sarkar and Sen, "Further Blood Group Investigations in Santal Parganas," 8.

24. Gerke, *Long Lives and Untimely Deaths: Life-Span Concepts and Longevity Practices among Tibetans in the Darjeeling Hills, India*, 132, 141, 152.

25. Gerke, *Long Lives and Untimely Deaths: Life-Span Concepts and Longevity Practices among Tibetans in the Darjeeling Hills, India*, 150–51.

26. Gerke, *Long Lives and Untimely Deaths: Life-Span Concepts and Longevity Practices among Tibetans in the Darjeeling Hills, India*, 141.

27. Gerke, *Long Lives and Untimely Deaths: Life-Span Concepts and Longevity Practices among Tibetans in the Darjeeling Hills, India*, 306–15.

28. Gerke, *Long Lives and Untimely Deaths: Life-Span Concepts and Longevity Practices among Tibetans in the Darjeeling Hills, India*, 152–54.

29. Gerke, *Long Lives and Untimely Deaths: Life-Span Concepts and Longevity Practices among Tibetans in the Darjeeling Hills, India*, 131.

30. Gerke, *Long Lives and Untimely Deaths: Life-Span Concepts and Longevity Practices among Tibetans in the Darjeeling Hills, India*, 131.

31. Gerke, *Long Lives and Untimely Deaths: Life-Span Concepts and Longevity Practices among Tibetans in the Darjeeling Hills, India*, 132.

32. Gerke, *Long Lives and Untimely Deaths: Life-Span Concepts and Longevity Practices among Tibetans in the Darjeeling Hills, India*, 141.

33. Büchi, "Blood Groups of Tibetans," 71.

34. Büchi, "Blood Groups of Tibetans," 71.

35. Macfarlane, "Eastern Himalayan Blood-Groups," 127.

36. The political and cultural sovereignty of Darjeeling and Kalimpong has also been a debated issue. Contested by the kingdoms of Sikkim and Bhutan in the eighteenth and nineteenth centuries, the areas were finally absorbed into the British empire in the 1860s. By the second quarter of the twentieth century, however, large-scale migration from Nepal had once again begun to reshape the cultural profile of the area. Later, in postcolonial India, a Nepalized political movement began to demand the separation of these regions from the state of West Bengal. Gerke, *Long Lives and Untimely Deaths: Life-Span Concepts and Longevity Practices among Tibetans in the Darjeeling Hills, India*, 45–56.

37. Gerke, "Of Matas, Jhakris, and other Healers: Fieldnotes on a Healing Event in Kalimpong, India," especially 240.

38. Sarkar, "Blood Grouping Investigations in India with Special Reference to Santal Perganas, Bihar," 90.

39. Sarkar, *The Aboriginal Races of India*, 37.

40. Sarkar, *The Aboriginal Races of India*, 37.

41. Sarkar, "The Social Institutions of the Mālpāhāriās," 26.

42. Nigam, "Disciplining and Policing the 'Criminals by Birth,' Part 1: The Making of a Colonial Stereotype—The Criminal Tribes and Castes of North India"; Nigam, "Disciplining and Policing the 'Criminals by Birth,' Part 2: The Development of a Disciplinary System, 1871–1900"; Brown, "Ethnology and Colonial Administration in Nineteenth-Century British India: The Question of Native Crime and Criminality"; Schwartz, *Constructing the Criminal Tribe in Colonial India: Acting Like a Thief*; Piliavsky, "The 'Criminal Tribe' in India before the British."

43. The phrase "ethnographic state" is used by Nicholas Dirks, *Castes of Mind: Colonialism and the Making of Modern India*.

44. Risley, "Of the Dravidian Tract: Santāl," 144.

45. Nichter, "The Layperson's Perception of Medicine as Perspective into the Utilization of Multiple Therapy Systems in the Indian Context," 231.

46. Datta-Majumder, *The Santals: A Study in Culture-Change*, 127.

47. Datta-Majumder, *The Santals: A Study in Culture-Change*, 103.

48. Bodding, *Santal Medicine*, 139.

49. Bodding, *Santal Medicine*, 144.

50. Dasgupta, "Reordering a World: The Tana Bhagat Movement, 1914–1919," 5–6.

51. Dasgupta, "Reordering a World: The Tana Bhagat Movement, 1914–1919," 28.

52. Lahiri, "Observations on the Medico-Legal Applications of Blood Grouping with a Note on Blood Groups in a Polyandrous Family," 971.

53. Vyas, Bhatia, and Sanghvi, "Three Cases of Weak B in an Indian Family."

54. Vyas, Bhatia, and Sanghvi, "Three Cases of Weak B in an Indian Family," 509.

55. Vyas, Bhatia, and Sanghvi, , "Three Cases of Weak B in an Indian Family," 510.

56. Karve, *Kinship Organization in India*, 84–86.

57. Karve, *Kinship Organization in India*, 63–67.

58. Karve, *Kinship Organization in India*, 132.

59. Karve, *Kinship Organization in India*, 132.

60. Sarkar, "Talking about Scandals: Religion, Law and Love in Late Nineteenth Century Bengal"; Sodhan, "Women in the Maharaj Libel Case."

61. Davis Jr., "Children: *Putr, Duhitr*," 158.

62. Schröder, "P. Matthias Hermanns SVD (1899–1972)."

63. Oomen, "P. Matthia Hermanns SVD (1899–1972," 120.

64. Hermanns, *The Evolution of Man: A Challenge to Darwinism through Human Biogenetics, Physical and Cultural Anthropology, Prehistory, and Palaeontology*, 27.

65. Hermanns, *The Evolution of Man: A Challenge to Darwinism through Human Biogenetics, Physical and Cultural Anthropology, Prehistory, and Palaeontology*, 27.

66. Hermanns, *The Evolution of Man: A Challenge to Darwinism through Human Biogenetics, Physical and Cultural Anthropology, Prehistory, and Palaeontology*, 28.

67. Hermanns, *The Evolution of Man: A Challenge to Darwinism through Human Biogenetics, Physical and Cultural Anthropology, Prehistory, and Palaeontology*, 6–8.

68. Hermanns, *The Evolution of Man: A Challenge to Darwinism through Human Biogenetics, Physical and Cultural Anthropology, Prehistory, and Palaeontology*, 7.

69. Hermanns, *The Evolution of Man: A Challenge to Darwinism through Human Biogenetics, Physical and Cultural Anthropology, Prehistory, and Palaeontology*, 8.

70. Hermanns, *The Evolution of Man: A Challenge to Darwinism through Human Biogenetics, Physical and Cultural Anthropology, Prehistory, and Palaeontology*, 15.

71. Hermanns, *The Evolution of Man: A Challenge to Darwinism through Human Biogenetics, Physical and Cultural Anthropology, Prehistory, and Palaeontology*, 33.

72. Hermanns, *The Evolution of Man: A Challenge to Darwinism through Human Biogenetics, Physical and Cultural Anthropology, Prehistory, and Palaeontology*, 17.

73. Hermanns, *The Evolution of Man: A Challenge to Darwinism through Human Biogenetics, Physical and Cultural Anthropology, Prehistory, and Palaeontology*, 56.

74. Hermanns, *The Evolution of Man: A Challenge to Darwinism through Human Biogenetics, Physical and Cultural Anthropology, Prehistory, and Palaeontology*, 54–57.

75. Chen, "A Non-Metaphysical Evaluation of Vitalism in the Early Twentieth Century," 51.

76. Chen, "A Non-Metaphysical Evaluation of Vitalism in the Early Twentieth Century," 55.

77. Reggiani, *God's Eugenicist: Alexis Carrel and the Sociobiology of Decline*.

78. Carrel, *Man, The Unknown: Complete and Unabridged*, 37.

79. Radin, *Life on Ice*, 22.

80. Radin, *Life on Ice*, 4.

81. Radin, *Life on Ice*, 131.

82. Brown and Young, "Indian Christians and Nehru's Nation-State."

83. Hardiman, *Coming of the Devi: Adivasi Assertion in Western India*; Baviskar, "Adivasi Encounters."

84. Sundar, *Subalterns and Sovereigns: An Anthropological History of Bastar, 1854–2006*.

85. Hermanns, *The Evolution of Man: A Challenge to Darwinism Through Human Biogenetics, Physical and Cultural Anthropology, Prehistory, and Palaeontology*, 136.

86. Ludden, "History outside Civilisation and Mobility of South Asia."

87. Beller, *Antisemitism: A Very Short Introduction*, 94.

88. Bayly, "The Ends of Liberalism and the Political Thought of Nehru's India."

89. On "reading against the grain" see Guha, "The Prose of Counter-Insurgency." On the use of the ethnographic archive, see Bhadra, "Four Rebels of Eighteen-Fifty-Seven." On the embrace of microhistory within the subalternist project, see Sarkar, "The Kalki-Avatar of Bikrampur: A Village Scandal in Early Twentieth Century Bengal." On the "defense of the fragment," see Pandey, "In Defense of the Fragment: Writing about Hindu-Muslim Riots in India Today."

90. Prakash, *Another Reason: Science and the Imagination of Modern India*.

91. Confino and Skaria, "The Local Life of Nationhood."

Chapter 7

1. Anon., "Preface," v.

2. Guha, "Application for Admission to Candidacy for a Degree in Arts or Philosophy, Harvard University" (Cambridge, Mass., 1922), 3.

3. Mahalanobis, "Analysis of Race Mixture in Bengal," 324.

4. Seal, "Meaning of Race, Tribe, Nation," 2.

5. Seal, "Meaning of Race, Tribe, Nation," 2.

6. Mukharji, "Profiling the Profiloscope."

7. Mukharji, "The Bengali Pharaoh."

8. Milam, *Looking for a Few Good Males: Female Choice in Evolutionary Biology*, 29–53.

9. Seal, "Meaning of Race, Tribe, Nation," 7.

10. Seal, "Meaning of Race, Tribe, Nation," 8.

11. Seal, "Meaning of Race, Tribe, Nation," 9.

12. Seal, "Meaning of Race, Tribe, Nation," 9.

13. Seal, "Meaning of Race, Tribe, Nation," 10.

14. Seal, "Meaning of Race, Tribe, Nation," 10.

15. Seal, "Meaning of Race, Tribe, Nation," 11.

16. Cohen, "The 'Social' De-Duplicated: On the Aadhaar Platform and the Engineering of Service."

17. Seal, "Meaning of Race, Tribe, Nation," 11.

18. Seal, "Meaning of Race, Tribe, Nation," 12–13.

19. Lewis, "Boas, Darwin, Science and Anthropology."

20. Sweeney, *"Fighting for the Good Cause": Reflections on Francis Galton's Legacy to American Hereditarian Psychology*, 81.

21. Anderson, *From Boas to Black Power: Racism, Liberalism, and American Anthropology*, 13.

22. Visweswaran, "Race and the Culture of Anthropology."

23. Arnold, "'An Ancient Race Outworn': Malaria and Race in Colonial Bengal, 1860–1930"; Gualtieri, *Between Arab and White: Race and Ethnicity in the Early Syrian American Diaspora*; Ergin, *"Is the Turk a White Man?" Race and Modernity in the Making of Turkish Identity*; Burton, *Genetic Crossroads: The Middle East and the Science of Human Heredity*; Hyun, "Blood Purity and Scientific Independence: Blood Science and Postcolonial Struggles in Korea, 1926–1975."

24. Arnold, "'An Ancient Race Outworn': Malaria and Race in Colonial Bengal, 1860–1930," 123.

25. Mukharji, "Profiling the Profiloscope."

26. Mukharji, "From Serosocial to Sanguinary Identities"; Mukharji, "The Bengali Pharaoh."

27. Mukharji, "Profiling the Profiloscope"; Mukharji, "The Bengali Pharaoh."

28. Basu et al., "Introduction," 3–5.

29. Sarkar, "The Place of Human Biology in Anthropology and Its Utility in the Service of the Nation," 1.

30. Seal, "Meaning of Race, Tribe, Nation," 1.

31. Basu et al., "Introduction," 10–11.

32. Bose, "Foreword."

33. Bose, "Foreword," 4.

34. Sarkar, "The Place of Human Biology in Anthropology and Its Utility in the Service of the Nation," 18–20.

35. Sarkar, "The Place of Human Biology in Anthropology and Its Utility in the Service of the Nation," 19.

36. Sarkar, "The Place of Human Biology in Anthropology and Its Utility in the Service of the Nation," 18.

37. Sarkar, *Indian Eugenics Society: Bulletin No. 1*, 22.

38. Chatterjee, "Foreword," i–iii.

39. Hodges, "South Asia's Eugenic Past," 5.

40. Sarkar, *Indian Eugenics Society: Bulletin No. 1*, 1.

41. On the history of laws regulating age of marriage in British India, see Pande, "Coming of Age: Law, Sex and Childhood in Late Colonial India."

42. Sarkar, "The Place of Human Biology in Anthropology and Its Utility in the Service of the Nation," 3–5.

43. Sarkar, *Indian Eugenics Society: Bulletin No. 1*, 7.

44. Sarkar, *Indian Eugenics Society: Bulletin No. 1*, 6–8.

45. Pande, *Sex, Law, and the Politics of Age: Child Marriage in India, 1891–1937*, 31–122. On the historicization of "age" and "aging," see also Cohen, *No Aging in India: Alzheimer's, the Bad Family and Other Modern Things*; Sivaramakrishnan, *As the World Ages: Rethinking a Demographic Crisis*.

46. Sarkar, "The Place of Human Biology in Anthropology and Its Utility in the Service of the Nation," 8.

47. Sarkar, *Indian Eugenics Society: Bulletin No. 1*, 14.

48. Sarkar, "The Place of Human Biology in Anthropology and Its Utility in the Service of the Nation," 12.

49. Sarkar, "The Place of Human Biology in Anthropology and Its Utility in the Service of the Nation," 12.

50. Sarkar, "The Place of Human Biology in Anthropology and Its Utility in the Service of the Nation," 13.

51. Sarkar, *Indian Eugenics Society: Bulletin No. 1*, 6.

52. Sarkar, "The Place of Human Biology in Anthropology and Its Utility in the Service of the Nation," 17–18.

53. Sarkar, "The Place of Human Biology in Anthropology and Its Utility in the Service of the Nation," 16.

54. Sarkar, "The Place of Human Biology in Anthropology and Its Utility in the Service of the Nation," 16–17.

55. Basu et al., "Introduction," 4.

56. Kevles, *In the Name of Eugenics: Genetics and the Uses of Human Heredity*, 164–75.

57. Penrose, "Report on Thesis, 'Studies on Twins' and Additional Papers by S. S. Sarkar" (London, July 15, 1948), 4.

58. The term *bhadralok* has wide currency in South Asian history but might be unfamiliar to non–South Asianists. It translates roughly as "genteel folk" and is commonly used in Bengali to designate middle- and upper-class people seen to be urbane and relatively elite. In scholarly works it is also often understood to be a collective term for the Bengali upper castes. The caste composition of the *bhadralok*, however, is more ambiguous since educated, middle-class, and relatively affluent people from lower castes or even non-Hindu Bengalis will frequently be included in the category. See Ghosh, *What Happened to the Bhadralok?*

59. Karve, "Racial Factor in Indian Social Life," 42.

60. Sundar, "In the Cause of Anthropology: The Life and Work of Irawati Karve."

61. Sundar, "In the Cause of Anthropology: The Life and Work of Irawati Karve," 393.

62. Graham, "Introduction," 4.

63. Knight, "Racism, Revolution and Indigenismo: Mexico, 1910–1940," 73.

64. Knight, "Racism, Revolution and Indigenismo: Mexico, 1910–1940," 85–86.

65. Wade et al., "Introduction: Genomics, Race Mixture, and Nation in Latin America," 14.

66. Miller, *Rise and Fall of the Cosmic Race: The Cult of Mestizaje in Latin America*, 44.

67. Wade et al., "Introduction: Genomics, Race Mixture, and Nation in Latin America," 14.

68. Miller, *Rise and Fall of the Cosmic Race: The Cult of Mestizaje in Latin America*, 43.

69. Souza and Ventura Santos, "The Emergence of Human Population Genetics and Narratives about the Formation of the Brazilian Nation (1950–1960)," 98.

70. Souza and Ventura Santos, "The Emergence of Human Population Genetics and Narratives about the Formation of the Brazilian Nation (1950–1960)," 101.

71. Souza and Ventura Santos, "The Emergence of Human Population Genetics and Narratives about the Formation of the Brazilian Nation (1950–1960)," 101–2.

72. Karve, "Racial Factor in Indian Social Life," 35.

73. Karve, "Racial Factor in Indian Social Life," 35.

74. Abu El-Haj, "The Genetic Reinscription of Race."

75. Reardon, *Race to the Finish: Identity and Governance in an Age of Genomics*, 17–45.

76. Gil-Riaño, "Relocating Anti-Racist Science: The UNESCO Statement on Race and Economic Development in the Global South," 303.

77. Sundar, "In the Cause of Anthropology: The Life and Work of Irawati Karve," 383.

78. Karve, "Racial Factor in Indian Social Life," 35.

79. Karve, "Racial Factor in Indian Social Life," 36–37.

80. Jaffrelot, "Sanskritization vs. Ethnicization in India: Changing Identities and Caste Politics before Mandal," 761–62.

81. Karve, "Racial Factor in Indian Social Life," 38.

82. Karve, "Racial Factor in Indian Social Life," 41.

83. Karve, "Racial Conflict," 44.

84. Karve, "Racial Conflict," 32.

85. Gil-Riaño, "Relocating Anti-Racist Science: The UNESCO Statement on Race and Economic Development in the Global South," 290.

86. Karve, "Racial Factor in Indian Social Life," 39.

87. Karve, "Racial Factor in Indian Social Life," 40.

88. Barbosa, "Making Human Differences in Berlin and Maharashtra: Considerations on the Production of Physical Anthropological Knowledge by Irawati Karve," 152.

89. Karve, "Racial Conflict," 49.

90. Sanghvi, "Perspectives for the Study of Racial Origins in India," 73.

91. Sanghvi, "Perspectives for the Study of Racial Origins in India," 72–73.

92. Sundar, "In the Cause of Anthropology: The Life and Work of Irawati Karve."

93. Sanghvi, "Genetic Adaptation in Man," 306.

94. Sanghvi, "Sanghvi, Labhshankar Dalichand."

95. Sanghvi, "Perspectives for the Study of Racial Origins in India," 70.

96. Sanghvi, "Preface," v.

97. Sanghvi, "Preface," v.

98. Sanghvi, "Inbreeding in India."

99. Sanghvi, "Preface," v.

100. Sanghvi, "Preface," v–vi.

101. Sanghvi, "Perspectives for the Study of Racial Origins in India," 72.

Conclusion

1. Subramaniam, *Holy Science: The Biopolitics of Hindu Nationalism*, 216.

2. Sarkar, *Lokgatha*.

3. Sarkar, *Lokgatha*, 1–3.

4. Singh, "The Sastri and the Air Pump: Experimental Fiction and Fictions of Experiment for Hindi Readers in Colonial North India."

5. Subramaniam, *Holy Science: The Biopolitics of Hindu Nationalism*, 221.

6. There is, of course, a longer genealogy to the "statistical frame" that dates from the late nineteenth century. See Samarendra, "Anthropological Knowledge and Statistical Frame: Caste in the Census in Colonial India." I will add, however, that the disciplinary formalism of this statistical frame is new in the period I have studied here.

7. Porter, *Trust in Numbers: The Pursuit of Objectivity in Science and Public Life*, ix.

8. Fanon, *Black Skin, White Masks*, 4.

9. Fanon, "Mental Alterations, Character Modifications, Psychic Disorders and Intellectual Deficit in Spinocerebellar Heredodegeneration: A Case of Friedreich's Ataxia with Delusions of Possession," 219.

10. Deleuze and Guattari, *Anti-Oedipus: Capitalism and Schizophrenia*, 96.

11. Arnold, "'An Ancient Race Outworn': Malaria and Race in Colonial Bengal, 1860–1930," 123.

12. Burton, *Genetic Crossroads: The Middle East and the Science of Human Heredity*, 38.

13. Fanon, *Black Skin, White Masks*, 42–43.

14. Gilroy, *Between Camps: Nations, Cultures and the Allure of Race*, 15.

15. Gilroy, *Between Camps: Nations, Cultures and the Allure of Race*, 8.

16. Gilroy, *Between Camps: Nations, Cultures and the Allure of Race*, 33.

17. Gilroy, *Between Camps: Nations, Cultures and the Allure of Race*, 11–53.

18. Gilroy, *Between Camps: Nations, Cultures and the Allure of Race*, 42.

19. Bashford, "Anticolonial Climates."

20. Guha, "Forget Not: In Search of a Nationalist Anthropology in India," 22.

21. Gilroy, *Between Camps: Nations, Cultures and the Allure of Race*, 48.

22. Fanon, *Black Skin, White Masks*, 91.

23. Gordon, "Requiem on a Life Well Lived: In Memory of Fanon," 21.

24. Fanon, *Black Skin, White Masks*, 36.

25. Fanon, *Black Skin, White Masks*, 134.

26. Fanon, *Black Skin, White Masks*, 145.

27. Fanon and Geronimi, "TAT in Muslim Women: Sociology of Perception and Imagination," 433.

28. For a critique of the position that sees nationalism merely as a "derivative discourse," see Chatterjee, *Nationalist Thought and the Colonial World: A Derivative Discourse*.

29. Mukharji, *Nationalizing the Body: The Medical Market, Print and Daktari Medicine*.

30. Mukharji, *Nationalizing the Body: The Medical Market, Print and Daktari Medicine*.146.

31. Mukharji, "The Bengali Pharaoh."

32. Gilroy, *Between Camps: Nations, Cultures and the Allure of Race*, 46.

33. Guha, *Bharater Jati Parichay* [India's Racial Identity], 6. The number of chromosomes mentioned by Guha was revised soon after and established to be 46, rather than 48. At the time, however, the number 48 was universally accepted in scientific circles. On the history of the recount, see Chadarevian, "Chromosome Photography and the Human Karyotype."

34. Guha, *Bharater Jati Parichay* [India's Racial Identity], 27. On linguistic nationalisms in India, see Partha Chatterjee, "On Religious and Linguistic Nationalisms: The Second Partition of Bengal"; Mitchell, *Language, Emotion and Politics in South India: The Making of a Mother Tongue*.

35. Guha, *Bharater Jati Parichay* [India's Racial Identity], 10–11.

36. Guha, *Bharater Jati Parichay* [India's Racial Identity], 27.

37. Honneth, "Foreword," vii–x.

38. Jaeggi, *Alienation*.

39. Jaeggi, *Alienation*, 3.

40. Jaeggi, *Alienation*, 90.

41. Sardar, "Foreword to the 2008 Edition," xvi.

42. Sardar, "Foreword to the 2008 Edition," xv. I am thankful to the anonymous referee for helping me recognize the difference between Fanon and Sardar.

43. Gilroy, *Between Camps: Nations, Cultures and the Allure of Race*, 20–21.

44. Here we do notice a clear difference between Gilroy's late-twentieth-century New Racism and the national raciology of seroanthropology. Gilroy is clear that modern consumer culture has produced a commoditized Black body that is positively racialized. National seroanthropology in mid-twentieth-century India had neither access to such a resourceful consumer culture nor a single body image that it sought to promote.

45. Gilroy, *Between Camps: Nations, Cultures and the Allure of Race*, 42.

46. Gilroy, *Between Camps: Nations, Cultures and the Allure of Race*, 103.

47. Roberts, *Fatal Invention: How Science, Politics, and Big Business Re-Create Race in the Twenty First Century*.

48. Gilroy, *Between Camps: Nations, Cultures and the Allure of Race*, 17.

49. Gilroy, *Between Camps: Nations, Cultures and the Allure of Race*, 273.

50. Gilroy, *Between Camps: Nations, Cultures and the Allure of Race*, 18.

51. Pande, *Medicine, Race and Liberalism in British Bengal*; Mehta, *Liberalism and Empire: A Study in Nineteenth-Century British Liberal Thought*.

52. Muñoz, *The Sense of Brown*, 2.

53. Muñoz, *The Sense of Brown*, 6.

54. Muñoz, *The Sense of Brown*, 17; Gilroy, *Between Camps: Nations, Cultures and the Allure of Race*.

55. Gilroy, *Between Camps: Nations, Cultures and the Allure of Race*, 25.

56. Khan, "Marginal Lives and the Microsociology of Overhearing in the Jamuna Chars."

57. Sarkar, *Lokgatha*, 1.

58. Sarkar, *Lokgatha*, Bhumika [preface].

59. Benjamin, "The Storyteller: Reflections on the Works of Nikolai Leskov," 373–74.

SOURCES FOR INTERCHAPTERS

Letter 1

Ray, Hemendrakumar. *Amanushik Manush* [Inhuman Humans]. In Gita Datta and Sukhamay Mukhopadhyay, eds., *Hemendrakumar Ray Rachanabali* [The Collected Works of Hemendrakumar Ray], *Vol.12*, 9–118. Calcutta: Asia Publishing Co., 1991.

Letter 2

Ray, Hemendrakumar. "Banabashi Abhishap" [The Forest-dwelling Curse]. In Gita Datta, ed., *Hemendrakumar Ray Rachanabali, Vol. 26*, 116–24. Kolkata: Asia Publishing Co., 2012.

———. "Bandi Atmar Kahini" [Story of the Imprisoned Soul]. In Ranjit Chattopadhyay, ed., *Nirbachita Bhuter Galpa*, 86–91. Calcutta: Suprakashani, 1963.

———. "Himalayer Bhayankar" [Terror of the Himalayas]. In Gita Datta and Sukhamay Mukhopadhyay, eds., *Hemendrakumar Ray Rachanabali, Vol. 10*, 169–236. Calcutta: Asia Publishing Co., 1987.

———. *Jorasankor Thakurbari* [The Tagore Family Home at Jorasanko]. In Gita Datta and Sukhamay Mukhopadhyay, eds., *Hemendrakumar Ray Rachanabali, Vol.1*, 239–45. Calcutta: Asia Publishing Co., 1974.

———. "Mandhatar Muluke" [In Mandhata's Country]. In Gita Datta and Sukhamay Mukhopadhyay, eds., *Hemendrakumar Ray Rachanabali, Vol. 13*, 113–84. Calcutta: Asia Publishing Co., 1992.

———. "Manusher Pratham Adventure" [Man's First Adventure]. In Gita Datta and Sukhamay Mukhopadhyay, eds., *Hemendrakumar Ray Rachanabali, Vol. 6*, 259–367. Calcutta: Asia Publishing Co., 1984.

———. "Nazrul Janmadin Smarane" [Remembering Nazrul's Birthday]. In Gita Datta and Sukhamay Mukhopadhyay, eds., *Hemendrakumar Ray Rachanabali, Vol.1*, 277–83. Calcutta: Asia Publishing Co., 1974.

Letter 3

Ray, Hemendrakumar. *Amanushik Manush* [Inhuman Humans]. In Gita Datta and Sukhamay Mukhopadhyay, eds., *Hemendrakumar Ray Rachanabali* [The Collected Works of Hemendrakumar Ray], *Vol.12*, 9–118. Calcutta: Asia Publishing Co., 1991.

———. "Banabashi Abhishap" [The Forest-dwelling Curse]. In Gita Datta, ed., *Hemendrakumar Ray Rachanabali, Vol. 26*, 116–24. Kolkata: Asia Publishing Co., 2012.

———. "Manush Pishash" [Human Ghoul]. In Gita Datta, ed., *Hemendrakumar Ray Rachanabali, Vol. 2*, 84–193. Kolkata: Asia Publishing Co., 1976.

———. "Manusher Pratham Adventure" [Man's First Adventure]. In Gita Datta and Sukhamay Mukhopadhyay, eds., *Hemendrakumar Ray Rachanabali, Vol. 6*, 259–367. Calcutta: Asia Publishing Co., 1984.

Letter 4

Ray, Hemendrakumar. "Kankal Sarathi" [Skeleton Charioteer]. In Asitabha Das, ed., *Sera Sera Bhuter Galpa* [Best of the Best Ghost Stories], 151–58. Calcutta: Arpita Prakashani, 2013.

———. "Nazrul Janmadin Smarane" [Remembering Nazrul's Birthday]. In Gita Datta and Sukhamay Mukhopadhyay, eds., *Hemendrakumar Ray Rachanabali, Vol.1*, 277–83. Calcutta: Asia Publishing Co., 1974.

Letter 5

"Kaji Nazrul Islamer Kali Puja" [Kazi Nazrul Islam's Kali Puja]. *Ei Samay*, 27 October 2019.

Ray, Hemendrakumar. "Mrs. Kumudini Choudhury." In Anish Deb, ed., *Sera 101 Bhoutik Aloukik* [101 Best Ghost and Supernatural (Stories)], 104–11. Calcutta: Patrabharati, 2018.

White, Luise. *Speaking with Vampires: Rumor and History in Colonial Africa*. Berkeley: University of California Press, 2000.

Letter 6

Choudhury, Amitabha. *Rabindranather Paralok Charcha* [Rabindranath's Practice of Spiritualism]. Saptarshi Prakashan, 2011.

Ray, Hemendrakumar. "Bhuter Raja" [The King of Ghosts]. In Gita Datta and Sukhamay Mukhopadhyay, eds., *Hemendrakumar Ray Rachanabali, Vol. 3*, 325–79. 3d ed. Calcutta: Asia Publishing Co., 1983.

Letter 7

Ray, Hemendrakumar. *Amanushik Manush* [Inhuman Humans]. In Gita Datta and Sukhamay Mukhopadhyay, eds., *Hemendrakumar Ray Rachanabali* [The Collected Works of Hemendrakumar Ray], *Vol.12*, 9–118. Calcutta: Asia Publishing Co., 1991.

Letter 8

Ghosh, Siddhartha. *Kaler Shahar Kolkata* [The City of Machines, Calcutta]. Calcutta: Ananda Publishers, 1991.

Ray, Hemendrakumar. "Chithi" [Letters]. In Gita Datta and Sukhamay Mukhopadhyay, eds., *Hemendrakumar Ray Rachanabali, Vol.1*, 377–80. Calcutta: Asia Publishing Co., 1974.

———. "Kalloler Dol" [The Kallol Group]. In Gita Datta and Sukhamay Mukhopadhyay, eds., *Hemendrakumar Ray Rachanabali, Vol.4*, 377–82. Calcutta: Asia Publishing Co., 1980.

———. *Rater Kolkata* [Calcutta by Night]. Kolkata: Urbi Prakashan, 2016.

BIBLIOGRAPHY

Abu El-Haj, Nadia. *The Genealogical Science: The Search for Jewish Origins and the Politics of Epistemology*. Chicago: University of Chicago Press, 2012.

———. "The Genetic Reinscription of Race." *Annual Review of Anthropology* 36 (2007): 283–300.

Adcock, C. S. "Sacred Cows and Secular History: Cow Protection Debates in Colonial North India." *Comparative Studies of South Asia, Africa and the Middle East* 30, no. 2 (2010): 297–311.

Agarwal, H. N. "ABO Blood Groups, P.T.C. Taste Sensitivity, Sickle Cell Trait, Middle Phalangeal Hairs, and Colour Blindness in the Coastal Nicobarese of Great Nicobar." *Acta Genetica et Statistica Medica* 18, no. 2 (1964): 147–54.

———. "A Study on ABO Blood Groups, P.T.C. Taste Sensitivity, Sickle Cell Trait and Middle Phalangeal Hairs among the Burmese Immigrants of Andaman Islands." *Eastern Anthropologist* 19 (1966): 107–16.

Ager, J. A. M., and H. Lehmann. "Haemoglobin K in an East Indian and His Family." *British Medical Journal* 1, no. 5033 (1957): 1449–50.

Ahmed, Rafiuddin. *The Bengal Muslims, 1871–1906: A Quest for Identity*. Delhi: Oxford University Press, 1998.

Alam, Eram. "The Care of Foreigners: A History of South Asian Physicians in the United States, 1965–2016." Diss., University of Pennsylvania, 2016.

Ambedkar, Bhimrao R. "Castes in India." *Indian Antiquary* 46 (1917): 81–95.

Amin, Shahid. *Conquest and Community: The Afterlife of Warrior Saint Ghazi Miyan*. Chicago: University of Chicago Press, 2015.

———. *Event, Metaphor, Memory, Chauri Chaura 1922–1992*. Berkeley: University of California Press, 1995.

Amrhein, Valentin, Sander Greenland, and Blake McShane. "Scientists Rise Up against Statistical Significance." *Nature* 567, no. 7748 (2019): 305–7.

Amrith, Sunil S. *Decolonizing International Health: India and Southeast Asia, 1930–65*. London: Palgrave Macmillan, 2006.

Anderson, Mark. *From Boas to Black Power: Racism, Liberalism, and American Anthropology*. Stanford: Stanford University Press, 2019.

Anderson, Warwick. "The Anomalous Blonds of the Maghreb: Carleton Coon Invents the African Nordics." In Martin Thomas and Amanda Harris, *Expedition-*

ary Anthropology: Teamwork, Travel and the "Science of Man," 150–74. New York: Berghahn, 2018.

———. *Cultivation of Whiteness: Science, Health, and Racial Destiny in Australia*. New York: Basic Books, 2003.

———. "From Racial Types to Aboriginal Clines: The Illustrative Career of Joseph B. Birdsell." *Historical Studies in the Natural Sciences* 50, no. 5 (2020): 498–524.

———. "Objectivity and Its Discontents." *Social Studies of Science* 43, no. 4 (2013): 557–76.

———. "Racial Hybridity, Physical Anthropology, and Human Biology in the Colonial Laboratories of the United States." *Current Anthropology* 53, no. S5 (2013): 95–107.

Anderson, Warwick, and M. Susan Lindee. "Pacific Biologies: How Humans Become Genetic." *Historical Studies in the Natural Sciences* 50, no. 5 (2020): 483–97.

Annandale, Nelson. "Introductory Note." *Records of the Indian Museum* 23 (1922): 1–4.

Anon. "Council of the Indian Anthropological Society, 1968–69." *Journal of the Indian Anthropological Society* 3, nos. 1–2 (1968): inside cover.

———. "Curriculum Vitae of Dr. Biraja Sankar Guha (a Sketch)." In R. K. Bhattacharya and Jayanta Sarkar, eds., *Anthropology of B. S. Guha (A Centenary Tribute)*, vii–viii. Calcutta: Anthropological Survey of India, 1996.

———. "Dhirendra Nath Majumdar: Vita." In T. N. Madan and Gopala Sarana, eds., *Indian Anthropology: Essays in Memory of D. N. Majumdar*, 11–13. New York: Asia Publishing House, 1962.

———. "[Entry on] Eileen M. Erlanson MacFarlane." *Michigan Alumnus* 10, no. 1 (1943): 76.

———. "How Government Dealt with Influenza Epidemic." *Amrita Bazar Patrika*, 26 January 1919.

———. "Life Sketch of Sudhir Ranjan Das." In Parasmani Dasgupta and Roland Hauspie, eds., *Perspectives in Human Growth, Development and Maturation*, 353–61. Dordrecht: Springer, 2001.

———. "Manifests of Passengers Arriving at St. Albans, VT, District through Canadian Pacific and Atlantic Ports, 1895–1954" (1919). Records of the Immigration and Naturalization Service, 1787–2004. RG 85. Washington, DC: National Archives.

———. "[Obituary of] Eileen Whitehead Erlanson MacFarlane." *Columbus Dispatch*, 19 February 2002.

———. "Preface." In G. [Gustav] Spiller, ed., *Papers on Inter-Racial Problems Communicated to the First Universal Races Congress, Held at the University of London*, v–vi. London: P. S. King & Sons, 1911.

———. "Scientific News." *Science* 74, no. 1930 (1931): 10–14.

———. "Sterilisation of the Unfit Bill." *Times of India*, 28 February 1969.

Appadurai, Arjun. "Number in the Colonial Imagination." In Carol A. Breckenridge and Peter van der Veer, eds., *Orientalism and the Postcolonial Predicament*, 314–40. Philadelphia: University of Pennsylvania Press, 1993.

Aquil, Raziuddin, and Partha Chatterjee, eds. *History in the Vernacular*. Bangalore: Permanent Black, 2008.

Arnold, David. "'An Ancient Race Outworn': Malaria and Race in Colonial Bengal, 1860–1930." In Waltraud Ernst and Bernard Harris, eds., *Race, Society and Medicine, 1700–1960*, 123–43. London: Routledge, 1999.

———. *Colonizing the Body: State Medicine and Epidemic Disease in Nineteenth-Century India*. Berkeley: University of California Press, 1993.

———. "Diabetes in the Tropics: Race, Place and Class in India, 1880–1965." *Social History of Medicine* 22, no. 2 (2009): 245–61.

———. "Nehruvian Science and Postcolonial India." *Isis* 104, no. 2 (2013): 360–70.

———. *Science, Technology and Medicine in Colonial India*. Cambridge: Cambridge University Press, 2000.

Aronowitz, Robert A. *Risky Medicine: Our Quest to Cure Fear and Uncertainty*. Chicago: University of Chicago Press, 2015.

Arya, Sunaina, and Akaash Singh Rathore, eds. *Dalit Feminist Theory: A Reader*. New York: Routledge, 2020.

Bailkin, Jordanna. "The Boot and the Spleen: When Was Murder Possible in British India?" *Comparative Studies in Society and History* 48, no. 2 (2006): 462–93.

Bald, Vivek. *Bengali Harlem and the Lost Histories of South Asian America*. Cambridge, MA: Harvard University Press, 2013.

Balibar, Étienne, and Immanuel Wallerstein. *Race, Nation, Class: Ambiguous Identities*. Trans. Chris Turner. London: Verso, 1991.

Ballantyne, Tony. *Orientalism and Race: Aryanism in the British Empire*. Basingstoke, UK: Palgrave Macmillan, 2007.

Ballhatchet, Kenneth. *Race, Sex, and Class under the Raj: Imperial Attitudes and Policies and Their Critics, 1793–1905*. London: Weidenfield & Nicholson, 1980.

Bandyopadhyay, Sekhar. *Caste, Culture and Hegemony: Social Domination in Colonial Bengal*. New Delhi: Sage, 2004.

———. "Caste, Social Reform and the Dilemmas of Indian Modernity: Reading Acharya Prafulla Chandra Ray." *Bengal Past & Present* 126, nos. 242–43 (2007): 31–51.

Banerjee, Prathama. "Between the Political and the Nonpolitical: The Vivekananda Moment and a Critique of the Social in Colonial Bengal, 1890s–1910s." *Social History* 39, no. 3 (2014): 323–39.

———. "Debt, Tome and Extravagance: Money and the Making of 'Primitives' in Colonial Bengal." *Indian Economic & Social History Review* 37, no. 4 (2000): 423–45.

Banerji, Amiya Kumar. *West Bengal District Gazetteers: Hooghly*. Calcutta: Durgadas Majumdar, 1972.

Bangham, Jenny. "Blood Groups and Human Groups: Collecting and Calibrating Genetic Data after World War Two." *Studies in History and Philosophy of Biological and Biomedical Sciences* 47, no. A (2014): 74–86.

———. *Blood Relations: Transfusion and the Making of Human Genetics*. Chicago: University of Chicago Press, 2020.

———. "Writing, Printing, Speaking: Rhesus Blood-Group Genetics and Nomenclatures in the Mid-Twentieth Century." *British Journal for the History of Science* 47, no. 2 (2014): 335–61.

Bansal, Inderjit Singh. "A Study of ABO Blood Groups of the People of Ladakh." *American Journal of Physical Anthropology* 27, no. 2 (1967): 211–12.

Barbosa, Thiago Pinto. "Making Human Differences in Berlin and Maharashtra: Considerations on the Production of Physical Anthropological Knowledge by Irawati Karve." *Südasien-Chronik/South Asia Chronicle* 8 (2018): 139–62.

Barnicot, N. A. "Taste Deficiency for Phenylthiourea in African Negroes and Chinese." *Annals of Eugenics* 15, no. 1 (1949): 248–54.

Bashford, Alison. "Anticolonial Climates: Physiology, Ecology, and Global Population, 1920-1950." *Bulletin of the History of Medicine* 86, no. 4 (2012): 596–626.

Basu, A. K. "J. B. Chatterjea: An Obituary." *Thrombosis Research* 1, no. 4 (1972): 281–82.

Basu, Amitabha, Alok Kumar Ghosh, Suhas Kumar Biswas, and Ramendra Ghosh. "Introduction." In Basu, Ghosh, Biswas, and Ghosh, eds., *Physical Anthropology and Its Extending Horizons: S. S. Sarkar Memorial Volume*, 1–17. Bombay: Orient Longman, 1973.

Basu, Arabinda, S. K. Bhattacharya, S. Chatterjee, and C. Piplai. "Physical Anthropological Research in the Survey: 1948–1990." In K. S. Singh, ed., *The History of the Anthropological Survey of India*, 64–78. Calcutta: Anthropological Survey of India, 1991.

Basu, Bani. *Bangla Sishushahitya: Granthapanji* [A Bibliography of Bengali Books for Children and Young Adults]. Calcutta: Bangiya Granthagar Parishad, 1965.

Bates, Crispin. "Race, Caste and Tribe in Central India: The Early Origins of Indian Anthropometry." In Peter Robb, ed., *The Concept of Race in South Asia*, 219–59. Delhi: Oxford University Press, 1995.

Baviskar, Amita. "Adivasi Encounters with Hindu Nationalism in MP." *Economic and Political Weekly* 40, no. 48 (2005): 5105–13.

Bayly, C. A. "The Ends of Liberalism and the Political Thought of Nehru's India." *Modern Intellectual History* 12, no. 3 (2015): 605–26.

Bayly, Susan. "Caste and 'Race' in the Colonial Ethnography of India." In Peter Robb, ed., *The Concept of Race in South Asia*, 165–218. Delhi: Oxford University Press, 1995.

———. *The New Cambridge History of India IV.3: Caste, Society and Politics in India from the Eighteenth Century to the Modern Age*. Cambridge: Cambridge University Press, 1999.

Beatty, John. "Genetics in the Atomic Age." In Keith Benson, Jane Maienschein, and Ronald Rainger, eds., *The Expansion of American Biology*, 284–324. New Brunswick, NJ: Rutgers University Press, 1991.

Beller, Steven. *Antisemitism: A Very Short Introduction*. Oxford: Oxford University Press, 2007.

Berenstein, Nadia. "There's No Voting on Matters of Taste: Phenylthiocarbamide and Genetics Education." http://nadiaberenstein.com/blog/tag/PTC.

Benjamin, Ruha. *Race after Technology: Abolitionist Tools for the New Jim Code*. Cambridge: Polity Press, 2019.

Benjamin, Walter. "The Storyteller: Reflections on the Works of Nikolai Leskov." In Dorothy J. Hale, ed., *The Novel: An Anthology of Criticism and Theory 1900–2000*. Malden, MA: Blackwell, 2006.

Berk, L., and G. M. Bull. "A Case of Sickle Cell Anaemia in an Indian Woman." *Clinical Proceedings* 2, no. 6 (1943): 147–52.

Béteille, André. "Race and Caste." In Sukhdeo Thorat and Umakant, eds., *Caste, Race and Discrimination: Discourses in International Context*, 49–52. New Delhi: Indian Institute of Dalit Studies, 2004.

———. "Race and Descent as Social Categories in India." *Daedalus* 96, no. 2 (1967): 444–63.

Bhadra, Gautam. "Four Rebels of Eighteen-Fifty-Seven." In Ranajit Guha and Gayatri Chakravorty Spivak, eds., *Selected Subaltern Studies*, 129–75. New York: Oxford University Press, 1988.

Bhalla, Vijender. "Caste as Evolutionary Force in Genetic Change." In L. P. Vidyarthi and B. K. Verma, eds., *Some Aspects of Applied Physical Anthropology*, 99–102. Ranchi: Council of Social & Cultural Research, 1963.

Bharati, Agehananda. "The Use of 'Superstition' as an Anti-Traditional Device in Urban Hinduism." *Contributions to Indian Sociology* 4, no. 1 (1970): 36–49.

Bhasin, M. K., and H. Walter. *Genetics of Castes and Tribes of India*. Delhi: Kamla-Raj Enterprises, 2001.

Bhattacharjee, P. N. "The Blood Groups (A1 A2 B O, MNS and Rh) of the Ladakhis." *Acta Genetica et Statistica Medica* 18, no. 1 (1968): 78–83.

———. "Distribution of the Blood Groups (A_1 A_2 B O, MNSs, Rh), and the Secretor Factor among the Muslims and the Pandits of Kashmir." *Zeitschrift für Morphologie und Anthropologie* 58, no. 1 (1966): 86–94.

———. "A Genetical Study of the Santals of the Santal Parganas." *Anthropologist*, special issue (1969): 93–103.

———. "A Genetic Survey in the Rarhi Brahmin and the Muslim of West Bengal." *Bulletin of the Department of Anthropology, Government of India* 5 (1956): 18–28.

Bhattacharjee, P. N., and N. Kumar. "A Blood Group Genetic Survey in the Dudh Kharias of the Ranchi District (Bihar, India)." *Human Heredity* 19, no. 4 (1969): 385–97.

Bhattacharjee, P. N., and Hirendra K. Rakshit. "Sero-Anthropology of the Indian Mongoloids." In *Bio-Anthropological Research in India: Proceedings of the Seminar in Physical Anthropology and Allied Disciplines*, 99–107. Calcutta: Anthropological Survey of India, 1970.

Bhattacharya, D. K. "A Study of ABO, RH-HR and MN Blood Groups of the Anglo-Indians of India." *Human Biology* 41, no. 1 (1969): 115–24.

———. "Tasting of P.T.C. among the Anglo-Indians of India." *Acta Geneticae Medicae et Gemellologiae* 13, no. 2 (1964): 159–66.

Bhattacharya, Nandini. *Contagion and Enclaves: Tropical Medicine in Colonial India*. Liverpool: Liverpool University Press, 2012.

Bhattacharya, R. K., and Jayanta Sarkar. "B. S. Guha: The Founder Director of the Anthropological Survey of India." In Bhattacharya and Sarkar, eds., *Anthropology of B. S. Guha*, 1–13. Calcutta: Anthropological Survey of India, 1996.

Bhisey, Rajani A. "Satyavati M. Sirsat (1925–2010)." *Current Science* 1010, no. 7 (2011): 964–65.

Binder, Stefan. "Magic Is Science: Atheist Conjuring and the Exposure of Superstition in South India." *HAU: Journal of Ethnographic Theory* 9, no. 2 (2019): 284–98.

Blakeslee, Albert F. "Genetics of Sensory Thresholds: Taste for Phenyl Thio Carbamide." *Proceedings of the National Academy of Sciences of the United States of America* 18, no. 1 (1932): 120–30.

Blakeslee, Albert F., and M. R. Salmon. "Odor and Taste Blindness." *Eugenical News* 16, no. 7 (n.d.): 105–8.

Bodding, Paul Olaf. *Santal Medicine*. Calcutta: Asiatic Society, 1927.

Bose, Nirmal Kumar. "Foreword." In Amitabha Basu, Alok Kumar Ghosh, Suhas Kumar Biswas, and Ramendra Ghosh, eds., *Physical Anthropology and Its Extending Horizons: S. S. Sarkar Memorial Volume*. Bombay: Orient Longman, 1973.

Bose, Shib Chunder. *The Hindoos as They Are: A Description of the Manners, Customs and Inner Life of Hindoo Society*. Calcutta: W. Newman & Co., 1881.

Bose, Uma. "Blood Groups of the Tribes of Travancore." *Bulletin of the Department of Anthropology, Government of India* 1, no. 1 (1952): 19–24.

Botha, M. C., and L. J. van Zyl. "Abnormal Haemoglobins in Cape Town." *South African Medical Journal* 40, no. 32 (1966): 752–56.

Bowler, Peter J., and John V. Pickstone. *The Cambridge History of Science Volume 6: The Modern Biological and Earth Sciences*. Cambridge: Cambridge University Press, 2009.

Boyd, William C. "Genetics and the Human Race." *Science* 140, no. 3571 (1963): 1057–64.

Boyd, William C., and Lyle G. Boyd. "Sexual and Racial Variations in Ability to Taste

Phenyl-Thio-Carbamide with Some Data on the Inheritance." *Annals of Eugenics* 8, no. 1 (1937): 46–51.

Breckenridge, Keith. *The Biometric State: The Global Politics of Identification and Surveillance in South Africa, 1850 to the Present*. Cambridge: Cambridge University Press, 2013.

Brimnes, Niels. "Variolation, Vaccination and Popular Resistance in Early Colonial South India." *Medical History* 48, no. 2 (2004): 199–228.

Brodkin, Karen. *How Jews Became White Folks & What That Says about America*. New Brunswick, NJ: Rutgers University Press, 1998.

Brown, Judith M., and Richard Fox Young. "Indian Christians and Nehru's Nation-State." In *India and the Indianness of Christianity: Essays on Understanding—Historical, Theological, and Bibliographical—in Honor of Robert Eric Frykenberg*, 217–34. Grand Rapids, MI: Wm. B. Eerdmans Publishing Co., 2009.

Brown, Mark. "Ethnology and Colonial Administration in Nineteenth-Century British India: The Question of Native Crime and Criminality." *British Journal for the History of Science* 36, no. 2 (2003): 201–19.

Büchi, Ernest C. "Blood Groups of Tibetans." *Bulletin of the Department of Anthropology, Government of India* 1, no. 2 (1952): 71–78.

———. "Blood Secretion and Taste among the Pallar—A South Indian Community." *Anthropologist* 2, no. 1 (1955): 1–8.

———. "Frequency of ABO-Blood Groups and Secretor Factor in Bengal." *Bulletin of the Department of Anthropology, Government of India* 2, no. 1 (1952): 49–54.

———. "Is Sickling a Weddid Trait?" *Anthropologist* 2 (1955): 25–29.

Burton, Elise K. *Genetic Crossroads: The Middle East and the Science of Human Heredity*. Stanford: Stanford University Press, 2020.

———. "Red Crescents: Race, Genetics, and Sickle Cell Disease in the Middle East." *ISIS: A Journal of the History of Science Society* 110, no. 2 (2019): 250–69.

Buxton, Hilary. "Imperial Amnesia: Race, Trauma and Indian Troops in the First World War." *Past & Present* 241, no. 1 (2018): 221–58.

Campbell, Chloe. *Race and Empire: Eugenics in Colonial Kenya*. Manchester, UK: Manchester University Press, 2007.

Cantor, David. "The Frustrations of Families: Henry Lynch, Heredity, and Cancer Control, 1962–1975." *Medical History* 50, no. 3 (2006): 279–302.

Caplan, Lionel. "Martial Gurkhas: The Persistence of a British Military Discourse on 'Race.'" In Peter Robb, ed., *The Concept of Race in South Asia*, 260–81. Delhi: Oxford University Press, 1995.

Carrel, Alexis. *Man, the Unknown: Complete and Unabridged*. Bombay: Wilco Publishing House, 1959.

Central Bureau of Health Intelligence. *Annual Report of the Directorate General of Health Services 1957*. New Delhi: Directorate General of Health Services, Ministry of Health, n.d.

———. *Annual Report of the Directorate General of Health Services 1960*. New Delhi: Directorate General of Health Services, Ministry of Health, n.d.

Chadarevian, Soraya de. "Chromosome Photography and the Human Karyotype." *Historical Studies in the Natural Sciences* 45, no.1 (2015): 115–46.

———. "Chromosome Surveys of Human Populations: Between Epidemiology and Anthropology." *Studies in History and Philosophy of Biological and Biomedical Sciences* 47, Part A (2014): 87–96.

———. "Following Molecules: Hemoglobin between the Clinic and the Laboratory." In Soraya de Chadarevian and Harmke Kamminga, eds., *Molecularizing Biology and Medicine New Practices and Alliances, 1910s–1970s*, 160–89. Amsterdam: Overseas Publishers Association, 1998.

Chakrabarti, Pratik. *Bacteriology in British India: Laboratory Medicine and the Tropics*. Rochester: Rochester University Press, 2012.

Chakrabarty, Dipesh. *The Calling of History: Sir Jadunath Sarkar and His Empire of Truth*. Chicago: University of Chicago Press, 2015.

Chakraborty, Mukul. "ABO Blood Groups of the Zeliang Naga." *Man in India* 45 (1965): 311–15.

Chakraborty, Ranjit. "A Theorem on Race Mixture." In M. R. Chakravartti, ed., *Proceedings of the International Symposium on Human Genetics (1971)*, 176–79. Waltair: Andhra University Press, 1973.

Chakravorty, Mayurika. "'Skeletons of History': Fact and Fiction in Rakhaldas Bandyopadhyay's Sasanka." *South Asia Research* 24, no. 2 (2004): 171–83.

Chandra, Shefali. "Whiteness on the Margins of Native Patriarchy: Race, Caste, Sexuality, and the Agenda of Transnational Studies." *Feminist Studies* 37, no. 1 (2011): 127–53.

Chatterjea, J. B. "Haemoglobinopathies, Glucose-6-Phosphate Dehydrogenase Deficiency and Allied Problems in the Indian Subcontinent." *Bulletin of the World Health Organization* 35, no. 6 (1966): 837–56.

———. "Haemoglobinopathy in India." In J. H. P. Jonxis and J. F. Delafresnaye, eds., *Abnormal Haemoglobins*, 322–39. Oxford: Blackwell Scientific Publications, 1959.

Chatterjee, B. K. "Foreword." In S. S Sarkar, ed., *Indian Eugenics Society: Bulletin No. 1*, i–iii. Calcutta: Indian Eugenics Society, 1941.

Chatterjee, Indrani. "Genealogy, History and Law: The Case of Tripura Rajamala." In Partha Chatterjee and Anjan Ghosh, eds., *History and the Present*, 108–43. Delhi: Permanent Black, 2002.

———. "Monastic 'Governmentality': Revisiting 'Community' and 'Communalism' in South Asia." *History Compass* 13, no. 10 (n.d.): 497–511.

Chatterjee, Partha. *Nationalist Thought and the Colonial World: A Derivative Discourse*. Minneapolis: University of Minnesota Press, 2008.

———. "On Religious and Linguistic Nationalisms: The Second Partition of Bengal." In Peter van der Veer and Hartmut Lehmann, eds., *Nation and Religion: Perspectives on Asia and Europe*, 112–28. Princeton: Princeton University Press, 1999.

Chatterjee, Piya. *A Time for Tea: Women, Labor, and Post/Colonial Politics*. Durham, NC: Duke University Press, 2001.

Chattopadhyay, Bodhisattva. "Hemendrakumar Ray and the Birth of Adventure Kalpabigyan." *Jadavpur University Essays and Studies* 27 (2013): 35–54.

Chattopadhyay, Debiprasad. *Payer Nokh theke Mathar Chul* [From Toe Nails to the Hair on the Head]. Calcutta: Eagle Publishing Co. Ltd., n.d.

Chattopadhyay, Prasanta Kumar. "Taste Sensitivity to Phenylthiocarbamide among the Jats." *Anthropologischer Anzeiger* 33, no. 1 (1971): 52–60.

Chaudhuri, S., J. Ghosh, B. Mukherjee, and A. K. Roychoudhury. "Study of Blood Groups and Haemoglobin Variants among the Santal Tribe in Midnapore District of West Bengal, India." *American Journal of Physical Anthropology* 26, no. 3 (1967): 307–11.

Chen, Bohang. "A Non-Metaphysical Evaluation of Vitalism in the Early Twentieth Century." *History & Philosophy of the Life Sciences* 40, no. 3 (2018): 50–72.

Cleminson, Richard, and Ricardo Roque. "Imagining the 'Biochemical Race': Sero-Anthropology and Concepts of Racial Purity in Portugal (1900s–1950s)." *European History Quarterly* 51, no. 3 (2021).

Clouser, Kirstin. "Chinese Whiteness: The Discourse of Race in Modern and Contemporary Chinese Culture." Honors thesis, Bucknell University, 2013.

Cohen, Lawrence. *No Aging in India: Alzheimer's, the Bad Family and Other Modern Things*. Berkeley: University of California Press, 2000.

———. "The Other Kidney: Biopolitics beyond Recognition." *Body & Society* 7, nos. 2–3 (2001): 9–29.

———. "The 'Social' De-Duplicated: On the Aadhaar Platform and the Engineering of Service." *South Asia: Journal of South Asian Studies* 42, no. 3 (2019): 482–500.

Confino, Alon, and Ajay Skaria. "The Local Life of Nationhood." *National Identities* 4, no. 1 (1 March 2002): 7–24.

Copeman, Jacob, and Dwaipayan Banerjee. *Hematologies: The Political Life of Blood in India*. Ithaca: Cornell University Press, 2019.

Coulson, Doug. *Race, Nation, and Refuge: The Rhetoric of Race in Asian American Citizenship Cases*. Albany: SUNY Press, 2017.

Cowles, Henry M. *The Scientific Method: An Evolution of Thinking from Darwin to Dewey*. Cambridge, MA: Harvard University Press, 2020.

Crawford, D. G. *A Brief History of the Hughli District*. Calcutta: Bengal Secretariat Press, 1902.

Crow, James F. "Some Possibilities for Measuring Selection Intensities in Man." *Human Biology* 30, no. 1 (1958): 1–13.

Dabassa, Anbessa, and Ketema Bacha. "The Prevalence and Antibiogram of Salmonella and Shigella Isolated from Abattoir, Jimma Town, South West Ethiopia." *International Journal of Pharmaceutical and Biological Research* 3, no. 4 (2012): 143–48.

Dacie, John. "Hermann Lehmann, 8 July 1910–13 July 1985." *Biographical Memoirs of Fellows of the Royal Society* 34 (1988): 406–49.

Das, S. R. "Application of Phenylthiocarbamide Taste Character in the Study of Racial Variation." *Journal of Indian Anthropological Society* 1, no. 1 (1966): 63–80.

———. "A Contribution to the Heredity of the P.T.C. Taste Character Based on a Study of 845 Sib-Pairs." *Annals of Human Genetics* 20, no. 4 (1956): 334–43.

———. "Inheritance of the P. T. C. Taste Characters in Man: An Analysis of 126 Rarhi Brahmin Families of West Bengal." *Annals of Human Genetics* 22, no. 3 (1958): 200–212.

Das, S. R., and P. N. Bhattacharjee. "Blood Groups (A1A2BO), ABH Secretion, Sickle-Cell, P.T.C. Taste and Colour Blindness in the Rajbanshi of Midnapur District, West Bengal." *Bulletin of the Anthropological Survey of India* 12, no. 1 (1963): 1–6.

Das, S. R., N. Kumar, P. N. Bhattacharjee, and D. B. Sastry. "Blood Groups (ABO, M-N and Rh), ABH Secretion, Sickle-Cell, P.T.C. Taste, and Colour Blindness in the Mahar of Nagpur." *Journal of the Royal Anthropological Institute of Great Britain and Ireland* 91, no. 2 (1961): 345–55.

Das, S. R., and D. P. Mukherjee. "Phenylthiocarbamide Taste Sensitivity Survey among the Pareng Gadaba, the Ollaro Gadabaand, and the Konda Paroja of Koraput District, Orissa." *Acta Genetica et Statistica Medica* 14, no. 2 (1964): 168–76.

Das, S. R., D. P. Mukherjee, and P. N. Bhattacharjee. "P.T.C. Taste Threshold Distribution in the Bado Gadaba and the Bareng Paroja of KoraputDistrict in Orissa." *Acta Genetica et Statistica Medica* 13, no. 4 (1963): 369–77.

Das, S. R., D. P. Mukherjee, and P. N. Bhattacharjee. "Survey of the Blood Groups and PTC Taste among the Rajbanshi Caste of West Bengal (ABO, MNS, Rh, Duffy and Diego)." *Acta Genetica et Statistica Medica* 17, no. 5 (1967): 433–45.

Das, S. R., D. B. Sastry, and D. P. Mukherjee. "Blood Groups (A_1A_2BO, M-N, Rh) and ABH Secretion in the Pareng Gadaba, the Ollaro Gadaba and the Konda Paroja of Koraput District in Orissa." *Acta Genetica et Statistica Medica* 16, no. 2 (1966): 169–83.

Dasgupta, Parasmani, and Roland Hauspie. "Editor's Note." In Dasgupta and Hauspie, eds., *Perspectives in Human Growth, Development and Maturation*, xv–xvi. Dordrecht: Springer, 2001.

Dasgupta, Sangeeta. "The Journey of an Anthropologist in Chhotanagpur." *Indian Economic & Social History Review* 41, no. 2 (2004): 165–98.

———. "Reordering a World: The Tana Bhagat Movement, 1914–1919." *Studies in History* 15, no. 1 (1999): 1–41.

Daston, Lorraine. "Objectivity and the Escape from Perspective." *Social Studies of Science* 22, no. 4 (1992): 597–618.

Daston, Lorraine, and Peter Galison. *Objectivity*. Zone Books, 2007.

Datta-Majumder, Nabendu. *The Santals: A Study in Culture-Change*. Calcutta: Manager of Publications—Delhi, 1956.

Davis, Donald R., Jr. "Children: *Putr, Duhitr*." In Patrick Olivelle and Donald R. Davis Jr., eds. *Hindu Law: A New History of Dharmaśāstra*, 151–63. Oxford: Oxford University Press, 2018.

Delafresnaye, J. F. "Foreword." In J. H. P. Jonxis and J. F. Delafresnaye, eds., *Abnormal Haemoglobins*, ix. Oxford: Blackwell Scientific Publications, 1959.

Deleuze, Gilles, and Félix Guattari. *Anti-Oedipus: Capitalism and Schizophrenia*. Trans. Robert Hurley, Mark Seem, and Helen Lane. Minneapolis: University of Minnesota Press, 1983.

Desai, Sapur F. "Place of Eugenics in a Democracy." *Times of India*, 20 September 1959.

Deshpande, Prachi. "Caste as Maratha: Social Categories, Colonial Policy and Identity in Early Twentieth-Century Maharashtra." *Indian Economic & Social History Review* 41, no. 1 (2004): 7–32.

Dey, Kanny Lall. *The Indigenous Drugs of India*. 2d ed. Calcutta: Thacker, Spink & Co., 1896.

Dikötter, Frank. *The Discourse of Race in Modern China*. Oxford: Oxford University Press, 2015.

Directorate General of Health Services. *Tri-Annual Reports of the Directorate General of Health Services, 1954–56*. New Delhi: Ministry of Health, n.d.

Dirks, Nicholas B. *Castes of Mind: Colonialism and the Making of Modern India*. Princeton: Princeton University Press, 2001.

Dobzhansky, Theodosius. "Commentary." *Current Anthropology* 3, no. 3 (1962): 279–80.

———. "Genetic Loads in Natural Populations." *Science* 126, no. 3266 (1957): 191–94.

———. "Mendelian Populations and Their Evolution." *American Naturalist* 84, no. 819 (1950): 401–18.

———. "Of Flies and Men." *American Psychologist* 22, no. 1 (1967): 41–48.

Dodson, Michael. *Orientalism, Empire, and National Culture: India, 1770–1880*. Basingstoke, UK: Palgrave Macmillan, 2007.

Dragsdahl, Rune-Christoffer. "The Practices of Indian Vegetarianism in a World of Limited Resources: The Case of Bengaluru." In *Food Consumption in the City: Practices and Patterns in Urban Asia and the Pacific*, 141–58. Abingdon, UK: Routledge, 2016.

Dronamraju, K. R. "Mating Systems of the Andhra Pradesh People." *Cold Spring Harbor Symposium on Quantitative Biology* 29 (1964): 81–84.

D'Souza, Dilip. "De-Notified Tribes: Still 'Criminal'?" *Economic and Political Weekly* 34, no. 51 (1999): 3576–78.

Duncan, Ian. "Dalits and the Raj: The Persistence of the Jatavs in the United Provinces." *Indian Economic & Social History Review* 56, no. 2 (2019): 119–45.

Dunlop, K. J., and U. K. Mozumder. "The Occurrence of Sickle Cell Anaemia among a Group of Tea Garden Labourers in Upper Assam." *Indian Medical Gazette* 87, no. 9 (1952): 387–91.

Duster, Troy. "A Post-Genomic Surprise: The Molecular Reinscription of Race in Science, Law and Medicine." *British Journal of Sociology* 66, no. 1 (2015): 1–27.

Dutta, P. C. "A Note on the Ear Lobe." *Acta Genetica et Statistica Medica* 13, no. 3 (1963): 290–94.

Edney, Matthew H. *Mapping an Empire: The Geographical Construction of British India, 1765–1843*. Chicago: University of Chicago Press, 1997.

Egorova, Yulia. "Castes of Genes? Representing Human Genetic Diversity in India." *Genomics, Society and Policy* 6, no. 3 (November 2010): 32–49.

———. "De/Geneticizing Caste: Population Genetic Research in South Asia." *Science as Culture* 18, no. 4 (2009): 417–34.

———. "The Substance That Empowers: DNA in South Asia." *Contemporary South Asia* 21, no. 3 (2013): 291–303.

Egorova, Yulia, and Tudor Parfitt. "Genetics, History, and Identity: The Case of the Bene Israel and the Lemba." *Culture, Medicine and Psychiatry* 29, no. 2 (2005): 193–224.

El Shakry, Omnia. *The Great Social Laboratory: Subjects of Knowledge in Colonial and Postcolonial Egypt*. Stanford: Stanford University Press, 2007.

Ergin, Murat. *"Is the Turk a White Man?": Race and Modernity in the Making of Turkish Identity*. Boston: Brill, 2017.

Ernst, Waltraud. "Colonial Policies, Racial Politics and the Development of Psychiatric Institutions in Early Nineteenth-Century British India." In Waltraud Ernst and Bernard Harris, eds., *Race, Science and Medicine, 1700–1960*, 80–100. London: Routledge, 1999.

———. "Colonial Psychiatry, Magic and Religion: The Case of Mesmerism in British India." *History of Psychiatry* 15, no. 57 (March 2004): 57–71.

Falconer, D. S. "Sensory Thresholds for Solutions of Phenyl-Thio-Carbamide." *Annals of Eugenics* 13, no. 1 (1946): 211–22.

Fanon, Frantz. *Black Skin, White Masks*. Trans. Charles Lam Markmann. London: Pluto Press, 1986.

———. "Mental Alterations, Character Modifications, Psychic Disorders and Intellectual Deficit in Spinocerebellar Heredodegeneration: A Case of Friedreich's Ataxia with Delusions of Possession." In Fanon, *Alienation and Freedom*, ed. Jean Khalfa and Robert J. C. Young, trans. Steven Corcoran, 203–75. London: Bloomsbury Academic, 2018.

———. "Parallel Hands." In Fanon, *Alienation and Freedom*, ed. Jean Khalfa and Robert J. C. Young, trans. Steven Corcoran, 113–64. London: Bloomsbury Academic, 2015.

———. "The Trials and Tribulations of National Consciousness." In *The Wretched of the Earth*, trans. Richard Philcox. New York: Grove Press, 2004.

Fanon, Frantz, and Charles Geronimi. "TAT in Muslim Women: Sociology of Perception and Imagination." In Fanon, *Alienation and Freedom*, ed. Jean Khalfa and Robert J. C. Young, trans. Steven Corcoran, 427–33. London: Bloomsbury Academic, 2018.

Fischer-Tiné, Harald. "From Brahmacharya to 'Conscious Race Culture': Victorian Discourses of 'Science' and Hindu Traditions in Early Indian Nationalism." In Crispin Bates, ed., *Beyond Representation: Colonial and Postcolonial Constructions of Indian Identity*, 241–69. Delhi: Oxford University Press, 2006.

———. "Hierarchies of Punishment in Colonial India: European Convicts and the Racial Dividend, c. 1860–1890." In Harald Fischer-Tiné and Susanne Gehrmann, eds., *Empires and Boundaries: Rethinking Race, Class, and Gender in Colonial Settings*, 41–65. New York: Routledge, 2009.

Fox, Arthur L. "The Relationship between Chemical Constitution and Taste." *Proceedings of the National Academy of Sciences of the United States of America* 18, no. 1 (1932): 115–20.

Fujimura, Joan H., Troy Duster, and Ramya Rajagopalan. "Introduction: Race, Genetics, and Disease: Questions of Evidence, Matters of Consequence." *Social Studies of Science* 38, no. 5 (2008): 643–56.

Fuller, C. J. "Anthropologists and Viceroys: Colonial Knowledge and Policy Making in India, 1871–1911." *Modern Asian Studies* 50, no. 1 (n.d.): 217–58.

Fullwiley, Duana. *The Enculturated Gene: Sickle Cell Health Politics and Biological Difference in West Africa*. Princeton: Princeton University Press, 2011.

Gannett, Lisa. "The Biological Reification of Race." *British Journal for the Philosophy of Science* 55, no. 2 (2004): 323–45.

———. "Racism and Human Genome Diversity Research: The Ethical Limits of 'Population Thinking.'" *Philosophy of Science* 68, no. 3 (2001): S479–92.

———. "Theodosius Dobzhansky and the Genetic Race Concept." *Studies in History and Philosophy of Biological and Biomedical Sciences* 44, no. 3 (2013): 251–60.

Gannett, Lisa, James R. Griesemer, Jean-Paul Gaudillière, and Hans-Jörg Rheinberger. "Classical Genetics and the Geography of Genes." In Jean-Paul Gaudillière, and Hans-Jörg Rheinberger, eds., *Classical Genetic Research and Its Legacy: The Mapping Cultures of Twentieth-Century Genetics*, 57–88. New York: Routledge, 2004.

Gates, R. Ruggles. "Blood Groups from the Andamans." *Man* 40 (1940): 55–57.

Gerke, Barbara. *Long Lives and Untimely Deaths: Life-Span Concepts and Longevity Practices among Tibetans in the Darjeeling Hills, India*. Leiden: Brill, 2011.

———. "Of Matas, Jhakris, and Other Healers: Fieldnotes on a Healing Event in Kalimpong,India." In Charles Ramble and Ulrike Roesler, eds., *Tibetan and Himalayan Healing. An Anthology for Anthony Aris*, 231–48. Kathmandu: Vajra Publications, 2015.

Germann, Pascal. "'Nature's Laboratories of Human Genetics': Alpine Isolates, Hereditary Diseases and Medical Genetic Fieldwork, 1920–1970." In H. Petermann, P. Harper, and S. Doetz, eds., *History of Human Genetics*, 145–66. Dordrecht: Springer, 2017.

Ghosh, Amitav. *In an Antique Land: History in the Guise of a Traveler's Tale*. New York: Vintage, 1992.

———. *Occasional Paper No. 125: The Slave of MS.H.6*. Calcutta: Centre for Studies in Social Sciences, 1990.

Ghosh, Kaushik. "A Market for Aboriginality: Primitivism and Race Classification in the Indentured Labour Market of Colonial India." In Gautam Bhadra, Gyan Prakash, and

Susie Tharu, eds., *Subaltern Studies 10: Writings on South Asian History and Society*, 8–48. Delhi: Oxford University Press, 1999.

Ghosh, Parimal. *What Happened to the Bhadralok*. New Delhi: Primus Books, 2016.

Ghoshal, Sayori. "Race in South Asia: Colonialism, Nationalism and Modern Science." *History Compass* 19, no. 2 (2021): 1–11.

Gilman, Sander. *The Jew's Body*. New York: Routledge, 1991.

Gil-Riaño, Sebastián. "Relocating Anti-Racist Science: The UNESCO Statement on Race and Economic Development in the Global South." *British Journal for the History of Science* 51, no. 2 (2018): 281–303.

Gilroy, Paul. *Between Camps: Nations, Cultures and the Allure of Race*. London: Penguin Books, 2000.

Gissis, Snait B. "When Is 'Race' a Race? 1946–2003." *Studies in the History and Philosophy of Biology and Biomedical Sciences* 39, no. 4 (2008): 437–50.

Giuffrida-Ruggeri, V. *The First Outlines of Systematic Anthropology of Asia*. Trans. Haranchandra Chakladar. Calcutta: Calcutta University Press, 1921.

Glass, Bentley, Milton S. Sacks, Elsa F. Jahn, and Charles Hess. "Genetic Drift in a Religious Isolate: An Analysis of the Causes of Variation in Blood Group and Other Gene Frequencies in a Small Population." *American Naturalist* 86, no. 828 (1 May 1952): 145–59.

Goldstein, Eric L. *The Price of Whiteness: Jews, Race, and American Identity*. Princeton: Princeton University Press, 2006.

Golinski, Jan. *Making Natural Knowledge: Constructivism and the History of Science*. Cambridge: Cambridge University Press, 1998.

Gordin, Michael D. *On the Fringe: Where Science Meets Pseudoscience*. New York: Oxford University Press, 2021.

———. *The Pseudoscience Wars: Immanuel Velikovsky and the Birth of the Modern Fringe*. Chicago: University of Chicago Press, 2013.

Gordon, Avery. *Ghostly Matters: Haunting and the Sociological Imagination*. Minneapolis: University of Minnesota Press, 2008.

Gordon, Lewis R. "Requiem on a Life Well Lived: In Memory of Fanon." In Nigel C. Gibson, ed., *Living Fanon: Global Perspectives*, 11–26. Basingstoke, UK: Palgrave Macmillan, 2011.

Gould, William. *Hindu Nationalism and the Language of Politics in Late Colonial India*. Cambridge: Cambridge University Press, 2011.

Graham, Richard. "Introduction." In Graham, ed., *The Idea of Race in Latin America, 1870–1940*, 1–7. Austin: University of Texas Press, 1990.

Gualtieri, Sarah. *Between Arab and White: Race and Ethnicity in the Early Syrian American Diaspora*. Berkeley: University of California Press, 2009.

Guha, Abhijit. "Forget Not: In Search of a Nationalist Anthropology in India." *Bulletin of the Ramakrishna Mission Institute of Culture* 71, no. 9 (2020): 21–28.

———. *In Search of Nationalist Trends in Indian Anthropology: Opening a New Discourse*. Kolkata: Institute of Development Studies, 2018.

Guha, Birajashankar. "Application for Admission to Candidacy for a Degree in Arts or Philosophy, Harvard University" (1922). B. S. Guha's Admissions File, Harvard University Archives, Cambridge, MA.

———. *Bharater Jati Parichay* [India's Racial Identity]. Vernacular Pamphlet No. 1. Calcutta: Department of Anthropology Publications, 1949.

———. *An Outline of Racial Ethnology*. Calcutta: Indian Science Congress Association, 1937.

———. "The Role of Social Sciences in Nation Building." *Sociological Bulletin* 7, no. 2 (1958): 148–51.

Guha, Ranajit. *Elementary Aspects of Peasant Insurgency in Colonial India*. Delhi: Oxford University Press, 2002.

———. "The Prose of Counter-Insurgency." In Ranajit Guha and Gayatri Chakravorty Spivak, eds., *Select Subaltern Studies*, 45–88. New York: Oxford University Press, 1988.

Guha, Sumit. "Lower Strata, Older Races, and Aboriginal Peoples: Racial Anthropology and Mythical History Past and Present." *Journal of Asian Studies* 57, no. 2 (1998): 423–41.

———. "The Politics of Identity and Enumeration in India, c. 1600–1990." *Comparative Studies in Society and History* 45, no. 1 (2003): 148–67.

Gupta, Dipankar. "Caste Is Not Race: But, Let's Go to the UN Forum Anyway." In Sukhdeo Thorat and Umakant, eds., *Caste, Race, and Discrimination: Discourses in International Context*, 53–56. New Delhi: Indian Institute of Dalit Studies, 2004.

Gupta, Meghnad. *Rater Kolkata* [Night Life of Calcutta]. Kolkata: Urbi Prakashan, 2016.

Gupta, Sanjukta. "The Domestication of a Goddess: Carana-Tirtha Kalighat, the Mahapitha of Kali." In Rachel Fell McDermott and Jeffrey J. Kripal, eds., *Encountering Kali: In the Margins, at the Center, in the West*, 60–79. Berkeley: University of California Press, 2003.

Hacking, Ian. "Making Up People." In T. L. Heller, M. Sosna, and D. E. Wellbery, eds., *Reconstructing Individualism*. Stanford: Stanford University Press, 1985.

Haddon, A. C. *The Races of Man and Their Distribution*. New York: Frederick A. Stokes Co., 1909.

Hage, Ghassan. "Recalling Anti-Racism." *Ethnic and Racial Studies* 39, no. 1 (n.d.): 123–33.

Hardiman, David. *Coming of the Devi: Adivasi Assertion in Western India*. Delhi: Oxford University Press, 1987.

Harris, H., and H. Kalmus. "The Measurement of Taste Sensitivity to Phenylthiourea (P.T.C.)." *Annals of Eugenics* 15, no. 1 (n.d.): 24–31.

Harrison, Mark. *Climates and Constitutions: Health, Race, Environment, and British Imperialism in India, 1600–1850*. New York: Oxford University Press, 2002.

———. "'The Tender Frame of Man': Disease, Climate, and Racial Difference in India and the West Indies, 1760–1860." *Bulletin of the History of Medicine* 70, no. 1 (1996): 68–93.

Hartman, Saidiya. "Venus in Two Acts." *Small Axe* 12, no. 2 (2008): 1–14.

Haughton, Graves C. *A Dictionary, Bengali and Sanskrit, Explained in English*. London: Parbury, Allen & Co., 1833.

Heidegger, Martin. "The Question concerning Technology." In Heidegger, *The Question concerning Technology and Other Essays*, trans. Lovitt, 3–35. New York: Garland Publishing, 1977.

Hellman-Rajanayagam, Dagmar. "Is There a Tamil Race?" In Peter Robb, ed., *The Concept of Race in South Asia*, 109–45. Delhi: Oxford University Press, 1995.

Hermanns, P. Matthias. *The Evolution of Man: A Challenge to Darwinism through Human Biogenetics, Physical and Cultural Anthropology, Prehistory, and Palaeontology*. Allahabad: Society of St. Paul, 1955.

Highmore, Ben. "Bitter after Taste: Affect, Food, and Social Aesthetics." In Melissa Gregg and Gregory J. Seigworth, eds., *The Affect Theory Reader*, 118–38. Durham, NC: Duke University Press, 2010.

Hirschfeld [Hirszfeld], Ludwik, and Hanka Hirschfeld [Hirszfeld]. "Serological Differences between the Blood of Different Races: The Result of Researches on the Macedonian Front." *Lancet* 194, no. 5016 (1919): 675–79.

Hirszfeld, Ludwik. *Ludwik Hirszfeld: The Story of One Life*. Ed. Marta A. Balinska and William H. Schneider, trans. Marta A. Balinska. Rochester: Rochester University Press, 2010.

Hodges, Sarah. "South Asia's Eugenic Past." In Alison Bashford and Philippa Levine, eds., *The Oxford Handbook of the History of Eugenics*, 228–42. Oxford: Oxford University Press, n.d.

Honneth, Axel. "Foreword." In Rahel Jaeggi, *Alienation*, ed. Frederick Neuhouser, trans. Neuhouser and Alan E. Smith, vii–x. New York: Columbia University Press, 2014.

Huxley, T. H. "On the Geographical Distribution of the Chief Modifications of Mankind." *Journal of the Ethnological Society of London* 2, no. 4 (1870): 404–12.

Hyun, Jaehwan. "Blood Purity and Scientific Independence: Blood Science and Postcolonial Struggles in Korea, 1926–1975." *Science in Context* 32, no. 3 (2019): 239–60.

Imy, Kate. *Faithful Fighters: Identity and Power in the British Indian Army*. Stanford: Stanford University Press, 2019.

Indian Genome Variation Consortium (IGVC). "The Indian Genome Variation Database (IGVdb): A Project Overview." *Human Genetics* 118, no. 1 (2005): 1–11.

Jaeggi, Rahel. *Alienation*. Ed. Frederick Neuhouser, trans. Neuhouser and Alan E. Smith. New York: Columbia University Press, 2014.

Jaffrelot, Christophe. "Sanskritization vs. Ethnicization in India: Changing Identities and Caste Politics before Mandal." *Asian Survery* 40, no. 5 (2000): 756–66.

Jakobsen, Kjetil A. "A Fragment of Future History: Gabriel Tarde's Archival Utopia." In Susi K. Frank and Kjetil A. Jakobsen, eds., *Arctic Archives: Ice, Memory and Entropy*, 107–30. Bielefeld, Germany: Transcript, 2019.

Kabat, Elvin A. *Blood Group Substances: Their Chemistry and Immunochemistry*. New York: Academic Press, 1956.

Kapoor, Shivani. "The Smell of Caste: Leatherwork and Scientific Knowledge in Colonial India." *South Asia: Journal of South Asian Studies* 44, no. 5 (2021): 983–99.

Karve, Irawati. *Kinship Organization in India*. 2d ed. Bombay: Asia Publishing House, 1965.

———. "Racial Conflict." In D. N. Majumdar and Irawati Karve, *Racial Problems in Asia*, 27–54. New Delhi: Indian Council of World Affairs, 1947.

———. "Racial Factor in Indian Social Life." In O. P. Bhatnagar, ed., *Studies in Social History (Modern India)*, 33–42. Allahabad: St. Paul's Press Training School, 1964.

Kaviraj, Narayan Das. *Dravyagun Darpan*. Ed. Abhaycharan Gupta. Calcutta: Anglo-Indian Union Yantra, 1865.

Kevles, Daniel J. *In the Name of Eugenics: Genetics and the Uses of Human Heredity*. Berkeley: University of California Press, 1985.

Khan, Naveeda. "Marginal Lives and the Microsociology of Overhearing in the Jamuna Chars." *Ethnos: Journal of Anthropology* (2021).

Khullar, P. "Taste Sensitivity to Phenylthiocarbamide among Sindhi Children of Delhi." *Anthropology* 12 (1965): 17–24.

Knight, Alan. "Racism, Revolution and Indigenismo: Mexico, 1910–1940." In Richard Graham, ed., *The Idea of Race in Latin America, 1870–1940*, 71–114. Austin: University of Texas Press, 1990.

Kolsky, Elizabeth. *Colonial Justice in British India: White Violence and the Rule of Law*. Cambridge: Cambridge University Press, 2010.

Kowal, Emma. "Orphan DNA: Indigenous Samples, Ethical Biovalue and Postcolonial Science." *Social Studies of Science* 43, no. 4 (2013): 577–97.

Kowal, Emma, Ashley Greenwood, and Rebekah E. McWhirter. "All in the Blood: A Review of Aboriginal Australians' Cultural Beliefs about Blood and Implications for Biospecimen Research." *Journal of Empirical Research on Human Research Ethics* 10, no. 4 (2015): 347–59.

Kowal, Emma, Joanna Radin, and Jenny Reardon. "Indigenous Body Parts, Mutating Temporalities, and the Half-Lives of Postcolonial Technoscience." *Social Studies of Science* 43, no. 4 (2013): 465–83.

Kuklick, Henrika. "After Ishmael: The Fieldwork Tradition and Its Future." In Akhil Gupta and James Ferguson, eds., *Anthropological Locations: Boundaries and Grounds of a Field Science*, 47–65. Berkeley: University of California Press, 1997.

Kumar, Narendra. "Blood Group and Secretor Frequency among the Galong." *Bulletin of the Department of Anthropology, Government of India* 3, no. 1 (1954): 55–64.

Kumar, Narendra, and Alok Kumar Ghosh. "ABO Blood Groups and Sickle-Cell Trait Investigations in Madhya Pradesh: Ujjain and Dewas Districts." *Acta Genetica et Statistica Medica* 17, no. 1 (1967): 55–61.

Kumar, Narendra, and D. B. Sastry. "A Genetic Survey among the Riang: A Mongoloid Tribe of Tripura (North East India)." *Zeitschrift für Morphologie und Anthropologie* 51, no. 3 (1961): 346–55.

Lahiri, M. N. "Observations on the Medico-Legal Applications of Blood Grouping with a Note on Blood Groups in a Polyandrous Family." *Indian Journal of Medical Research* 16 (1929): 969–72.

Lal, Chaman. *Hindu America: Revealing the Story of the Romance of the Surya Vanshi Hindus and Depicting the Imprints of Hindu Culture on the Two Americas*. Bombay: New Book Co., 1940.

Leavitt, Judith Walzer. "'Typhoid Mary' Strikes Back: Bacteriological Theory and Practice in Early Twentieth-Century Public Health." *ISIS: A Journal of the History of Science Society* 83, no. 4 (1992): 608–29.

Lehmann, H., and Marie Cutbush. "Sickle-Cell Trait in Southern India." *British Medical Journal* 1, no. 4755 (1952): 404–5.

Lehmann, H., and P. K. Sukumaran. "Examination of 146 South Indian Aboriginals for Haemoglobin Variants." *Man* 56, nos. 96–97 (1956): 95–96.

Lewis, Herbert S. "Boas, Darwin, Science and Anthropology." *Cultural Anthropology* 42, no. 3 (2001): 381–406.

Lewis, Walter H., Anton A. Reznicek, and Richard K. Rabeler. "Identifications and Typifications of Rosa (Rosaceae) Taxa in North America Described or Used by E. W. Erlanson, 1925–1934." *Novon: A Journal for Botanical Nomenclature* 22, no. 1 (2012): 32–42.

Lindee, M. Susan. *American Science and the Survivors at Hiroshima*. Chicago: University of Chicago Press, 1994.

———. "Provenance and the Pedigree: Victor McKusick's Fieldwork with the Old Order Amish." In Alan H. Goodman, Deborah Heath, and M. Susan Lindee, eds., *Genetic Nature Culture: Anthropology and Science beyond the Two-Culture Divide*, 41–57. Berkeley: University of California Press, 2003.

Lindee, Susan, and Ricardo Ventura Santos. "The Biological Anthropology of Living Human Populations: World Histories, National Styles, and International Networks: An Introduction to Supplement 5." *Current Anthropology* 53, suppl. 5 (2012): S3–16.

Lipphardt, Veronika. "'Geographical Distribution Patterns of Various Genes': Genetic

Studies of Human Variation after 1945." *SHPSC: Studies in History and Philosophy of Biological & Biomedical Sciences: Part A* 47 (2014): 50–61.
———. "Isolates and Crosses in Human Population Genetics; or, A Contextualization of German Race Science." *Current Anthropology* 53, suppl. 5 (2012): S69–82.
———. "The Jewish Community of Rome: An Isolated Population? Sampling Procedures and Bio-Historical Narratives in Genetic Analysis in the 1950s." *BioSocieties* 5, no. 3 (2010): 306–29.
Lipphardt, Veronika, and Jörg Niewöhner. "Producing Difference in an Age of Biosociality: Biohistorical Narratives, Standardisation and Resistance as Translations." *Science, Technology & Innovation Studies* 3, no. 1 (2007): 45–65.
Lourdusamy, J. *Science and National Consciousness in Bengal: 1870–1930*. New Delhi: Orient Longman, 2004.
Ludden, David. "History outside Civilisation and Mobility of South Asia." *South Asia: Journal of South Asian Studies* 17, no. 1 (1994): 1–23.
———. "Orientalist Empiricism: Transformations of Colonial Knowledge." In Carol A. Breckenridge and Peter van der Veer, eds., *Orientalism and the Postcolonial Predicament: Perspectives on South Asia*. Philadelphia: University of Pennsylvania Press, 1993.
Lugg, J. W. H. "Taste-Thresholds for Phenylthiocarbamide of Some Population Groups: The Threshold of Two Uncivilized Ethnic Groups Living in Malaya." *Annals of Human Genetics* 22, no. 3 (1957): 244–53.
Lugg, J. W. H., and J. M. Whyte. "Taste Thresholds for Phenylthiocarbamide of Some Population Groups: The Thresholds of Some Civilized Ethnic Groups Living in Malaya." *Annals of Human Genetics* 19, no. 4 (1955): 290–311.
Luzzato, L. "In Memoriam: P. K. Sukumaran." *Hemoglobin: International Journal of Hemoglobin Research* 15, no. 5 (1991): v–vii.
Macfarlane, Eileen. "Blood Group Distribution in India with Special Reference to Bengal." *Journal of Genetics* 36 (1938): 225–37.
———. "Blood Grouping in the Deccan and the Eastern Ghats." *Journal of the Royal Asiatic Society of Bengal, Science* 6, no. 5 (1940): 39–53.
———. "Eastern Himalayan Blood-Groups." *Man* 37 (1937): 127–29.
———. "Eileen Macfarlane: Lecturer, Traveller, Scientist" (flyer, 1946). Jack Schultz Papers, American Philosophical Society, Philadelphia.
———. "Genetics." *Current Science,* special issue (1938): 1–3.
———. "Letter to J. B. S. Haldane," 12 July 1942. Special Collections, University College London.
———. "Preliminary Note on the Blood Groups of Some Cochin Castes." *Current Science* 4 (1936): 653–54.
———. "The Racial Affinities of the Jews of Cochin." *Journal of the Asiatic Society of Bengal* 3 (1938): 1–24.
———. Untitled Communication on Bagdi Blood Groups. *Current Science* 5 (1937): 284.
Macfarlane, Eileen, and S. S Sarkar. "Blood Groups in India." *American Journal of Physical Anthropology* 28, no. 4 (1941): 397–410.
Madan, T. N., and Gopala Sarana. "Introduction." In Madan and Sarana, eds., *Indian Anthropology: Essays in Memory of D. N. Majumdar*, 3–7. New York: Asia Publishing House, 1962.
Mahalanobis, P. C. "Analysis of Race Mixture in Bengal." In *Proceedings of the Twelfth Indian Science Congress*, 301–33. Benares: Indian Science Congress, 1925.
———. "A Revision of Risley's Anthropometric Data Relating to the Chittagong Hill Tribes." *Sankhya: The Indian Journal of Statistics* 1, nos. 2–3 (1934): 267–76.

———. "A Revision of Risley's Anthropometric Data Relating to the Tribes and Castes of Bengal." *Sankhya: The Indian Journal of Statistics* 1, no. 1 (1933): 76–105.

Mahalanobis, P. C., D. N. Majumdar, M. W. M. Yeatts, and C. Radhakrishna Rao. "Anthropometric Survey of the United Provinces, 1941: A Statistical Study." *Sankhya: The Indian Journal of Statistics* 9, nos. 2–3 (1949): 89–324.

Mahapatra, Usha Deka, and Priyabala Das. "Taste Threshold for Phenylthiocarbamide in Some Endogamous Groups of Assam." *Anthropologist* 15 (1968): 25–32.

Majeed, Javed. "Race and Pan-Islam in Iqbal's Thought." In Peter Robb, ed., *The Concept of Race in South Asia*, 304–26. Delhi: Oxford University Press, 1995.

Major, Andrew J. "State and Criminal Tribes in Colonial Punjab: Surveillance, Control and Reclamation of the 'Dangerous Classes.'" *Modern Asian Studies* 33, no. 3 (1999): 657–88.

Majumdar, D. N. "Blood Groups among the Makranis of Western Khandesh." *Current Science* 14, no. 5 (1945): 128–29.

———. "Blood Groups of the Doms." *Current Science* 11, no. 4 (1942): 153–54.

———. "Blood Groups of Tribes and Castes of the U.P. with Special Reference to the Korwas." *Journal of the Royal Asiatic Society of Bengal, Science* 9 (1943): 81–94.

———. "Muslim Blood Groups with Particular Reference to the U.P." *Current Science* 12, no. 10 (1943): 269–70.

———. *Race Realities in Cultural Gujarat: Report on the Anthropometric, Serological and Health Survey of Maha Gujarat*. Bombay: Gujarat Research Society, 1950.

———. "The Tharus and the Blood Groups." *Journal of the Royal Asiatic Society of Bengal, Science* 8 (1942): 25–37.

Majumdar, D. N., and Irawati Karve. *Racial Problems in Asia*. New Delhi: Indian Council of World Affairs, 1947.

Majumdar, D. N., and Kunwar Kishen. "Blood Group Distribution in the United Provinces: Report on the Serological Survey of the United Provinces." *Eastern Anthropologist* 1, nos. 1–2 (1947): 8–15.

Majumdar, D. N., and C. R. Rao. "Bengal Anthropometric Survey, 1945: A Statistical Study." *Sankhya: The Indian Journal of Statistics* 19, nos. 3–4 (1958): 201–402.

Majumdar, Rochona. "Feluda on Feluda: A Letter to Topshe." *South Asian History & Culture* 8, no. 2 (2017): 233–44.

Majumder, Partha P. "People of India: Biological Diversity and Affinities." In D. Balasubramanian and N. Appaji Rao, eds., *The Indian Human Heritage*, 45–59. Hyderabad: Universities Press, 1998.

———. "People of India: Biological Diversity and Affinities." *Evolutionary Biology* 6, no. 3 (1998): 100–110.

Malone, R. H., and M. N. Lahiri. "The Distribution of the Blood-Groups of Certain Races and Castes in India." *Indian Journal of Medical Research* 16 (1929): 963–68.

Maluf, N. S. R. "History of Blood Transfusion." *Journal of the History of Medicine and Allied Sciences* 9, no. 1 (1954): 59–107.

Mantena, Karuna. *Alibis of Empire: Henry Maine and the Ends of Liberal Imperialism*. Princeton: Princeton University Press, 2010.

Martin, Emily. "The Potentiality of Ethnography and the Limits of Affect Theory." *Current Anthropology* 54, suppl. 7 (2013): S149–59.

Matson, G. Albin. "Blood Groups and Ageusia in Indians of Montana and Alberta." *American Journal of Physical Anthropology* 24, no. 1 (1938): 81–84.

Mavhunga, Clapperton Chakanetsa. *Transient Workspaces: Technologies of Everyday Innovation in Zimbabwe*. Cambridge, MA: MIT Press, 2014.

McGonigle, Ian. *Genomic Citizenship: The Molecularization of Identity in the Contemporary Middle East.* Cambridge, MA: MIT Press, 2021.

McGranahan, Carole. "Theorizing Refusal: An Introduction." *Cultural Anthropology* 31, no. 3 (2016): 319–25.

McKennan, Robert A. "The Physical Anthropology of Two Alaskan Athapaskan Groups." *American Journal of Physical Anthropology* 22 (1964): 43–52.

McLeod, John. "Marriage and Identity among the Siddis of Janjira and Sachin." In John C. Hawley, ed., *India in Africa, Africa in India: Indian Ocean Cosmopolitanisms*, 253–72. Bloomington: Indiana University Press, 2008.

McMahon, Richard. "The History of Transdisciplinary Race Classification: Methods, Politics and Institutions, 1840s–1940s." *British Journal for the History of Science* 51, no. 1 (2018): 41–67.

Mehta, Uday Singh. *Liberalism and Empire: A Study in Nineteenth-Century British Liberal Thought.* Chicago: University of Chicago Press, 1999.

Melleuish, Greg, Konstantin Sheiko, and Stephen Brown. "Pseudo History/Weird History: Nationalism and the Internet." *History Compass* 7, no. 6 (2009): 1484–95.

Mikanowski, Jacob. "Dr. Hirszfeld's War: Tropical Medicine and the Invention of Sero-Anthropology on the Macedonian Front." *Social History of Medicine* 25, no. 1 (2012): 103–21.

Miki, Toshiyuki, Takashi Tanaka, and Tanemoto Furuhata. "On the Distribution of the ABO Blood Groups and the Taste Ability for Phenyl-Thio-Carbamide (P.T.C.) of the Lepchas and the Khasis." *Proceedings of the Japan Academy* 36, no. 2 (1960): 78–80.

Milam, Erika L. *Looking for a Few Good Males: Female Choice in Evolutionary Biology.* Baltimore: Johns Hopkins University Press, 2010.

Miller, Marilyn Grace. *Rise and Fall of the Cosmic Race: The Cult of Mestizaje in Latin America.* Austin: University of Texas Press, 2004.

Ministry of Food and Agriculture, Government of India. *Report of the Ad-Hoc Committee on Slaughter-Houses and Meat Inspection Practices.* Delhi: Manager of Publications, 1958.

Mishra, Saurabh. "Of Poisoners, Tanners and the British Raj: Redefining Chamar Identity in Colonial North India, 1850–90." *Indian Economic & Social History Review* 48, no. 3 (2011): 317–38.

Mital, M. S., J. G. Parekh, P. K. Sukumaran, R. S. Sharma, and P. J. Dave. "A Focus of Sickle Cell Gene near Bombay." *Acta Genetica et Statistica Medica* 27, no. 5 (1962): 257–67.

Mitchell, Lisa. *Language, Emotion and Politics in South India: The Making of a Mother Tongue.* Bloomington: Indiana University Press, 2009.

Mitra, Durba. "'Surplus Woman': Female Sexuality and the Concept of Endogamy." *Journal of Asian History* 80, no. 1 (2021): 3–26.

Mitter, Gopee Kissen. *Bengali and English Dictionary, for the Use of Schools.* Calcutta: Calcutta School Book Society, 1868.

Mizutani, Satoshi. *The Meaning of White: Race, Class, and the "Domiciled Community" in British India, 1858–1930.* Oxford: Oxford University Press, 2011.

M'Lennan, John F. *Primitive Marriage: An Inquiry into the Origins of the Form of Capture in Marriage Ceremonies.* Edinburgh: Adam & Charles Black, 1865.

Mohr, Jan. "Taste Sensitivity to Phenylthiourea in Denmark." *Annals of Eugenics* 16, no. 1 (1951): 282–86.

Mol, Annemarie. *The Body Multiple: Ontology in Medical Practice*. Durham, NC: Duke University Press, 2002.

Molina, Santiago José. "Amerindians, Europeans, Makiritare, Mestizos, Puerto Rican, and Quechua: Categorical Heterogeneity in Latin American Human Biology." *Perspectives on Science* 25, no. 5 (2017): 655–79.

Morton, Newton E., James F. Crow, and H. J. Muller. "An Estimate of the Mutational Damage in Man from Data on Consanguineous Marriages." *Proceedings of the National Academy of Sciences of the United States of America* 42, no. 11 (1956): 855–63.

Mukerji, D. P. "Majumdar: Scholar and Friend," in T. N. Madan and Gopala Sarana, eds., *Indian Anthropology: Essays in Memory of D. N. Majumdar*, 8–11. New York: Asia Publishing House, 1962.

Mukharji, Projit Bihari. "The Bengali Pharaoh: Upper-Caste Aryanism, Pan-Egyptianism, and the Contested History of Biometric Nationalism in Twentieth-Century Bengal." *Comparative Studies in Society and History* 59, no. 02 (2017): 446–76.

———. "Bloodworlds: A Hematology of the 1952 Indo-Australian Genetical Survey of the Chenchus." *Historical Studies in the Natural Sciences* 50, no. 5 (2020): 525–53.

———. "Cat and Mouse: Animal Technologies, Trans-Imperial Networks and Public Health from Below, British India, c. 1907–1918." *Social History of Medicine* 31, no. 3 (1 August 2018): 510–32.

———. *Doctoring Traditions: Ayurveda, Small Technologies, and Braided Sciences*. Chicago: University of Chicago Press, 2016.

———. "From Serosocial to Sanguinary Identities: Caste, Transnational Race Science and the Shifting Metonymies of Blood Group B, India c. 1918–1960." *Indian Economic & Social History Review* 51, no. 2 (2014): 143–76.

———. "Historicizing 'Indian Systems of Knowledge': Ayurveda, Exotic Foods, and Contemporary Antihistorical Holisms." *Osiris* 35 (2020): 228–48.

———. "Hylozoic Anticolonialism: Archaic Modernity, Internationalism, and Electromagnetism in British Bengal, 1909–1940." *Osiris* 34, no. 1 (2019): 101–20.

———. *Nationalizing the Body: The Medical Market, Print and Daktari Medicine*. London: Anthem Press, 2009.

———. "Profiling the Profiloscope: Facialization of Race Technologies and the Rise of Biometric Nationalism in Inter-War British India." *History and Technology* 31, no. 4 (2016): 376–96.

———. "Vernacularizing the Body: Informational Egalitarianism, Hindu Divine Design, and Race in Physiology Schoolbooks, Bengal 1859–1877." *Bulletin of the History of Medicine* 91, no. 3 (2017): 554–85.

Mukherji, M. "Cooley's Anaemia (Erythroblastic or Mediterranean Anaemia)." *Indian Journal of Pediatrics* 5, no. 17 (1938): 1–7.

Mumtaz, Zubia, Sarah Bowen, and Rubina Mumtaz. "Meanings of Blood, Bleeding and Blood Donations in Pakistan: Implications for National vs Global Safe Blood Supply Policies." *Health Policy and Planning* 27, no. 2 (2012): 147–55.

Muñoz, José Esteban. *The Sense of Brown*. Durham, NC: Duke University Press, 2020.

Nanda, Meera. *Prophets Facing Backwards: Postmodern Critiques of Science and Hindu Nationalism in India*. New Brunswick, NJ: Rutgers University Press, 2003.

Nandy, Ashis. *Alternative Sciences: Creativity and Authenticity in Two Indian Scientists*. Delhi: Oxford University Press, 2001.

———. *The Intimate Enemy: Loss and Recovery of Self under Colonialism*. New Delhi: Oxford University Press, 1988.

Negi, R. S. "The Incidence of Sickle-Cell Trait in Two Bastar Tribes, I." *Man* 62, no. 142 (1962): 84–86.

———. "The Incidence of Sickle-Cell Trait in Bastar, II." *Man* 63, nos. 21–22 (1963): 19–21.

———. "New Incidence of Sickle-Cell Trait in Bastar, III." *Man* 64, no. 214 (1964): 171–74.

Newman, William R. *Atoms and Alchemy: Chymistry and the Experimental Origins of the Scientific Revolution*. Chicago: University of Chicago Press, 2006.

Newman, William R., and Lawrence M. Principe. *Alchemy Tried in the Fire: Starkley, Boyle, and the Fate of Helmotian Chymistry*. Chicago: University of Chicago Press, 2002.

Nichter, Mark. "The Layperson's Perception of Medicine as Perspective into the Utilization of Multiple Therapy Systems in the Indian Context." *Social Science & Medicine* 14, no. 4 (1980): 225–33.

Nigam, Sanjay. "Disciplining and Policing the 'Criminals by Birth', Part 1: The Making of a Colonial Stereotype—The Criminal Tribes and Castes of North India." *Indian Economic & Social History Review* 27, no. 2 (1990): 131–64.

———. "Disciplining and Policing the 'Criminals by Birth', Part 2: The Development of a Disciplinary System, 1871–1900." *Indian Economic & Social History Review* 27, no. 3 (1990): 257–87.

Obeng, Pashington. "Religion and Empire: Belief and Identity among African Indians of Karnataka, South India." In John C. Hawley, ed., *India in Africa, Africa in India: Indian Ocean Cosmopolitanisms*, 231–52. Bloomington: Indiana University Press, 2008.

O'Brien, Alison D., and Randall K. Holmes. "Shiga and Shiga-like Toxins." *Microbiological Reviews* 51, no. 2 (1987): 206–20.

Oka, Rahul C., and Chapurukha M. Kusimba. "Siddi as Mercenary or as African Success Story on the West Coast of India." In John C. Hawley, ed., *India in Africa, Africa in India: Indian Ocean Cosmopolitanisms*, 203–30. Bloomington: Indiana University Press, 2008.

Oommen, T. K. "P. Matthias Hermanns SVD (1899–1972)." *Sociological Bulletin* 22, no. 1 (1973): 120.

———. "Race and Caste: Anthropological and Sociological Perspectives." In Sukhdeo Thorat and Umakant, eds., *Caste, Race, and Discrimination: Discourses in International Context*, 97–109. New Delhi: Indian Institute of Dalit Studies, 2004.

Owen, G. E. *Bihar and Orissa in 1921*. Patna: Superintendent of Government Printing, Bihar and Orissa, 1922.

Paik, Shailaja. "Amchya Jalmachi Chittarkatha (The Bioscope of Our Lives): Who Is My Ally?" *Economic and Political Weekly* 44, no. 40 (2009): 39–47.

———. *Dalit Women's Education in Modern India: Double Discrimination*. Abingdon, UK: Routledge, 2014.

———. "Mangala Bansode and the Social Life of Tamasha: Caste, Sexuality, and Discrimination in Modern Maharashtra." *Biography* 40, no. 1 (2017): 170–98.

Pande, Ishita. "Coming of Age: Law, Sex and Childhood in Late Colonial India." *Gender & History* 24, no.1 (2012): 205–30.

———. *Medicine, Race and Liberalism in British Bengal: Symptoms of Empire*. London: Routledge, 2012.

———. *Sex, Law, and the Politics of Age: Child Marriage in India, 1891–1937*. Cambridge: Cambridge University Press, 2020.

Pandey, Gyanendra. "In Defense of the Fragment: Writing about Hindu-Muslim Riots in India Today." *Representations* 37 (1992): 27–55.

Pandey, Sanjay Kumar. *Role of the Modulating Factors on the Phenotype of Sickle Cell Disease*. Raleigh, NC: Lulu Publication, 2019.

Parekh, J. G. "Hereditary Haemolytic Anaemias." *Journal of the J.J. Group of Hospitals and Grant Medical College* 2, no. 1 (1957): 1–24.

Parfionovitch, Yuri, Gyurmed Dorje, and Fernand Meyer. *Tibetan Medical Paintings: Illustrations to the Blue Beryl Treatise of Sangye Gyatso (1653–1705)*. Ed. Anthony Aris. London: Serindia, 1992.

Parfitt, Tudor, and Yulia Egorova. *Genetics, Mass Media and Identity: A Case of Genetic Research on the Lemba*. New York: Routledge, 2006.

Parikh, N. P., A. J. Baxi, and H. I. Jhala. "Blood Groups, Abnormal Haemoglobins and Other Genetical Characters in Three Gujarati-Speaking Groups." *Human Heredity* 19, no. 5 (1969): 486–98.

Parmar, P. K. "Taste Sensitivity to Phenyl Thio Carbamide (P.T.C.) in Gorkhas of Dhauladhar Range (Himachal Pradesh)." *Eastern Anthropologist* 21 (1968): 267–77.

Paul, Diane B. "'Our Load of Mutations' Revisited." *Journal of the History of Biology* 20, no. 3 (1987): 321–35.

Peabody, Norbert. "Cents, Sense, Census: Human Inventories in Late Precolonial and Early Colonial India." *Comparative Studies in Society and History* 43, no. 4 (2001): 819–50.

Pemberton, Stephen. "Canine Technologies, Model Patients: The Historical Production of Hemophiliac Dogs in American Biomedicine." In Susan Schrepfer and Philip Scranton, eds., *Industrializing Organisms: Introducing Evolutionary History*, 191–213. New York: Routledge, 2004.

Penrose, Lionel. "Report on Thesis, 'Studies on Twins' and Additional Papers by S. S. Sarkar." London, 15 July 1948. Penrose Correspondence, University College London.

Pickstone, John V. *Ways of Knowing: A New History of Science, Technology and Medicine*. Chicago: University of Chicago Press, 2001.

Piliavsky, Anastasia. "The 'Criminal Tribe' in India before the British." *Comparative Studies in Society and History* 57, no. 2 (2015): 323–54.

Pols, Hans, and Warwick Anderson. "The Mestizos of Kisar: An Insular Racial Laboratory in the Malay Archipelago." *Journal of Southeast Asian Studies* 49, no. 3 (2018): 445–63.

Porter, Theodore M. *Trust in Numbers: The Pursuit of Objectivity in Science and Public Life*. Princeton: Princeton University Press, 1996.

Prakash, Gyan. *Another Reason: Science and the Imagination of Modern India*. New York: Oxford University Press, 2000.

Prasad, Amit. "Burdens of the Scientific Revolution: Euro/West-Centrism, Black Boxed Machines, and the (Post) Colonial Present." *Technology & Culture* 60, no. 4 (2019): 1059–82.

Proctor, Robert N., and Londa Schiebinger, eds. *Agnotology: The Making and Unmaking of Ignorance*. Stanford: Stanford University Press, 2008.

Quack, Johannes. *Disenchanting India: Organized Rationalism and Criticism of Religion in India*. New York: Oxford University Press, 2012.

Race, R. R., and Ruth Sanger. *Blood Groups in Man*. 2d ed. Oxford: Blackwell Scientific Publications, 1954.

Radin, Joanna. *Life on Ice: A History of New Uses for Cold Blood*. Chicago: University of Chicago Press, 2017.

Ragab, Ahmed. "Islam Intensified: Snapshot Historiography and the Making of Muslim Identities." *Postcolonial Studies* 22, no. 2 (2019): 203–19.
Rai, Rohini. "From Colonial 'Mongoloid' to Neoliberal 'Northeastern': Theorising 'Race,' Racialization, and Racism in Contemporary India." *Asian Ethnicity* (2021). https://doi.org/10.1080/14631369.2020.1869518.
Raina, B. L. "Family Planning with Special Reference to Medical Aspects." In *Aspects of Population Policy in India*, 148–59. New Delhi: Council for Social Development, 1971.
Raj, Kapil. "When Human Travellers Become Instruments: The Indo-British Exploration of Central Asia in the Nineteenth Century." In Marie-Noëlle Bourguet, Christian Licoppe, and H. Otto Sibum, eds., *Instruments, Travel and Science: Itineraries of Precision from the Seventeenth to the Twentieth Century*, 156–88. New York: Routledge, 2002.
Rajagopal, Arvind. "Hindu Nationalism in the US: Changing Configurations of Political Practice." *Ethnic and Racial Studies* 23, no. 3 (2000): 467–96.
Rajan, Kaushik Sunder. *Biocapital: The Constitution of Postgenomic Life*. Durham, NC: Duke University Press, 2006.
Rakshit, Hirendra K., ed. *Bio-Anthropological Research in India: Proceedings of the Seminar in Physical Anthropology and Allied Disciplines*. Calcutta: Anthropological Survey of India, 1970.
Ramaswamy, Sumathi. "Remains of the Race: Archaeology, Nationalism, and the Yearning for Civilization in the Indus Valley." *Indian Economic & Social History Review* 38, no. 2 (2001): 105–45.
Ramberg, Lucinda. *Given to the Goddess: South Indian Devadasis and the Sexuality of Religion*. Durham, NC: Duke University Press, 2014.
Rand, Gavin, and Kim A. Wagner. "Recruiting the 'Martial Races': Identities and Military Service in Colonial India." *Patterns of Prejudice* 46, nos. 3–4 (n.d.): 232–54.
Ranganathan, Shubha. "Healing Temples, the Anti-Superstition Discourse and Global Mental Health: Some Questions from Mahanubhav Temples in India." *South Asia: Journal of South Asian Studies* 37, no. 4 (2014): 625–39.
Rao, Anupama. *The Caste Question: Dalits and the Politics of Modern India*. Berkeley: University of California Press, 2009.
Raper, Alan B. "Unusual Haemoglobin Variant in a Gujerati Indian." *British Medical Journal* 1, no. 5030 (1957): 1285–86.
Ray [Roy], Hemendrakumar. *Amanushik Manush* [Inhuman Humans]. In Gita Dutta and Sukhomoy Mukhopadhyay, eds., *Hemendrakumar Ray [Roy] Rachanabali* [The Collected Works of Hemendrakumar Ray], 12:9–118. Calcutta: Asia Publishing Co., 1991.
Ray [Roy], Hemendrakumar. "Najruler Janmadin Smarane" [Remembering Nazrul's Birthday]. In *Hemendrakumar [Ray] Roy Rachanabali* [Collected Works of Hemendrakumar Ray], 1:277–83. Calcutta: Asia Publishing Company, 1978.
Ray, Subhadeepta. "Studying Laboratories: A Sociological Inquiry into Research Practices in Genetics." In Mahuya Bandyopadhyay and Ritambhara Hebbar, eds., *Towards a New Sociology in India*. Hyderabad: Orient Blackswan, 2016.
Reardon, Jenny. *Race to the Finish: Identity and Governance in an Age of Genomics*. Princeton: Princeton University Press, 2005.
Reardon, Jenny, and Kim TallBear. "'Your DNA Is Our History': Genomics, Anthropology, and the Construction of Whiteness as Property." *Current Anthropology* 53, no. 55 (2012): S233–45.
Rees, Amanda. "Doing 'Deep Big History': Race, Landscape and the Humanity of H. J. Fleure (1877–1969)." *History of the Human Sciences* 32, no. 1 (2019): 99–120.

Reggiani, Andrés Horacio. *God's Eugenicist: Alexis Carrel and the Sociobiology of Decline.* New York: Berghahn Books, 2007.

Risley, H. H. "Of the Dravidian Tract: Santāl." In Risley, ed., *Census of India, 1901: Volume I—Ethnographic Appendices*, 143–48. Calcutta: Office of the Superintendent of Government Printing, India, 1903.

———. *Tribes and Castes of Bengal: Ethnographic Glossary, Volumes 1 and 2.* Calcutta: Bengal Secretariat Press, 1892.

Robbins, Kenneth X., and John McLeod. *African Elites in India: Habshi Amarat.* Ahmedabad: Mapin, 2006.

Robbins, Kenneth X., and Pushkar Sohoni. *Jewish Heritage of the Deccan, Mumbai, the Northern Konkan, Pune.* Ahmedabad: Jaico Publishing House, 2015.

Roberts, Dorothy E. *Fatal Invention: How Science, Politics, and Big Business Re-Create Race in the Twenty First Century.* New York: New Press, 2011.

Roberts, Lissa. "The Death of the Sensuous Chemist: The 'New' Chemistry and the Transformation of Sensuous Technology." *Studies in History and Philosophy of Science Part A* 26, no. 4 (1 December 1995): 503–29.

Rocher, Ludo. "The Aurasa Son." In Rocher, *Studies in Hindu Law and Dharmasastra*, ed. Donald R. Davis, 613–23. London: Anthem Press, 2012.

Rogaski, Ruth. *Hygienic Modernity: Meanings of Health and Disease in Treaty-Port China.* Berkeley: University of California Press, 2014.

Roos, Anna Marie. *The Salt of the Earth: Natural Philosophy, Medicine, and Chymistry in England, 1650–1750.* Boston: Brill, 2007.

Roque, Ricardo. "Bleeding Languages: Blood Types and Linguistic Groups in the Timor Anthropological Mission." *Current Anthropology*, forthcoming.

———. "The Blood That Remains: Card Collections from the Colonial Anthropological Missions." *BJHS Themes* 4 (2009): 29–53.

Rose, H. A. "Legitimisation and Adoption in Hindu Law." *Man* 22, nos. 94–96 (1922): 166–69.

Roy, D. N., and S. K. Roy Chaudhuri. "Sickle-Cell Trait in the Tribal Population in Madhya Pradesh and Orissa (India)." *Journal of the Indian Medical Association* 49, no. 3 (1967): 107–12.

Roy, Kaushik. "Race and Recruitment in the Indian Army: 1800–1918." *Modern Asian Studies* 47, no. 4 (2013): 1310–47.

Roy, Rohan Deb. *Malarial Subjects: Empire, Medicine and Nonhumans in British India, 1820–1909.* Cambridge: Cambridge University Press, 2017.

Roy, Tapti. "Tracking the Ephemeral: Elokeshi-Nabin-Mohanto Episode and the History of Print in Colonial Bengal." In Ishita Banerjee-Dube and Sarvani Gooptu, eds., *On Modern Indian Sensibilities: Culture, Politics, Histories* (Abingdon, UK: Routledge, 2018: 217–33.

Samanta, Arabinda. *Living with Epidemics in Colonial Bengal.* London: Routledge, 2017.

Samanta, Suchitra. "The 'Self-Animal' and Divine Digestion: Goat Sacrifice to the Goddess Kali in Bengal." *Journal of Asian Studies* 53, no. 3 (1994): 779–803.

Samarendra, Padmanabh. "Anthropological Knowledge and Statistical Frame: Caste in the Census in Colonial India." In Sumit Sarkar and Tanika Sarkar, eds., *Caste in Modern India: A Reader*, 255–96. Ranikhet: Permanent Black, 2014.

———. "Census in Colonial India and the Birth of Caste." *Economic and Political Weekly* 46, no. 33 (2011): 51–58

Sanghvi, L. D. "Comparison of Genetical and Morphological Methods for a Study of

Biological Differences." *American Journal of Physical Anthropology* 11, no. 3 (1953): 385–404.
———. "The Concept of Genetic Load: A Critique." *American Journal of Human Genetics* 15, no. 3 (1963): 298–309.
———. "Genetic Adaptation in Man." In P. T. Baker and J. S. Weiner, eds., *Biology of Human Heredity*, 305–28. Oxford: Clarendon Press, 1966.
———. "Genetics of Caste in India." In Amitabha Basu, Alok Kumar Ghosh, Suhas Kumar Biswas, and Ramendra Ghosh, eds., *Physical Anthropology and Its Extending Horizons: S. S. Sarkar Memorial Volume*, 175–88. Bombay: Orient Longman, 1973.
———. "Inbreeding in India." *Eugenics Quarterly* 13, no. 4 (1966): 291–301.
———. "Mystery That Is Heredity." *Journal of Family Welfare* 14, no. 3 (1968): 1–8.
———. "Mystery That Is Heredity (Part II)." *Journal of Family Welfare* 14, no. 4 (1968): 23–30.
———. "Perspectives for the Study of Racial Origins in India." *Anthropologist*, special issue (1969): 69–73.
———. "Preface." In L. D. Sanghvi, V. Balakrishnan, and Irawati Karve, eds., *Biology of the People of Tamil Nadu*, v–vi. Calcutta: Anthropological Society and Indian Society of Human Genetics, Pune, 1981.
———. "Sanghvi, Labhshankar Dalichand" (curriculum vitae, 1964). Penrose Papers, University College London.
Sanghvi, L. D., and V. R. Khanolkar. "Data Relating to Seven Genetical Characters in Six Endogamous Groups in Bombay." *Annals of Eugenics* 15, no. 1 (1949): 52–64.
Sanghvi, L. D., P. K. Sukumaran, and H. Lehmann. "Haemoglobin J Trait in Two Indian Women: Associated with Thalassaemia in One." *British Medical Journal* 2, no. 5100 (1958): 828–30.
Santos, Boaventura de Sousa. *The End of the Cognitive Empire: The Coming of Age of the Epistemologies of the South*. Durham, NC: Duke University Press, 2018.
Sarbadhikary, Sukanya. "The Leftover Touch: Sensing Caste in the Modern Urban Lives of a Devotional Instrument." In Supriya Chaudhuri, ed., *Religion and the City in India*, 130–45. London: Routledge, 2021.
Sardar, Ziauddin. "Foreword to the 2008 Edition." In Frantz Fanon, *Black Skin, White Masks*, vi–xx. London: Pluto Press, 2008.
Sarkar, Jayanta. *The Jarawa*. Calcutta: Seagull Books/Anthropological Survey of India, 1990.
Sarkar, Sasanka Sekhar. *The Aboriginal Races of India*. Calcutta: Bookland, 1954.
———. "Analysis of Indian Blood Group Data with Special Reference to the Oraons." *Transations of the Bose Research Institute* 15 (1942): 1–15.
———. "Blood Grouping Investigations in India with Special Reference to Santal Perganas, Bihar." *Transations of the Bose Research Institute* 12 (1937): 89–101.
———. "Blood Groups from the Andaman and Nicobar Islands." *Bulletin of the Department of Anthropology, Government of India* 1, no. 1 (1952): 25–30.
———. *Indian Eugenics Society: Bulletin No. 1*. Calcutta: Indian Eugenics Society, 1941.
———. "The Jarawa of the Andaman Islands." *Anthropos* 57, no. 3/6 (1962): 670–77.
———. *Lokgatha*. 2d ed. Calcutta: S. S. Sarkar Society for Human Science, 1998.
———. "The Place of Human Biology in Anthropology and Its Utility in the Service of the Nation." *Man in India* 31, no. 1 (1951): 1–22.
———. "Race and Race Movement in India." *Bulletin of the Ramakrishna Mission Institute of Culture* 18, no. 11 (1967): 371–75.

———. "The Social Institutions of the Mālpāhāriās." *Journal of the Royal Asiatic Society of Bengal, Science* 3, no. 1 (1932): 25–32.

Sarkar, Sasanka Sekhar, and Dilip Kumar Sen. "Further Blood Group Investigations in Santal Parganas." *Bulletin of the Department of Anthropology, Government of India* 1, no. 1 (1952): 8–13.

Sarkar, Sumit. "The Kalki-Avatar of Bikrampur: A Village Scandal in Early Twentieth Century Bengal." In *Subaltern Studies No. VI: Writings on South Asian History and Society*, edited by Ranajit Guha, 1–53. Delhi: Oxford University Press, 1989.

———. *Modern India, 1885–1947*. New York: St. Martin's Press, 1988.

Sarkar, Tanika. "Talking about Scandals: Religion, Law and Love in Late Nineteenth Century Bengal." *Studies in History* 13, no. 1 (1997): 63–95.

Savary, Luzia. *Evolution, Race and Public Sphere in India: Vernacular Concepts and Sciences (1860–1930)*. Abingdon, UK: Routledge, 2019.

———. "Vernacular Eugenics? Santati-Sastra in Popular Hindi Advisory Literature (1900–1940)." *South Asia: Journal of South Asian Studies* 37, no. 3 (2014): 381–97.

Schär, Bernard C. "From Batticaloa via Basel to Berlin: Transimperial Science in Ceylon and Beyond around 1900." *Journal of Imperial and Commonwealth History* 48, no. 2 (2020): 230–62.

Schiebinger, Londa. *Plants and Empire: Colonial Bioprospecting in the Atlantic World*. Cambridge, MA: Harvard University Press, 2004.

Schiff, Fritz, and William C. Boyd. *Blood Grouping Technic: A Manual for Clinicians, Serologists, Anthropologists, and Students of Legal and Military Medicine*. New York: Interscience Publishers, 1942.

Schneider, William H. "Blood Group Research in Great Britain, France, and the United States between the World Wars." *Yearbook of Physical Anthropology* 38 (1995): 87–114.

———. *The History of Blood Transfusion in Sub-Saharan Africa*. Athens: Ohio University Press, 2013.

Schneider, William H., and Ernest Drucker. "Blood Transfusions in the Early History of AIDS in Sub-Saharan Africa." *American Journal of Public Health* 96 (2006): 984–94.

Schröder, Dominik. "P. Matthias Hermanns SVD (1899–1972)." *Anthropos* 67, nos. 1–2 (1972): 1–8.

Schwartz, Henry. *Constructing the Criminal Tribe in Colonial India: Acting Like a Thief*. Chichester, UK: Wiley-Blackwell, 2010.

Seal, Brajendranath. "Meaning of Race, Tribe, Nation." In Gustav Spiller, ed., *Papers on Inter-Racial Problems Communicated to the First Universal Races Congress, Held at the University of London*, 1–13. London: P. S. King & Sons, 1911.

Segal, J. B. "White and Black Jews at Cochin, the Story of a Controversy." *Journal of the Royal Asiatic Society of Great Britain and Ireland* 2 (1983): 228–52.

Sen, Abala Kanta. *Comprehensive Anglo-Bengali Dictionary*. Calcutta: School Book Press, 1892.

Sen, Dwaipayan. "'No Matter How, Jogendranath Had to Be Defeated': The Scheduled Castes Federation and the Making of Partition in Bengal, 1945–1947." *Indian Economic & Social History Review* 49, no. 3 (2012): 321–64.

Sen, Sudipta. "Confessions of the Unfriendly Spleen: Medicine, Violence, and the Mysterious Organ of Colonial India." In Rohan Deb Roy and Guy N. A. Attewell, eds., *Locating the Medical: Explorations in South Asian History*, 71–100. New Delhi: Oxford University Press, 2018.

Sen-Gupta, Nares C. "The Early History of Sonship in Ancient India." *Man* 24 (1924): 40–43.

Seth, Praveen Kumar. "PTC Taste Distribution among the Betel Chewers, Non-Vegetarians and Smokers." *Eastern Anthropologist* 15, no. 5 (1962): 36–49.

Seth, Suman. *Difference and Disease: Medicine, Race, and the Eighteenth-Century British Empire*. Cambridge: Cambridge University Press, 2020.

Shahan, Maurice S., and Ward T. Huffman. *Diseases of Sheep and Goats: Farmer's Bulletin No. 1943*. Washington: US Department of Agriculture, 1943.

Shapin, Steven. *The Scientific Life: A Moral History of a Late Modern Vocation*. Chicago: University of Chicago Press, 2008.

Sharafi, Mitra. "Abortion in South Asia, 1860–1947: A Medico-Legal History." *Modern Asian Studies* 55, no. 2 (2021): 371–428.

———. "The Imperial Serologist and Punitive Self-Harm: Bloodstains and Legal Pluralism in British India." In Ian A. Burney and Christopher Hamlin, eds., *Global Forensic Cultures: Making Fact and Justice in the Modern Era*, 60–85. Baltimore: Johns Hopkins University Press, 2019.

Sharma, J. C. "Blood and P.T.C. Taste Studies in Punjabis and the Effects of Age and Certain Eating Habits on Taste Thresholds." *Anthropologist* 6 (1959): 40–46.

———. "Taste Sensitivity to Phenylthiocarbamide among Three Mongoloid Populations of the Indian Borders." *Acta Geneticae Medicae et Gemellologiae* 16, no. 3 (1967): 317–24.

Sharma, Jayeeta. *Empire's Garden: Assam and the Making of India*. Durham, NC: Duke University Press, 2011.

Shodhan, Amrita. "Women in the Maharaj Libel Case: A Re-Examination." *Indian Journal of Gender Studies* 4, no. 2 (1997): 123–39.

Shukla, R. N., and B. R. Solanki. "Sickle-Cell Trait in Central India." *Lancet* 1, no. 7015 (1958): 297–98.

Shukla, R. N., B. R. Solanki, and A. S. Parande. "Sickle Cell Disease in India." *Blood* 13, no. 6 (1958): 552–58.

Sibum, Heinz Otto. "Reworking the Mechanical Value of Heat: Instruments of Precision and Gestures of Accuracy in Early Victorian England." *Studies in History and Philosophy of Science Part A* 26, no. 1 (1995): 73–106.

Simmons, Roy T., J. J. Graydon, N. M. Semple, and G. W. L. D'Sena. "A Genetical Survey in Chenchu, South India: Blood, Taste and Secretion." *Medical Journal of Australia* 1, no. 15 (1953): 497–503.

Simpson, Audra. "On Ethnographic Refusal: Indigeneity, 'Voice' and Colonial Citizenship." *Junctures: The Journal for Thematic Dialogue* 9 (2007): 67–80.

Singh, Charu. "The Shastri and the Air-Pump: Experimental Fiction and Fictions of Experiment for Hindi Readers, 1915–1919." *History of Science* (January 2021). https://doi.org/10.1177/0073275320987421.

Singh, Indera P. "History of Development of Physical Anthropology in India." *Anthropologist*, special issue (1969): 217–22.

Singh, K. S. "Introduction." In *The History of the Anthropological Survey of India: Proceedings of a Seminar*, 1–4. Calcutta: Anthropological Survey of India, 1991.

Singh, Om Prakash. "Evolution of Dalit Identity: History of Adi Hindu Movement in United Province (1900–1950)." *Proceedings of the Indian History Congress* 70 (2010): 574–85.

Singha, Radhika. *A Despotism of the Law: Crime and Justice in Early Colonial India*. New Delhi: Oxford University Press, 2000.

Sinha, Mrinalini. "Britishness, Clubbability, and the Colonial Public Sphere: The Genealogy of an Imperial Institution in Colonial India." *Journal of British Studies* 40, no. 4 (2001): 489–521.

———. *Colonial Masculinity: The "Manly Englishman" and the "Effeminate Bengali" in the Late Nineteenth Century*. Manchester, UK: Manchester University Press, 1995.

———. "Totaram Sanadhya's *Fiji Mein Mere Ekkis Varsh*: A History of Empire and Nation in a Minor Key." In Antoinette Burton and Isabel Hofmeyr, eds., *Ten Books That Shaped the British Empire: Creating an Imperial Commons*, 168–89. Durham, NC: Duke University Press, 2014.

Sirsat, Satyavati M. "Effects of Migration on Some Genetical Characters in Six Endogamous Groups in India." *Annals of Human Genetics* 21, no. 1 (1956): 145–54.

———. "Exploring Nature's Secrets." In Rohini Godbole and Ram Ramswamy, eds., *Lilavati's Daughters: The Women Scientists of India*, 310–13. Bangalore: Indian Academy of Sciences, 2008.

Sivaramakrishnan, Kavita. *As the World Ages: Rethinking the Demographic Crisis*. Cambridge, MA: Harvard University Press, 2018.

Sivasundaram, Sujit. "Imperial Transgressions: The Animal and Human in the Idea of Race." *Comparative Studies of South Asia, Africa and the Middle East* 35, no. 1 (2015): 156–72.

Skaria, Ajay. "Shades of Wildness: Tribe, Caste, and Gender in Western India." *Journal of Asian History* 56, no. 3 (1997): 726–45.

Skude, G. "Complexities of Human Taste Variation." *Journal of Heredity* 51 (1960): 259–63.

———. "Saliva and Sweet Taste Perception for Phenylthiourea (P.T.C.)." *Acta Geneticae Medicae et Gemellologiae* 10 (1961): 316–20.

———. "Studies in Sweet Taste Perception for Phenylthiourea (P.T.C.)." *Hereditas* 50 (1963): 196–202.

———. "Sweet Taste Perception for Phenylthiourea (P.T.C.)." *Hereditas* 45 (1959): 597–622.

Smith, Linda Tuhiwai. *Decolonizing Methodologies: Research and Indigenous Peoples*. London: Zed Books, 1999.

Smith, Mark M. *How Race Is Made: Slavery, Segregation and the Senses*. Chapel Hill: University of North Carolina Press, 2006.

Snow, Jennifer. "The Civilization of White Men: The Race of the Hindu in *United States v. Bhagat Singh Thind*." In Henry Goldschmidt and Elizabeth McAlister, eds., *Race, Nation, and Religion in the Americas*, 259–82. New York: Oxford University Press, 2004.

Solomon, Harris. *Metabolic Living: Food, Fat, and the Absorption of Illness in India*. Durham, N.C.; London: Duke University Press, 2016.

Soneji, Davesh. *Unfinished Gestures: Devadasis, Memory, and Modernity in South India*. Chicago: University of Chicago Press, 2012.

Souza, Vanderlei Sebastião de, and Ricardo Ventura Santos. "The Emergence of Human Population Genetics and Narratives about the Formation of the Brazilian Nation (1950–1960)." *Studies in History and Philosophy of Biological and Biomedical Sciences* 47 (2014): 97–107.

Srivastava, R. P. "Blood Groups in the Tharus of Uttar Pradesh and Their Bearing on the Ethnic and Genetic Relationships." *Human Biology* 37, no. 1 (1965): 1–12.

———. "Frequency of Non-Tasters among the Danguria Tharu of Uttar Pradesh." *Eastern Anthropologist* 14, no. 3 (1961): 258–59.

———. "Further Data on Non-Tasters among the Tharus of Uttar Pradesh." *Eastern Anthropologist* 17 (1964): 19–22.

———. "Measurement of Taste Sensitivity to Phenylthiourel (P.T.C.) in Uttar Pradesh." *Eastern Anthropologist* 12 (1959): 267–72.

Stepan, Nancy Leys. *The Hour of Eugenics: Race, Gender and Nation in Latin America.* Ithaca, NY: Cornell University Press, 1991.

Strasser, Bruno J., and Soraya de Chadarevian. "The Comparative and the Exmplary: Revisiting the Early History of Molecular Biology." *History of Science* 49, no. 3 (2011): 317–36.

Subramaniam, Banu. *Holy Science: The Biopolitics of Hindu Nationalism.* Seattle: University of Washington Press, 2019.

Subramaniam, Samanth. *A Dominant Character: The Radical Science and Restless Politics of JBS Haldane.* New York: W. W. Norton, 2020.

Sukumaran, P. K., L. D. Sanghvi, J. A. M. Ager, and H. Lehmann. "Haemoglobin L in Bombay: Findings in Three Gujarati Speaking Lohana Families." *Acta Genetica et Statistica Medica* 9, no. 3 (1959): 202–6.

Sukumaran, P. K., L. D. Sanghvi, and G. N. Vyas. "Sickle-Cell Trait in Some Tribes of Western India." *Current Science* 25, no. 9 (1956): 290–91.

Summers, William C. "Cholera and Plague in India: The Bacteriophage Inquiry of 1927–1936." *Journal of the History of Medicine and Allied Sciences* 48 (1993): 275–301.

Sundar, Nandini. "In the Cause of Anthropology: The Life and Work of Irawati Karve." In Patricia Uberoi, Nandini Sundar, and Satish Deshpande, eds., *Anthropology in the East: Founders of Indian Sociology and Anthropology*, 360–416. Ranikhet: Permanent Black, 2007.

———. *Subalterns and Sovereigns: An Anthropological History of Bastar, 1854–2006.* New Delhi: Oxford University Press, 2011.

Sunseri, Thaddeus. "Blood Trials: Transfusions, Injections, and Experiments in Africa, 1890–1920." *Journal of the History of Medicine and Allied Sciences* 71, no. 3 (2016): 293–321.

Sur, Atulkrishna. *Bangalir Nrittatwik Parichay.* Calcutta: Jigyasa, 1977.

Swarup Mitra, Sushiela, and Bharati Dutta. "Jyoti Bhusan Chatterjea." In *Biographical Memoirs of the Fellows of the Indian National Science Academy*, 4:154–70. New Delhi: Indian National Science Academy, 1976.

Sweeney, Gerald. *"Fighting for the Good Cause": Reflections on Francis Galton's Legacy to American Hereditarian Psychology.* Philadelphia: American Philosophical Society, 2001.

Sysling, Fenneke. "Measurement, Self-Tracking and the History of Science: An Introduction." *History of Science* 58, no. 2 (2020): 103–16.

———. *Racial Science and Human Diversity in Colonial Indonesia.* Singapore: National University of Singapore Press, 2016.

TallBear, Kim. "Genomic Articulations of Indigeneity." *Social Studies of Science* 43, no. 4 (2013): 509–33.

———. *Native American DNA: Tribal Belonging and the False Promise of Genetic Science.* Minneapolis: University of Minnesota Press, 2013.

Tapper, Melbourne. *In the Blood: Sickle Cell Anemia and the Politics of Race.* Philadelphia: University of Pennsylvania Press, 1999.

Tarde, Gabriel. *Underground Man.* Trans. Cloudsley Brereton. London: Duckworth & Co., 1905.

Thomas, C. T. "Reminiscences in the Service of the Anthropological Survey of India." In K.S. Singh, ed., *History of the Anthropological Survey of India: Proceedings of a Seminar*, 20–46. Calcutta: Anthropological Survey of India, 1991.

TOI News Service. "Minister Explains Bill to Rationalise Abortion Law." *Times of India*, 4 December 1969.

Tracy, Sarah E. "Delicious Molecules: Big Food Science, the Chemosenses, and Umami." *Senses and Society* 13, no. 1 (2018): 89–107.

Trautmann, Thomas R. *Aryans and British India*. New Delhi: Yoda Press, 2008.

———. "Discovering Aryan and Dravidian: A Tale of Two Cities." *Historiographia Linguistica* 31, no. 1 (2004): 33–58.

———. *Languages and Nations: The Dravidian Proof in Colonial Madras*. Berkeley: University of California Press, 2006.

Tripathy, K. C. "PTC Taste Sensitivity in Some Orissan Castes." *Man in India* 49, no. 1 (1969): 64–70.

Trofa, Andrew F., Hannah Uneo-Olsen, Ruiko Oiwa, and Masanosuke Yoshikawa. "Dr. Kiyoshi Shiga: Discoverer of the Dysentery Bacillus." *Clinical Infectious Diseases* 29 (1999): 1303–6.

Tyagi, D. "Taste Sensitivity to Phenyl-Thio-Urea (P. T. C.) among Oraons and Mundas of Ranchi (India)." *Journal of the Anthropological Society of Nippon* 77, nos. 5–6 (1969): 195–200.

Uberoi, Patricia, Nandini Sundar, and Satish Deshpande, eds. *Anthropology in the East: Founders of Indian Sociology and Anthropology*. Ranikhet: Permanent Black, 2007.

Undevia, J. V., Aparna Bhagwe, S. Naik N., and M. B. Agarwal. "Dr. L. D. Sanghvi." In *Indian Legends in Haematology*, 87–89. Mumbai: Indian Society of Haematology and Blood Transfusion, 2013.

University Grants Commission. *University Development in India: Basic Facts and Figures, 1964–65*. New Delhi: University Grants Commission, 1966.

Vargas, Eduardo Viana, Bruno Latour, Bruno Karsenti, and Frédérique Aït-Touati. "The Debate between Tarde and Durkheim." *Environment and Planning D: Society and Space* 26, no. 5 (2008): 761–77.

Vella, F., and P. L. de V. Hart. "Sickle-Cell Anaemia in an Indian Family in Malaya." *Medical Journal of Malaya* 14, no. 2 (1959): 144–50.

Visvanathan, Shiv. *A Carnival of Science: Essays on Science, Technology and Development*. Delhi: Oxford University Press, 1997.

Visweswaran, Kamala. "Race and the Culture of Anthropology." *American Anthropologist* 100, no. 1 (1998): 70–83.

Vyas, G. N., H. M. Bhatia, D. D. Banker, and N. M. Purandare. "Study of Blood Groups and Other Genetical Characters in Six Endogamous Groups in Western India." *Annals of Human Genetics* 22, no. 3 (1958): 185–99.

Vyas, G. N., H. M. Bhatia, and L. D. Sanghvi. "Three Cases of Weak B in an Indian Family." *Vox Sanguinis* 5, no. 5 (1960): 509–16.

Wade, Peter, Carlos López Beltrán, Eduardo Restrepo, and Ricardo Ventura Santos. "Introduction: Genomics, Race Mixture, and Nation in Latin America." In Wade, López Beltrán, Restrepo, and Ventura Santos, eds., *Mestizo Genomics: Race Mixture, Nation, and Science in Latin America*, 1–32. Durham, NC: Duke University Press, 2014.

Wagner, Kim A. "Confessions of a Skull: Phrenology and Colonial Knowledge in Early Nineteenth-Century India." *History Workshop Journal* 69 (2010): 28–51.

Wailoo, Keith. *Drawing Blood: Technology and Disease Identity in Twentieth-Century America*. Baltimore: TJohns Hopkins University Press, 1997.

———. *Dying in the City of Blues: Sickle Cell Anemia and the Politics of Race and Health*. Chapel Hill: University of North Carolina Press, 2001.

———. "Who Am I? Genes and the Problem of Historical Identity." In Keith Wailoo, Alondra Nelson, and Catherine Lee, eds., *Genetics and the Unsettled Past: The Collision of DNA, Race and Hsitory*, 13–19. New Brunswick, NJ: Rutgers University Press, 2012.

Wald, Erica. *Vice in the Barracks: Medicine, the Military and the Making of Colonial India, 1780–1868*. New York: Palgrave Macmillan, 2014.

Wallace, Bruce. *Genetic Load: Its Biological and Conceptual Aspects*. Englewood Cliffs, NJ: Prentice-Hall, 1970.

Watt, Elizabeth, Emma Kowal, and Carmen Cummings. "Traditional Laws Meet Emerging Biotechnologies: The Impact of Genetic Genealogy on Indigenous Land Title in Australia." *Human Organization* 79, no. 2 (2020): 140–49.

Weitzman, Steven. *The Origin of the Jews: The Quest for Roots in a Rootless Age.*: Princeton University Press, 2017.

White, Luise. *Speaking with Vampires: Rumor and History in Colonial Africa*. Berkeley: University of California Press, 2000.

Widmer, Alexandra. "Making Blood 'Melanesian': Fieldwork and Isolating Techniques in Genetic Epidemiology." *Studies in History and Philosophy of Biological and Biomedical Sciences* 47, no. A (2014): 118–29.

World Health Organization. "Research on Genetics in Psychiatry: Report of a WHO Scientific Group." Geneva: World Health Organization, 1966.

Yang, Anand A. "Sacred Symbol and Sacred Space in Rural India: Community Mobilization in the 'Anti-Cow Killing' Riot of 1893." *Comparative Studies in Society and History* 22, no. 4 (1980): 576–96.

INDEX

Page numbers followed by "f" refer to figures.

CPSIA information can be obtained
at www.ICGtesting.com
Printed in the USA
LVHW041326270123
738090LV00008B/15